T0226550

Advances in
MARINE BIOLOGY

VOLUME 30

FRONTISPIECE

Examples of invertebrate aggregations. (A) School of spawning squid, *Loligo opalescens*. (B) Swarm of mysids in shallow coastal water, south east Tasmania, Australia. (C) School of mysids, *Paramesopodopsis rufa*. (A, courtesy of Planet Earth Pictures, photograph by Norbert Wu; B and C, courtesy of Jon Bryan.)

Advances in
MARINE BIOLOGY

Edited by

J.H.S. BLAXTER

Dunstaffnage Marine Research Laboratory, Oban, Scotland

and

A.J. SOUTHWARD

Marine Biological Association, The Laboratory, Citadel Hill, Plymouth, England

ACADEMIC PRESS

Harcourt Brace & Company, Publishers
London   San Diego   New York   Boston
Sydney   Tokyo   Toronto

ACADEMIC PRESS LIMITED
24/28 Oval Road
LONDON NW1 7DX

United States Edition published by
ACADEMIC PRESS INC.
San Diego CA 92101

A catalogue record for this book is available from the British Library

ISBN 0–12–026130–8

Typeset by Keyset Composition, Colchester, Essex
Transferred to digital print 2007
Printed and bound by CPI Antony Rowe, Eastbourne

CONTRIBUTORS TO VOLUME 30

B.J. BETT, *Institute of Oceanographic Sciences Deacon Laboratory, Brook Road, Wormley, Godalming, Surrey GU8 5UB, UK*

A.C. BROWN, *Department of Zoology, University of Cape Town, South Africa 7700*

A. DINET, *Laboratoire Arago, F 66650 Banyuls sur Mer, France*

T. FERRERO, *Department of Zoology, The Natural History Museum, Cromwell Road, London SW7 5BD, UK*

A. FERRON, *Department of Biology, McGill University, Montréal, Quebec, Canada, H3A 1B1*

A.J. GOODAY, *Institute of Oceanographic Sciences Deacon Laboratory, Brook Road, Wormley, Godalming, Surrey GU8 5UB, UK*

P.J.D. LAMBSHEAD, *Department of Zoology, The Natural History Museum, Cromwell Road, London SW7 5BD, UK*

W.C. LEGGETT, *Department of Biology, McGill University, Montréal, Quebec, Canada, H3A 1B1*

F.J. ODENDAAL, *Department of Zoology, University of Cape Town, South Africa 7700*

O. PFANNKUCHE, *Forschungzentrum für Marine Geowissenschaften, GEO MAR Abt. Marine Umweltgeologie, Universität Kiel, Wischhofstr. 1–3, Kiel, Germany*

D.A. RITZ, *Zoology Department, University of Tasmania, Box 252C, GPO, Hobart, Tasmania 7001, Australia*

A.D. ROGERS, *Marine Biological Association of the United Kingdom, The Laboratory, Citadel Hill, Plymouth, PL1 2PB, United Kingdom*

T. SOLTWEDEL, *Institut für Hydrobiologie und Fischereiwissenschaft, Universität Hamburg, Zeiseweg 9, 22765 Hamburg, Germany*

A. VANREUSEL, *University of Gent, Zoology Institute, Marine Biology Section, K.L. Ledeganckstraat 35, B 9000 Gent, Belgium*

M. VINCX, *University of Gent, Zoology Institute, Marine Biology Section, K.L. Ledeganckstraat 35, B 9000 Gent, Belgium*

CONTENTS

Meiobenthos of the Deep Northeast Atlantic

M. Vincx, B.J. Bett, A. Dinet, T. Ferrero, A.J. Gooday, P.J.D. Lambshead, O. Pfannkuche, T. Soltwedel and A. Vanreusel

The Biology of Oniscid Isopoda of the Genus *Tylos*

A.C. Brown and F.J. Odendaal

Social Aggregation in Pelagic Invertebrates

D.A. Ritz

An Appraisal of Condition Measures for Marine Fish Larvae

A. Ferron and W.C. Leggett

The Biology of Seamounts

A.D. Rogers

Meiobenthos of the Deep Northeast Atlantic

M. Vincx,[1] B.J. Bett,[2] A. Dinet,[3] T. Ferrero,[4] A.J. Gooday,[2] P.J.D. Lambshead,[4] O. Pfannkuche,[6] T. Soltwedel[5] and A. Vanreusel[1]

[1]*University of Gent, Zoology Institute, Marine Biology Section, K.L. Ledeganckstraat 35, B 9000 Gent, Belgium.*
[2]*Institute of Oceanographic Sciences Deacon Laboratory, Brook Road, Wormley, Godalming, Surrey GU8 5UB, UK.*
[3]*Laboratoire Arago, F 66650 Banyuls sur Mer, France.*
[4]*Department of Zoology, The Natural History Museum, Cromwell Road, London SW7 5BD, UK.*
[5]*Institut für Hydrobiologie und Fischereiwissenschaft, Universität Hamburg, Zeiseweg 9, 22765, Germany.*
[6]*Forschungzentrum für Marine Geowissenschaften, GEO MAR Abt. Marine Umweltgeologie, Universität Kiel, Wischhofstr. 1–3, Kiel, Germany*

ADVANCES IN MARINE BIOLOGY VOL 30
ISBN 0–12–026130–8

1. INTRODUCTION

Although the first ecological investigations on the meiobenthic communities of the deep northeast Atlantic were carried out 20 years ago (Thiel, 1972b), it is only recently (1990) that co-operative research has been initiated by the European Community under the EC MAST I (Marine Science and Technology I 1990–1992) programme: "Natural variability and the prediction of change in marine benthic ecosystems". The general objectives of this EC programme, which continues in a MAST II project (1993–1996), are (i) to describe the natural structure and variability of offshore benthic populations in the northeast Atlantic, (ii) to relate the structure and variability to processes in the physical, chemical and biological environment, (iii) to describe the trophic network in the benthic boundary layer and to estimate the organic carbon flux through the deep-sea benthic ecosystem, and (iv) to attempt to predict the changes that are likely to be associated with natural and anthropogenic disturbance. An important component of benthic ecosystems, particularly in the deep sea (Thiel, 1975, 1983), is the meiobenthos, generally considered to include organisms in the size range 31–500 μm. The combined efforts of five laboratories in four of the countries participating in the MAST project have highlighted the gaps that exist in our knowledge of the meiobenthos of the northeast Atlantic and have prompted this review. Our main purpose is to summarize literature data and new results from an area lying between 15°N and 53°N and extending from the continental margin of western Europe and northwest Africa to the Mid-Atlantic Ridge (Figure 1).

Since the first quantitative investigation by Wigley and McIntyre (1964), data on deep-sea meiobenthos have been gathered from all oceans and attempts made to relate the broad geographical patterns observed to various environmental factors. On a planetary scale, one of the major environmental gradients is created by the slope of the ocean floor, a gradient which has important effects on benthic communities. As in the case of macrobenthos (Lampitt *et al.*, 1986), the data available on meiobenthic densities in deep-sea environments also show trends which can be related to the amount and nature of organic matter reaching the seafloor (Thiel, 1983; Shirayama, 1983; Pfannkuche, 1985; Pfannkuche and Thiel, 1987). The distribution patterns of deep-sea organisms are

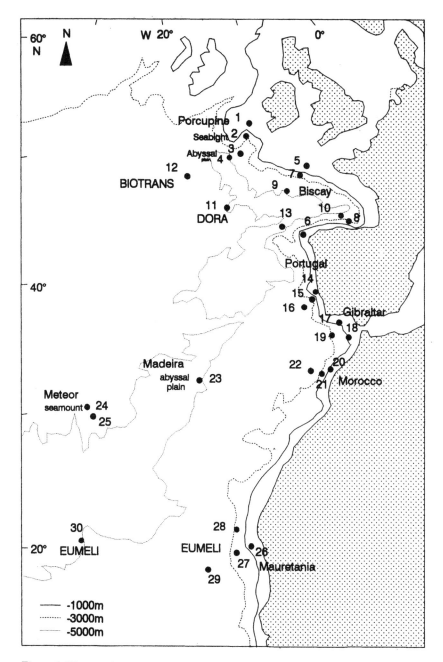

Figure 1 The northeast Atlantic showing the positions of the 30 sampling *areas*.

influenced by other variables such as sediment type, bottom currents and bottom water masses. Local topographic and hydrodynamic features, such as canyons, seamounts and deep boundary currents, are also important. In addition, the ever-improving resolution of the physical structure of the deep sea, and technical advances in sampling gear and surface navigation, have permitted biologists to address small-scale variability, on scales ranging from centimetres to kilometres.

In this review we consider first the nature and scope of meiofaunal research in the northeast Atlantic and then discuss the environmental parameters which are believed to influence meiofaunal organisms. Next, we discuss the various types and scales of pattern observed among meiofaunal populations within our study area, progressing from the large-scale bathymetric and latitudinal trends and then to small-scale horizontal patterns within particular areas. Faunal densities and faunal composition are considered separately and compared with data from other regions. Finally, we discuss the distribution of meiofauna within sediment profiles and the temporal variability of populations. Our approach differs, therefore, from that adopted in Tietjen's (1992) recent review of deep-sea meiofauna which focused mainly on abundance and biomass data from different oceans and on the relationship between the biomass of the meiofauna and that of other faunal components.

2. MEIOBENTHOS IN THE NORTHEAST ATLANTIC

2.1. Physiographic Setting

The area under investigation consists of a series of deep basins separated by ridges. Basin depth tends to increase from north to south, with depths in excess of 5000 m occurring in basins to the west and northwest of the Cape Verde Islands.

A number of physiographic zones can be recognized within this region: continental shelf, continental slope, continental rise, abyssal apron and abyssal plain (Emery and Uchupi, 1984; see also Rona, 1980; Udintsev *et al.*, 1989–1990). Secondary features include the zone of abyssal hills which separates the continental rise from the Mid-Atlantic Ridge and a number of major seamounts and volcanic islands. Notable aspects include the abyssal aprons (sediment masses deposited by geostrophic bottom currents) of the northwest African margin and around and to the west of the Rockall Trough (Hill, 1987) and the series of abyssal plains (from north to south the Porcupine, Biscay, Iberian, Tagus, Horseshoe, Seine, Madeira, Cape Verde, Gambia) which lie seaward of the continental rise.

The area consists of different biogeochemical provinces of plankton productivity, such as upwelling (NW-Africa), trade wind regime, subtropical gyre, etc., which are of great consequence to the supply of food to the seabed and ultimately for the sediment type where pelagic sedimentation prevails.

2.2. Historical Background

The study of some meiobenthic taxa, particularly foraminifera, living in this region has a long history (e.g. Parker & Jones, 1856; Brady, 1884). However, sampling for meiobenthos was incidental until the 1960s and 1970s when the first quantitative meiobenthic samples were collected from the German research vessel *Meteor*; numerical abundance data from these samples were published by Thiel (1972a, b, 1975, 1978, 1983) and Rachor (1975). Another quantitative investigation which included the meiobenthos was the BIOGAS programme, carried out during the 1970s in the Bay of Biscay (Dinet and Vivier, 1977; see also Dinet *et al.*, 1985). More recent papers devoted partly or exclusively to the meiobenthos are those of Pfannkuche *et al.* (1990), Pfannkuche (1992, 1993b) in the BIOTRANS area and Rutgers van der Loeff and Lavaleye (1986) at the DORA site.

Other studies have focused on particular aspects of the meiobenthos. Some authors have considered just the nematodes (Riemann, 1974; Dinet and Vivier, 1979, 1981). Desbruyères *et al.* (1985) evaluated meiobenthic taxa as part of a recolonization experiment in the Bay of Biscay. Another approach has been to look for correlations between meiobenthic densities and environmental parameters such as bathymetric depth and the amount of organic matter in the sediment (Thiel, 1979b, 1983; Dinet and Khripounoff, 1980; Sibuet *et al.*, 1989; Vanreusel *et al.*, 1992). Although taxonomic studies of deep-sea meiobenthos are fairly rare in our area, some new taxa have been described among the harpacticoids (Bodin, 1968; Dinet 1977), nematodes (Decraemer, 1983), ostracods (Kornicker, 1989, van Harten, 1990) and tardigrades (Renaud-Mornant, 1989). Some of these investigations have dealt exclusively with the metazoans while others have included foraminifera within the scope of the meiobenthos. Several papers by Gooday (1986a, b, 1988), Gooday and Lambshead (1989), Lambshead and Gooday (1990) have described the foraminiferal meiobenthos in the northeast Atlantic. Gooday and Turley (1990) presented some additional data and Gooday (1990) established a new, ecologically important allogromiid species. The numerous geologically orientated studies of modern deep-sea foraminifera in the northwest Atlantic (Murray, 1991) deal only with the hard-shelled taxa and are not considered further.

2.3. Sampling Areas

Nine of the papers cited above provide data on the density and composition of metazoan and foraminiferal meiobenthos in the area under consideration and are further treated in a general data analysis (Thiel, 1972a, 1975; Dinet and Vivier, 1977; Pfannkuche *et al.*, 1983, 1990; Pfannkuche, 1985, 1992; Rutgers van der Loeff and Lavaleye, 1986; Vanreusel *et al.*, 1992). Information on foraminifera published by Gooday (1986a, b, 1988) and Gooday and Lambshead (1989) is not considered in the general approach because no total meiobenthic data are given in these articles. However, this information is discussed in the sections on foraminifera. For reasons explained below, we exclude the data of Rachor (1975) from this survey. Additional unpublished results are available from the Porcupine Seabight (Gooday, IOSDL benthic biology programme), Porcupine and Madeira Abyssal Plains (Gooday and Ferrero, IOSDL DEEPSEAS and EC MAST Programmes), Bay of Biscay (Vincx and Vanreusel, EC MAST programme), BIOTRANS site (Soltwedel and Gooday, BIOTRANS Programme) and from off northwest Africa (Dinet, EUMELI and EC MAST programmes).

In order to recognize general trends among the meiobenthos, we have grouped all the stations sampled during these published and unpublished studies into 30 *areas* (identified by number in Table 1) on the basis of geographical and bathymetric proximity. In what follows, density data and other relevant abiotic information on *areas* will always be an average value of the sampling stations situated within one of the 30 *areas* as defined in Table 1. References to *area* numbers in the following text refer always to one of the *area* numbers shown in Figure 1. Bathymetric proximity is arbitrarily defined with limits of following depth zones: < 1000 m, 1000–3000 m, 3000–4500 m and > 4500 m. Inevitably, most of these *areas* are broader in their areal and bathymetric extent than the stations described in the original publications. *Area* locations and numbers are summarized in Figure 1. The original data from all sampling stations are summarized in Table 2.

2.4. Collection and Processing

Methods for the collection and processing of meiobenthic samples are discussed by Thiel (1983), Fleeger *et al.* (1988) and Pfannkuche and Thiel (1988). The data reviewed in the present chapter were obtained from samples collected with various kinds of coring devices (Table 1). A Reineck box corer was used in some of the early studies (Thiel, 1972b, 1975; Dinet and Vivier, 1977) but most investigators have used either an

USNEL box corer or multiple corer (Hessler and Jumars, 1974; Barnett *et al.*, 1984). Blomqvist (1991) has identified four factors which bias the sampling of soft bottom sediments, namely, the loss of superficial sediment, the distribution, resuspension and loss of enclosed sediments, core shortening, and repenetration. He urges extreme caution when evaluating results based on grab samples but considers box corers to be fairly reliable if used with a supporting stand and mechanism which secures the lids during retrieval through the water column. Blomqvist (1991) concludes that the Barnett Watson (SMBA) multiple corer (Barnett *et al.*, 1984) "seems to be the best device for general sampling of open-sea, soft bottom sediments at present". It is equipped with a battery of up to 12 coring tubes and the penetration of these into the sediment is slowed by a hydraulic damping system. The multiple corer collects samples in which the sediment–water interface is virtually undisturbed; it is the only remotely operated coring device capable of reliably recovering phytodetritus. Sampling bias, within the data-set considered here, is investigated in detail in Bett *et al.* (1994).

Box core and multiple core samples have been subsampled in a variety of ways. Early German studies used a "meiostecher" with a cross-sectional area of $25\,cm^2$ or $10\,cm^2$ (Thiel, 1983). Dinet and Vivier (1977) and Rutgers van der Loeff and Lavaleye (1986) used transparent tubes with a cross-sectional area of 10 and $25\,cm^2$ respectively. More recently, smaller subcores have been taken from both box and multiple corers using modified 20 or 50-ml syringes (3.46 and $5.31\,cm^2$ across sectional area respectively). Some complete multiple corer samples have been sorted for metazoan and foraminiferal meiobenthos. Gooday and Lambshead (1989) and Lambshead and Gooday (1990) examined only the upper 1 cm of sediment for foraminifera analysis. However, most authors have followed the lead of Thiel (1966) by examining 1-cm thick sediment slices down to a depth of 4–7 cm. Samples have generally been fixed and preserved using 4% formaldehyde, buffered with borax.

Similar methods have been used to extract meiobenthos from the fixed samples (cf. Pfannkuche and Thiel, 1988). The sediment is passed through a sequence of sieves, the smallest mesh size used varying from 50 to 31 μm. The sieve residues are stained with Rose Bengal with added phenol and hand sorted under a binocular microscope. Rutgers van der Loeff and Lavaleye (1986) used elutriation to concentrate the meiobenthic organisms before sieving on a 50- or 31-μm mesh sieve. Rachor (1975) shook and decanted his samples at least eight times before preservation and sorting, a procedure which must account for the low densities of meiobenthos in his samples (Thiel, 1983). Because of these methodological difficulties, the data of Rachor (1975) are not included in our survey. In order to distinguish dead foraminiferal tests from those

Table 1 Source data on meiofauna of the northeast Atlantic (nr, *area* number, cf. Figure 1; st, number of sampling stations per *area*; sa, total number of samples per *area*).

nr	Author	Area of study	Depth (m)	st	sa	Corer
1	Gooday, unpubl.	Porcupine Seabight	398	1	1	Multi
	Pfannkuche, 1985	Porcupine Seabight	500–960	3	3	Multi
2	Gooday, unpubl.	Porcupine Seabight	1340	2	2	Multi
	Pfannkuche, 1985	Porcupine Seabight	1492–2785	4	4	Multi
3	Gooday, unpubl.	Porcupine Seabight	4090–4495	2	2	Multi
	Pfannkuche, 1985	Porcupine Seabight	3567–4167	2	2	Multi
4	Lambshead and Ferrero, unpubl.[a]	Porcupine Abyssal plain	4850	1	6	Multi
	Pfannkuche, 1985	Porcupine Abyssal plain	4500–4850	2	2	Multi
5	Vanreusel and Vincx, unpubl.	Bay of Biscay N	70–170	3	6	Box
6	Vanreusel et al., 1992	Bay of Biscay S	190–325	2	4	Box
7	Dinet and Vivier, 1977	Bay of Biscay N	2035–3370	20	20	Reineck
8	Dinet and Vivier, 1977	Bay of Biscay S	1912–2480	7	7	Reineck
9	Dinet and Vivier, 1977	Bay of Biscay N	4097–4725	12	12	Reineck
10	Dinet and Vivier, 1977	Bay of Biscay S	4315–4460	3	3	Reineck
11	Rutgers van der Loeff and Lavaleye, 1986	DORA	3958–4800	15	15	Box

No.	Reference	Location	Depth			Gear
12	Pfannkuche et al., 1990[b]	BIOTRANS	3900–4300	2	7	Multi
	Soltwedel, unpubl.	BIOTRANS	4560	1	3	Multi
	Gooday, unpubl.	BIOTRANS	4560	–	1	Multi
	Pfannkuche et al., 1990[b]	BIOTRANS	4550	1	18	Multi
13	Thiel, 1972a	Iberic sea	5272–5340	11	11	Box
14	Thiel, 1975[b]	Portugal	250–1250	2	2	Box
15	Thiel, 1975[b]	Portugal	1250–2250	2	2	Box
16	Thiel, 1975[b]	Portugal	3250–5250	2	2	Box
17	Thiel, 1975[b]	Gibraltar	250–1250	2	2	Box
18	Pfannkuche et al., 1983[b]	NW Africa	131–818	5	5	Box
19	Pfannkuche et al., 1983[b]	NW Africa	1163–3093	5	5	Box
20	Thiel, 1975[b]	Morocco	250–1250	2	2	Box
21	Thiel, 1975[b]	Morocco	1250–2250	2	2	Box
22	Thiel, 1975[b]	Morocco	4250–4750	1	1	Box
23	Gooday, unpubl.	Madeira Abyssal plain	4856–5120	2	2	Multi
	Lambshead and Ferrero, unpubl.[a]	Madeira Abyssal plain	4856	–	7	Multi
24	Thiel, 1975[b]	Great Meteor seamount	250–750	1	1	Box
25	Thiel, 1975[b]	Great Meteor seamount	1250–1750	1	1	Box
26	Thiel, 1975[b]	Mauretania	190–1250	3	3	Box
27	Thiel, 1975[b]	Mauretania	1750–3250	2	2	Box
28	Dinet, unpubl.	Eumeli	1543–2041	6	12	Box
29	Dinet, unpubl.	Eumeli	3107–3137	9	18	Box
30	Dinet, unpubl.	Eumeli	4568–4652	8	16	Box

[a] Only nematode densities available.
[b] Only total metazoan meiofauna densities available.

Table 2 Meiofaunal densities for samples collected in the northeast Atlantic (ar.nr, area number; Nem, Nematoda; Cop, harpacticoid copepods + nauplii; Turb, Turbellaria; Pol, Polychaeta; Oli, Oligochaeta; Ostr, Ostracoda; Hydr, Hydrozoa; Gas, Gastrotricha; Tar, Tardigrada; Kin, Kinorhyncha; Amp, Amphipoda; Hal, Halacaroidea; Lor, Loricifera; Iso, Isopoda; Biv, Bivalvia; Tan, Tanaidacea; Fora, Foraminifera; Meio, total meiofauna; Metaz, metazoan meiofauna).

Author	Area	ar.nr	Depth (m)	Sample	Nem	Cop	Turb	Pol	Oli	Ostr	Hydr
Dinet,	Eumeli	28	1970	KGS44 a	476.5	9.4		18.8		0.0	
unpubl.		28		b	256.1	30.1		28.2		1.9	
		28	1543	KGS38 a	849.3	97.9		1.9		1.9	
		28		b	764.6	58.4		24.5		0.0	
		28	1590	KGS41 a	1235.4	71.6		45.2		9.4	
		28		b	1033.9	75.3		13.2		1.9	
		28	2041	KGS43 a	1182.7	47.1		18.8		1.9	
		28		b	1367.2	94.2		32.0		1.9	
		28	1970	KGS45 a	862.5	16.9		13.2		3.8	
		28		b	779.7	30.1		5.6		0.0	
		28	1618	KGS39 a	1389.8	37.7		33.9		0.0	
		28		b	1133.7	26.4		15.1		1.9	
		29	3120	KGS10 a	333.3	16.9		7.5		1.9	
		29		b	337.1	30.1		11.3		1.9	
		29	3137	KGS13 a	256.1	30.1		20.7		0.0	
		29		b	421.8	16.9		5.6		0.0	
		29	3136	KGS14 a	474.6	43.3		15.1		0.0	
		29		b	655.4	32.0		11.3		0.0	
		29	3124	KGS11 a	455.7	33.9		3.8		0.0	
		29		b	559.3	16.9		13.2		0.0	
		29	3107	KGS12 a	551.8	13.2		33.9		5.6	
		29		b	448.2	11.3		13.2		5.6	
		29	3118	KGS18 a	572.5	13.2		11.3		0.0	
		29		b	540.5	11.3		9.4		0.0	
		29	3130	KGS15 a	745.8	33.9		13.2		0.0	
		29		b	476.5	33.9		3.8		1.9	
		29	3128	KGS16 a	1122.4	20.7		18.8		0.0	
		29		b	587.6	24.5		15.1		1.9	
		29	3118	KGS21 a	201.5	9.4		3.8		0.0	
		29		b	113.0	24.5		3.8		0.0	
		30	4629	KGS02 a	167.6	0.0		3.8		0.0	
		30		b	154.4	5.6		1.9		0.0	
		30	4582	KGS03 a	120.5	1.9		1.9		0.0	
		30		b	99.8	11.3		0.0		0.0	
		30	4590	KGS09 a	131.8	0.0		7.5		0.0	
		30		b	286.3	5.6		1.9		0.0	
		30	4582	KGS04 a	96.0	13.2		3.8		0.0	
		30		b	152.5	1.9		3.8		0.0	
		30	4568	KGS05 a	49.0	5.6		3.8		0.0	
		30		b	81.0	7.5		3.8		0.0	
		30	4580	KGS07 a	96.0	5.6		0.0		0.0	
		30		b	105.5	7.5		0.0		0.0	
		30	4652	KGS08 a	64.0	1.9		0.0		0.0	
		30		b	113.0	3.8		0.0		1.9	
		30	4569	KGS06 a	60.3	16.9		0.0		0.0	
		30		b	111.1	7.5		0.0		0.0	
Ferrero,	Porcupine	4	4850	16(5)	331.0						
unpubl.	Abyssal	4		41(1)	220.9						
	Plain	4		5(11)	374.0						
		4		24(7)	285.7						
		4		26(10)	603.2						
		4		26(11)	516.6						
Gooday,	Porcupine	1	398	51620	695.5	84.0	0.0	0.0	0.0	11.6	0.0
unpubl.	Seabight	2	1340	51502	1211.0	118.0	0.0	11.8	0.0	15.0	0.0
		2	1340	51615	1026.0	162.0	0.0	7.5	0.0	4.6	0.0
		3	4090	51504	215.6	38.0	0.0	3.6	0.0	0.7	0.0
		3	4495	51606	372.8	29.0	0.0	2.9	0.0	3.0	0.0
Gooday,	Madeira	23	5120	354	40.5	1.3	0.0	0.0	0.0	0.0	0.0
unpubl.	Abyssal	23	4856	12174(88)	179.6	19.0	0.0	2.4	0.0	0.4	0.0
Ferrero,	Plain	23	4856	12174(93)	79.5						
unpubl.		23	4856–4950	12174(94)	106.9						
Soltwedel,		23		12174(88)	59.9						
unpubl.		23		12174(24)	69.0						
		23		12174(26)	55.0						
		23		12174(86)	43.0						
Vanreusel	Bay of	5	70	1a	809.0	166.0	49.0	35.0	9.0	7.0	7.0
and Vincx,	Biscay	5		1b	658.0	72.0	48.0	25.0	5.0	2.0	4.0
unpubl.		5	170	3a	193.0	66.0	73.0	20.0	0.0	14.0	3.0
Vanreusel		5		3b	172.0	34.0	188.0	13.0	1.0	0.0	1.0
et al., 1992		5	148	S11a	255.0	145.0	10.0	27.0	0.0	20.0	3.0
		5		S11b	171.0	64.0	24.0	29.0	0.0	8.0	2.0
		6	123	6a	1031.0	45.0	10.0	18.0	0.0	2.0	0.0
		6		6b	773.0	46.0	32.0	21.0	0.0	3.0	0.0
		6	300	8a	419.0	30.0	4.0	10.0	0.0	0.0	0.0
		6		8b	479.0	15.0	0.0	9.0	0.0	0.0	0.0

Gas	Tar	Kin	Amp	Hal	Lor	Iso	Biv	Tan	Others	Fora	Meio	Metaz	Nem (%)	Cop (%)
	0.0	0.0					0.0	0.0	0.0			504.7	94.4	1.9
	0.0	1.9					0.0	3.8	3.8			325.8	78.6	9.2
	0.0	3.8					3.8	0.0	5.6			964.2	88.1	10.2
	0.0	3.8					0.0	0.0	7.5			858.8	89.0	6.8
	0.0	7.5					0.0	0.0	0.0			1369.1	90.2	5.2
	0.0	1.9					0.0	1.9	0.0			1128.1	91.7	6.7
	0.0	3.8					0.0	0.0	0.0			1254.2	94.3	3.8
	0.0	5.6					0.0	0.0	1.9			1502.8	91.0	6.3
	0.0	5.6					0.0	0.0	1.9			904.0	95.4	1.9
	0.0	1.9					0.0	0.0	0.0			817.3	95.4	3.7
	0.0	3.8					0.0	0.0	1.9			1467.0	94.7	2.6
	0.0	0.0					0.0	0.0	0.0			1177.0	96.3	2.2
	0.0	0.0					0.0	0.0	1.9			361.6	92.2	4.7
	0.0	0.0					1.9	0.0	0.0			382.3	88.2	7.9
	1.9	0.0					1.9	0.0	0.0			310.7	82.4	9.7
	0.0	0.0					0.0	0.0	0.0			444.4	94.9	3.8
	0.0	1.9					0.0	0.0	0.0			534.8	88.7	8.1
	0.0	0.0					1.9	0.0	0.0			700.6	93.5	4.6
	0.0	0.0					0.0	0.0	0.0			493.4	92.4	6.9
	0.0	0.0					1.9	0.0	0.0			591.3	94.6	2.9
	0.0	0.0					0.0	0.0	1.9			606.4	91.0	2.2
	0.0	1.9					0.0	0.0	0.0			482.1	93.0	2.3
	0.0	0.0					0.0	0.0	0.0			597.0	95.9	2.2
	0.0	0.0					0.0	0.0	1.9			563.1	96.0	2.0
	1.9	0.0					0.0	3.8	0.0			792.8	94.1	4.3
	0.0	0.0					0.0	0.0	0.0			521.7	91.3	6.5
	0.0	0.0					0.0	0.0	0.0			1162.0	96.6	1.8
	0.0	0.0					0.0	0.0	0.0			629.0	93.4	3.9
	0.0	0.0					0.0	0.0	0.0			214.7	93.9	4.4
	1.9	0.0					0.0	0.0	0.0			141.2	80.0	17.3
	0.0	1.9					0.0	0.0	0.0			173.3	96.7	0.0
	0.0	0.0					0.0	0.0	0.0			163.8	94.3	3.4
	0.0	0.0					0.0	0.0	0.0			124.3	97.0	1.5
	0.0	0.0					0.0	0.0	0.0			111.1	89.8	10.2
	0.0	1.9					0.0	0.0	0.0			139.4	94.6	0.0
	0.0	0.0					0.0	0.0	0.0			295.7	96.8	1.9
	0.0	1.9					0.0	1.9	0.0			114.9	83.6	11.5
	0.0	0.0					0.0	0.0	0.0			160.1	95.3	1.2
	0.0	0.0					0.0	0.0	0.0			58.4	83.9	9.7
	0.0	0.0					0.0	0.0	0.0			92.3	87.8	8.2
	0.0	0.0					0.0	0.0	0.0			101.7	94.4	5.6
	0.0	0.0					0.0	0.0	0.0			113.0	93.3	6.7
	0.0	0.0					0.0	0.0	0.0			65.9	97.1	2.9
	0.0	0.0					0.0	0.0	0.0			118.6	95.2	3.2
	0.0	0.0					0.0	0.0	0.0			77.2	78.0	22.0
	0.0	0.0					0.0	0.0	0.0			118.6	93.7	6.3

Gas	Tar	Kin	Amp	Hal	Lor	Iso	Biv	Tan	Others	Fora	Meio	Metaz	Nem (%)	Cop (%)
0.0	0.0	0.0	0.0	0.0	0.0	0.0	0.0	0.0	0.0	1098.3	1890.3	792.1	87.9	10.6
0.0	4.9	4.9	0.0	0.0	0.0	0.0	3.8	0.0	9.0	1835.2	3213.6	1378.4	87.9	8.6
0.0	9.2	3.8	0.0	0.0	0.0	0.0	1.7	0.0	9.2	2060.7	3284.8	1224.1	83.8	13.2
0.0	0.1	0.7	0.0	0.0	0.0	0.0	0.1	0.0	0.0		258.9	258.9	83.3	14.7
0.0	0.0	3.0	0.0	0.0	0.0	0.0	3.0	0.0	3.0	956.3	1373.0	416.7	89.5	7.0
0.0	0.0	0.0	0.0	0.0	0.0	0.0	0.0	0.0	0.0	34.4	76.2	41.8	96.9	3.1
0.0	0.0	0.8	0.0	0.0	0.0	0.4	0.0	0.0	0.0	234.0	436.6	202.6	88.6	9.4

Gas	Tar	Kin	Amp	Hal	Lor	Iso	Biv	Tan	Others	Fora	Meio	Metaz	Nem (%)	Cop (%)
24.0	14.0	0.0	0.0	0.0	0.0	0.0	0.0	0.0	0.0			1120.0	72.2	14.8
5.0	4.0	0.0	0.0	0.0	0.0	0.0	0.0	0.0	0.0			823.0	80.0	8.7
5.0	0.0	0.0	0.0	0.0	0.0	0.0	0.0	0.0	0.0			374.0	51.6	17.6
0.0	1.0	0.0	7.0	1.0	0.0	0.0	0.0	0.0	0.0			418.0	41.1	8.1
8.0	2.0	0.0	0.0	0.0	0.0	0.0	0.0	0.0	0.0			470.0	54.3	30.9
1.0	2.0	1.0	0.0	2.0	0.0	0.0	0.0	0.0	0.0			304.0	56.3	21.1
0.0	0.0	6.0	0.0	0.0	0.0	0.0	0.0	0.0	0.0			1112.0	92.7	4.0
0.0	1.0	4.0	0.0	0.0	0.0	0.0	0.0	0.0	0.0			880.0	87.8	5.2
0.0	0.0	0.0	0.0	0.0	0.0	0.0	0.0	0.0	0.0			463.0	90.5	6.5
0.0	0.0	0.0	0.0	0.0	0.0	0.0	0.0	0.0	0.0			503.0	95.2	3.0

Author	Area	ar.nr	Depth (m)	Sample	Nem	Cop	Turb	Pol	Oli	Ostr	Hydr
Dinet and	Bay of	7	2110	12-1	394.0	70.0	0.0	7.0	0.0	1.0	0.0
Vivler, 1977	Biscay	7	2200	16-1	912.0	49.0	0.0	4.0	0.0	2.0	0.0
		7	2035	17-1	396.0	35.0	0.0	2.0	0.0	2.0	0.0
		7	2150	37-1	766.0	122.0	0.0	6.0	0.0	1.0	0.0
		7	2235	38-1	278.0	27.0	0.0	11.0	0.0	0.0	0.0
		7	2125	39-1	280.0	6.0	0.0	9.0	0.0	0.0	0.0
		7	2136	40-1	228.0	25.0	.0.0	1.0	0.0	0.0	0.0
		7	2091	41-1	228.0	28.0	0.0	9.0	0.0	4.0	0.0
		7	2111	42-1	317.0	54.0	0.0	3.0	0.0	1.0	0.0
		7	2080	44-1	506.0	147.0	0.0	9.0	0.0	2.0	0.0
		7	2288	02-2	507.0	32.0	0.0	5.0	0.0	1.0	0.0
		7	3039	03-2	250.0	28.0	0.0	3.0	0.0	1.0	0.0
		7	2726	11-2	428.0	40.0	0.0	7.0	0.0	0.0	0.0
		7	2690	18-2	323.0	59.0	0.0	4.0	0.0	0.0	0.0
		7	3370	19-2	291.0	42.0	0.0	7.0	0.0	3.0	0.0
		7	2835	34-2	140.0	0.0	0.0	1.0	0.0	0.0	0.0
		7	2864	45-2	246.0	26.0	0.0	7.0	0.0	0.0	0.0
		7	2920	47-2	306.0	59.0	0.0	7.0	0.0	0.0	0.0
		7	2765	58-2	296.0	17.0	0.0	4.0	0.0	0.0	0.0
		7	2853	59-2	742.0	102.0	0.0	5.0	0.0	1.0	0.0
		8	1920	24-6	306.0	63.0	0.0	3.0	0.0	1.0	0.0
		8	2480	26-6	527.0	29.0	0.0	7.0	0.0	7.0	0.0
		8	1913	27-6	451.0	28.0	0.0	3.0	0.0	0.0	0.0
		8	1960	53-6	403.0	37.0	0.0	2.0	0.0	0.0	0.0
		8	1960	54-6	435.0	37.0	0.0	7.0	0.0	0.0	0.0
		8	1957	70-6	748.0	33.0	0.0	16.0	0.0	1.0	0.0
		8	1920	71-6	543.0	45.0	0.0	8.0	0.0	0.0	0.0
		9	4150	20-3	356.0	28.0	0.0	6.0	0.0	2.0	0.0
		9	4097	31-3	155.0	7.0	0.0	3.0	0.0	0.0	0.0
		9	4130	32-3	309.0	7.0	0.0	3.0	0.0	6.0	0.0
		9	4096	35-3	261.0	20.0	0.0	2.0	0.0	1.0	0.0
		9	4300	48-3	149.0	13.0	0.0	0.0	0.0	0.0	0.0
		9	4220	16-3	287.0	19.0	0.0	5.0	0.0	0.0	0.0
		9	4225	61-3	383.0	48.0	0.0	5.0	0.0	0.0	0.0
		9	4202	62-3	330.0	36.0	0.0	0.0	0.0	1.0	0.0
		9	4590	22-4	192.0	19.0	0.0	1.0	0.0	0.0	0.0
		9	4550	51-4	12.0	3.0	0.0	1.0	0.0	0.0	0.0
		9	4725	56-4	215.0	13.0	0.0	0.0	0.0	0.0	0.0
		9	4700	66-4	280.0	12.0	0.0	0.0	0.0	0.0	0.0
		10	4415	09-5	152.0	15.0	0.0	2.0	0.0	1.0	0.0
		10	4315	23-5	86.0	16.0	0.0	1.0	0.0	1.0	0.0
		10	4460	69-5	5.0	1.0	0.0	1.0	0.0	0.0	0.0
Pfannkuche,	Porcupine	1	500	51507	2382.0	98.0	0.0	3.0	0.0	18.0	0.0
1985	Seabight	1	510	51112	1676.0	121.0	0.0	26.0	0.0	9.0	0.0
		1	960	51103	1429.0	81.0	0.0	25.0	0.0	5.0	0.0
		2	1492	51104	820.0	75.0	0.0	11.0	0.0	7.0	0.0
		2	2000	51105	702.0	101.0	0.0	6.0	0.0	4.0	0.0
		2	2510	51106	658.0	66.0	0.0	6.0	0.0	2.0	0.0
		2	2785	51110	717.0	151.0	0.0	5.0	0.0	8.0	0.0
		3	3567	51108	595.0	54.0	0.0	4.0	0.0	1.0	0.0
		3	4167	51109	462.0	48.0	0.0	1.0	0.0	4.0	0.0
		4	4500	51505	300.0	52.0	0.0	1.0	0.0	2.0	0.0
		4	4850	51506	272.0	35.0	0.0	1.0	0.0	1.0	0.0
Pfannkuche	Morocco	18	131-133								
et al., 1983		18	191-207								
		18	381-407								
		18	598-607								
		18	800-818								
		19	1163								
		19	2014-2064								
		19	1463-1478								
		19	2435								
		19	2999-3093								
Pfannkuche	Biotrans	12									
et al. 1990	peak	12	3900								
		12									
		12									
		12									
	Slope	12	4300								
		12									
	Plain	12	4550								
		12									
		12									
		12									
		12									
		12									
		12									
		12									
		12									
		12									
		12									
		12									
		12									
		12									

Gas	Tar	Kin	Amp	Hal	Lor	Iso	Biv	Tan	Others	Fora	Meio	Metaz	Nem (%)	Cop (%)
0.0	1.0	1.0	0.0	1.0	0.0	0.0	0.0	0.0	1.0			478.0	82.4	14.6
0.0	2.0	2.0	0.0	1.0	0.0	0.0	0.0	2.0	4.0			978.0	93.3	5.0
0.0	2.0	1.0	0.0	0.0	0.0	1.0	0.0	0.0	1.0			440.0	90.0	8.0
0.0	4.0	2.0	0.0	0.0	0.0	0.0	0.0	0.0	5.0			907.0	84.5	13.5
0.0	2.0	0.0	0.0	0.0	0.0	0.0	0.0	0.0	0.0			320.0	86.9	8.4
0.0	0.0	0.0	0.0	0.0	0.0	1.0	0.0	0.0	8.0			304.0	92.1	2.0
0.0	3.0	0.0	0.0	0.0	0.0	0.0	0.0	0.0	0.0			257.0	88.7	9.7
0.0	0.0	1.0	0.0	0.0	0.0	0.0	0.0	0.0	0.0			270.0	84.4	10.4
0.0	2.0	0.0	0.0	0.0	0.0	0.0	0.0	0.0	3.0			380.0	83.4	14.2
0.0	2.0	3.0	0.0	0.0	0.0	2.0	0.0	0.0	0.0			671.0	75.4	21.9
0.0	0.0	1.0	0.0	1.0	0.0	1.0	0.0	0.0	1.0			549.0	92.3	5.8
0.0	2.0	1.0	0.0	0.0	0.0	0.0	0.0	0.0	2.0			288.0	86.8	9.7
0.0	1.0	1.0	0.0	1.0	0.0	0.0	0.0	0.0	3.0			536.0	79.9	7.5
0.0	0.0	1.0	0.0	0.0	0.0	0.0	0.0	1.0	0.0			387.0	83.5	15.2
0.0	0.0	1.0	0.0	1.0	0.0	1.0	0.0	0.0	0.0			346.0	84.1	12.1
0.0	0.0	0.0	0.0	1.0	0.0	0.0	0.0	0.0	0.0			142.0	98.6	0.0
0.0	1.0	0.0	0.0	0.0	0.0	0.0	0.0	0.0	0.0			281.0	87.5	9.3
0.0	0.0	1.0	0.0	0.0	0.0	0.0	0.0	0.0	0.0			374.0	81.8	15.8
0.0	0.0	0.0	0.0	0.0	0.0	0.0	0.0	1.0	0.0			315.0	94.0	5.4
0.0	0.0	3.0	0.0	0.0	0.0	0.0	0.0	0.0	0.0			853.0	87.0	12.0
0.0	1.0	1.0	0.0	0.0	0.0	1.0	0.0	0.0	3.0			379.0	80.7	16.6
0.0	0.0	2.0	0.0	0.0	0.0	0.0	0.0	0.0	7.0			579.0	91.0	5.0
0.0	0.0	0.0	0.0	0.0	0.0	0.0	0.0	0.0	0.0			482.0	93.6	5.8
0.0	0.0	1.0	0.0	0.0	0.0	0.0	0.0	0.0	0.0			445.0	90.6	8.3
0.0	0.0	2.0	0.0	0.0	0.0	0.0	0.0	1.0	1.0			483.0	90.1	7.7
0.0	1.0	5.0	0.0	0.0	0.0	0.0	0.0	0.0	1.0			808.0	92.6	4.1
0.0	4.0	1.0	0.0	0.0	0.0	0.0	0.0	0.0	1.0			603.0	90.0	7.5
0.0	0.0	1.0	0.0	1.0	0.0	0.0	1.0	0.0	0.0			395.0	90.1	7.1
0.0	0.0	0.0	0.0	0.0	0.0	0.0	0.0	0.0	0.0			165.0	93.9	4.2
0.0	0.0	0.0	0.0	0.0	0.0	0.0	0.0	0.0	0.0			320.0	96.6	2.2
0.0	0.0	1.0	0.0	0.0	0.0	0.0	0.0	0.0	2.0			287.0	90.9	7.0
0.0	0.0	0.0	0.0	0.0	0.0	0.0	0.0	0.0	0.0			62.0	240.3	21.0
0.0	0.0	1.0	0.0	0.0	0.0	0.0	0.0	0.0	0.0			311.0	92.3	6.1
0.0	0.0	2.0	0.0	0.0	0.0	0.0	0.0	0.0	0.0			437.0	87.6	11.0
0.0	0.0	0.0	0.0	0.0	0.0	1.0	0.0	0.0	0.0			369.0	89.4	9.8
0.0	0.0	0.0	0.0	0.0	0.0	0.0	0.0	0.0	0.0			213.0	90.1	8.9
0.0	0.0	1.0	0.0	0.0	0.0	0.0	0.0	0.0	1.0			16.0	75.0	18.8
0.0	0.0	0.0	0.0	0.0	0.0	0.0	0.0	0.0	1.0			230.0	93.5	5.7
0.0	0.0	0.0	0.0	0.0	0.0	1.0	0.0	0.0	0.0			297.0	94.3	4.0
0.0	0.0	0.0	0.0	0.0	0.0	1.0	0.0	0.0	0.0			171.0	88.9	8.8
0.0	0.0	0.0	0.0	0.0	0.0	0.0	0.0	0.0	0.0			105.0	81.9	15.2
20.0	0.0	8.0	0.0	2.0	0.0	0.0	62.0	0.0	1.0			7.0	71.4	14.3
81.0	0.0	6.0	0.0	3.0	0.0	0.0	12.0	0.0	29.0			2604.0	91.5	3.8
33.0	3.0	4.0	0.0	3.0	0.0	0.0	8.0	0.0	2.0			1963.0	85.4	6.2
8.0	3.0	2.0	0.0	7.0	0.0	0.0	4.0	0.0	5.0			1593.0	89.7	5.1
7.0	0.0	4.0	0.0	1.0	0.0	0.0	1.0	0.0	2.0			943.0	87.0	8.0
5.0	1.0	1.0	0.0	1.0	0.0	0.0	1.0	0.0	3.0			828.0	84.8	12.2
12.0	0.0	5.0	0.0	8.0	0.0	0.0	0.0	0.0	1.0			744.0	88.4	8.9
7.0	0.0	0.0	0.0	0.0	0.0	0.0	0.0	0.0	2.0			900.0	79.7	16.8
4.0	0.0	2.0	0.0	5.0	0.0	0.0	1.0	0.0	1.0			663.0	89.7	8.1
4.0	0.0	2.0	0.0	0.0	0.0	0.0	0.0	0.0	1.0			528.0	87.5	9.1
4.0	1.0	0.0	0.0	1.0	0.0	0.0	0.0	0.0	1.0			362.0	82.9	14.4
												315.0	86.3	11.1
												2656.0		
												1778.0		
												2480.0		
												1465.0		
												1175.0		
												631.0		
												570.0		
												620.0		
												550.0		
												557.0		
												364.0		
												376.0		
												352.0		
												308.0		
												386.0		
												302.0		
												302.0		
												316.0		
												475.0		
												434.0		
												270.0		
												306.0		
												397.0		
												281.0		
												307.0		
												375.0		
												315.0		
												409.0		
												429.0		
												315.0		
												434.0		
												348.0		
												450.0		
												457.0		
												326.0		

Author	Area	ar.nr	Depth (m)	Sample	Nem	Cop	Turb	Pol	Oli	Ostr	Hydr
Soltwedel, unpubl.	Biotrans	12	4560	113	124.8	20.0	0.0	0.0	0.0	0.0	0.0
		12		114	229.0	40.0	0.0	5.8	0.0	2.0	0.0
		12		115	120.4	15.0	0.0	2.9	0.0	0.0	0.0
Gooday, unpubl.		12		179(2)	760.1	53.0	0.0	0.0	0.0	0.0	0.0
Rutgers and Lavaleye, 1986	DORA	11	4787	21	315.0	10.5	0.0	1.2	0.0	0.6	0.0
		11	4325	5	386.0	16.7	0.0	1.2	0.0	2.0	0.0
		11	4333	23	534.0	14.7	0.0	3.2	0.0	2.0	0.0
		11	4310	22	403.0	3.9	0.0	0.8	0.0	1.2	0.0
		11	3958	24	560.0	19.3	0.0	1.4	0.0	0.8	0.0
		11	4723	25	622.0	16.1	0.0	1.8	0.0	0.6	0.0
		11	4540	26	720.0	12.0	0.0	2.0	0.0	0.6	0.0
		11	4800	1	322.0	5.7	0.0	0.0	0.0	0.4	0.0
		11	4200	2	101.0	9.7	0.0	0.4	0.0	0.8	0.0
		11	4300	4	246.0	11.7	0.0	0.0	0.0	0.8	0.0
		11	4570	6	300.0	15.4	0.0	0.0	0.0	0.4	0.0
		11	4725	8	373.0	13.0	0.0	0.0	0.0	0.4	0.0
		11	4725	11	401.0	12.6	0.0	0.8	0.0	0.8	0.0
		11	4700	13	595.0	32.1	0.0	1.2	0.0	0.4	0.0
		11	4000	15	985.0	48.7	0.0	5.7	0.0	2.4	0.0
Thiel, 1972	Iberic sea	13	5325	12-89-1	156.0	3.0	0.0	0.4	0.0	0.0	0.0
		13	5325	12-89-2	246.0	6.0	0.0	0.4	0.0	1.2	0.0
		13	5335	18-91-1	192.0	5.6	0.0	0.4	0.0	1.2	0.0
		13	5340	21-92-1	172.0	3.6	0.0	0.0	0.0	0.8	0.0
		13	5340	21-92-2	80.0	2.0	0.0	0.0	0.0	1.6	0.0
		13	5305	29-93-1	242.0	1.6	0.0	0.4	0.0	0.0	0.0
		13	5305	29-93-2	200.0	4.4	0.0	0.4	0.0	0.4	0.0
		13	5320	31-94-1	256.0	5.6	0.0	0.0	0.0	2.0	0.0
		13	5320	31-94-2	176.0	2.4	0.0	0.8	0.0	1.2	0.0
		13	5272	35-95-1	302.0	10.4	0.0	1.2	0.0	2.2	0.0
		13	5272	35-95-2	257.0	4.8	0.0	1.2	0.0	2.2	0.0
Thiel 1975	Mauretania	26	190-250								
	Mauretania	26	250-750								
	Gr. Met. seam.	24									
	Morocco	20									
	Gibraltar	17									
	Portugal	14									
	Mauretania	26	750-1250								
	Morocco	20									
	Gibraltar	17									
	Portugal	14									
	Gr. Met. seam.	25	1250-1750								
	Morocco	21									
	Portugal	15									
	Mauretania	27	1750-2250								
	Morocco	21									
	Portugal	15									
	Mauretania	27	2750-3250								
	Portugal	16	3750-4250								
	Morocco	22	4250-4750								
	Portugal	16	4750-5250								

which were living when collected, Gooday (1986a, b, 1988) and Gooday and Lambshead (1989) mounted stained specimens in glycerol and examined them under a compound microscope to ensure that the stained material was foraminiferal protoplasm. Only specimens with a convincing protoplasmic mass were regarded as living.

A controversial issue in benthic research is the standardization of size groups. During the Sixth Deep-Sea Biology Symposium in Copenhagen (1991), a workshop was organized by Hjalmar Thiel on the "Standardization of methods for benthos studies and biochemical measurements in sediments" (Thiel, 1993). For meiobenthos, this workshop recommended a lower limit of 31 μm and an upper limit of 1 mm for all benthic work in the deep sea. In fact, these mesh-size based categories have little ecological and even less taxonomic justification in the deep sea because a

Gas	Tar	Kin	Amp	Hal	Lor	Iso	Biv	Tan	Others	Fora	Meio	Metaz	Nem (%)	Cop (%)
0.0	0.0	0.0	0.0	0.0	0.0	0.0	0.0	0.0	2.6	196.0	343.4	147.4	84.7	13.6
0.0	0.0	1.0	0.0	0.0	0.0	0.0	0.0	0.0	2.0	319.0	598.8	279.8	81.8	14.3
0.0	0.0	0.0	0.0	0.0	0.0	0.0	0.0	0.0	0.0	464.3	602.6	138.3	87.1	10.8
0.0	0.0	0.0	0.0	0.0	0.0	0.0	0.0	0.0	0.0	864.2	1677.3	813.1	93.5	6.5
0.0	0.0	0.2	0.0	0.0	0.0	0.0	0.0	0.0	4.8	12.6	374.2	361.6	87.1	2.9
0.0	0.0	0.0	0.0	0.0	0.0	0.0	0.0	0.0	18.7	37.8	462.1	424.3	91.0	3.9
0.0	0.6	1.0	0.0	0.0	0.0	0.0	0.0	0.0	4.6	22.8	667.3	644.5	82.9	2.3
0.0	0.0	0.0	0.0	0.0	0.0	0.0	0.2	0.0	2.2	4.6	453.6	449.0	89.8	0.9
0.0	0.0	0.2	0.0	0.0	0.0	0.0	0.2	0.0	5.4	6.8	630.3	623.5	89.8	3.1
0.0	0.0	0.0	0.0	0.0	0.0	0.0	0.2	0.0	8.8	4.6	705.3	700.7	88.8	2.3
0.0	0.0	0.3	0.0	0.0	0.0	0.0	0.0	0.2	8.4	11.2	831.4	820.2	87.8	1.5
0.0	0.0	0.0	0.0	0.0	0.0	0.0	0.0	0.0	2.4	22.8	352.9	330.1	97.5	1.7
0.0	0.0	0.0	0.0	0.0	0.0	0.0	0.0	0.0	2.9	6.1	130.6	124.5	81.1	7.8
0.0	0.8	0.0	0.0	0.0	0.0	0.0	0.4	0.0	1.2	1.2	262.5	261.3	94.1	4.5
0.0	0.4	0.0	0.0	0.0	0.0	0.0	0.0	0.0	11.0	13.8	341.4	327.6	91.6	4.7
0.0	0.0	0.8	0.0	0.0	0.0	0.0	0.0	0.0	8.1	19.1	415.7	396.6	94.0	3.3
0.0	0.0	0.0	0.0	0.0	0.0	0.0	0.0	0.0	2.4	7.3	429.1	421.8	95.1	3.0
0.0	0.0	0.0	0.0	0.0	0.0	0.0	0.0	0.0	16.7	7.3	669.5	662.2	89.9	4.8
0.0	0.4	1.2	0.0	0.0	0.0	0.0	0.8	0.0	6.1	34.1	1120.9	1086.8	90.6	4.5
0.0	0.0	0.0	0.0	0.0	0.0	0.0	0.8	0.0	0.0			162.0	96.3	1.9
0.0	0.0	0.0	0.0	0.0	0.0	0.0	0.0	0.0	0.0			257.0	95.7	2.3
0.0	0.0	0.0	0.0	0.0	0.0	0.0	0.0	0.0	0.0			200.0	96.0	2.8
0.0	0.0	0.0	0.0	0.0	0.0	0.0	0.0	0.0	0.0			176.0	97.7	2.0
0.0	0.0	0.0	0.0	0.0	0.0	0.4	0.0	0.0	0.0			84.0	95.2	2.4
0.0	0.0	0.0	0.0	0.0	0.0	0.0	0.0	0.0	0.0			245.0	98.8	0.7
0.0	0.0	0.0	0.0	0.4	0.0	0.0	0.0	0.0	0.0			210.0	95.2	2.1
0.0	0.0	0.0	0.0	0.4	0.0	0.0	0.0	0.4	0.0			264.0	97.0	2.1
0.0	0.0	0.0	0.0	0.0	0.0	0.0	0.0	0.0	0.0			180.0	97.8	1.3
0.0	0.0	0.0	0.0	0.0	0.0	0.0	0.0	0.0	0.0			317.0	95.3	3.3
0.0	0.0	0.0	0.0	0.0	0.0	0.0	0.0	0.0	0.0			366.0	70.2	1.3
												156.0		
												417.0		
												246.0		
												629.0		
												370.0		
												1387.0		
												1082.0		
												331.0		
												298.0		
												1250.0		
												48.0		
												306.0		
												521.0		
												315.0		
												140.0		
												474.0		
												1099.0		
												192.0		
												130.0		
												123.0		

trend towards miniaturization has occurred among the benthos above the size of nanobenthos (Thiel, 1975). This trend has not affected the larger macrofauna and megafauna in which, for some taxa, a reverse trend towards gigantism has occurred.

Therefore, as suggested by Hessler and Jumars (1974), the use of the term "meiobenthic taxa" is more appropriate in the context of deep-sea habitats. These taxa consist of metazoan animals traditionally regarded as "meiobenthos", and larger protozoans, almost exclusively foraminifera. Nearly all meiobenthic phyla have been found in deep-sea sediments. The numerical dominance of a few taxa, notably foraminifera, nematodes and copepods, however, is more pronounced in comparison with shallower areas.

3. ENVIRONMENTAL VARIABLES

3.1. Sediment Type

Bottom sediments in the northeast Atlantic originate from terrestrial or volcanic sources, from turbidity currents and related catastrophic phenomena, from bottom currents and from pelagic sedimentation. In general, coarse sediments (sands and gravels) of terrestrial or biogenic origin are restricted to the continental shelf and upper slope and become progressively finer (silty muds and muds) with increasing bathymetric depth and distance from land (Emery and Uchupi, 1984; Auffret, 1985; Lampitt *et al.*, 1986). However, in some areas, for example off the western approaches to the English Channel (Mart *et al.*, 1979; Auffret, 1985; Weston, 1985), the continental slope is dissected by active submarine canyons which channel coarse sediments onto the continental slope and rise. Coal, clinker (derived from steamships) and ice-rafted debris are also widespread in the northeast Atlantic (Kidd and Huggett, 1981) and provide a substratum for sessile organisms.

Over much of the northeast Atlantic the superficial sediments consist of pelagic calcareous oozes with a calcium carbonate content exceeding 30 or 50% and a mean particle size of <0.01 mm (Apostolescu *et al.*, 1978; Emery and Uchupi, 1984; Auffret, 1985; Lampitt *et al.*, 1986; Udintsev *et al.*, 1989–1990). Clay particles (<0.01 mm) generally make up 50% or more of the sediment (Udintsev *et al.*, 1989–1990). The carbonate compensation depth (CCD) in the northeast Atlantic exceeds 5000 m (Berger, 1975; Biscaye *et al.*, 1976) and hence areas of very fine grained red clay (from which the carbonate has been removed by dissolution) are restricted to the deepest basins, located west and northwest of the Cape Verde Islands.

Many continental slopes in the northeast Atlantic are characterized by widespread slope failure leading to turbidity currents, debris flows and sediment slides which have deposited sediments across huge areas of the adjacent rise and abyssal plain (Emery and Uchupi, 1984; Stein, 1991). Sediment slides and debris flows have been most fully described off the northwest African coast where some cover a great area (Embley, 1976; Jacobi, 1976; Kidd *et al.*, 1986; Masson *et al.*, 1994). Turbidity currents originating from slope failure may travel considerable distances across the ocean floor, depositing progressively finer sediments as they do so. Distal turbidites are typically fine grained and carbonate rich and have a higher total organic carbon content (1–3%) than sediments of pelagic origin (Wilson and Wallace, 1990; Stein, 1991). Such deposits are known to blanket the Madeira Abyssal Plain where the Quaternary succession consists of thick turbidite units separated by thin pelagic layers (Weaver and Kuijpers, 1983; Weaver *et al.*, 1986). They clearly influence the abyssal biota since areas of the Madeira Abyssal Plain underlain by turbiditic and pelagic sediments have quite different assemblages of animal traces (Huggett, 1986).

Turbidity currents must have devastated benthic life over large tracts of ocean floor and controlled the nature of recolonizing communities. However, such events have been rare along the northwest African margin with only one major turbidite being deposited, on average, every 30 000 years during the last 730 000 years (Masson *et al.*, 1994). Hence, all but the most recent turbidites are blanketed by pelagic sediments. Bottom current deposits (contourites) are well developed on the abyssal aprons, for example around the Rockall Plateau (Hill, 1987).

Detailed information on sediment grain size composition of the *areas* (Figure 1) reviewed in this chapter are only available for *area* 6 (Vanreusel *et al.*, 1992), *areas* 7–10 (Dinet and Vivier, 1977) and *area* 13 (Thiel, 1972a). Some *areas* less than 1000 m deep have coarser sediments, while most of the other *areas* are characterized by fine silty clays.

3.2. Oxygen

Unlike those in shallow water, deep-sea sediments are usually well oxygenated. Oxygen profiles for abyssal plain sediments in the northeast Atlantic are given by Sorensen and Wilson (1984), Wilson *et al.* (1985, 1986), Wallace *et al.* (1988), Rutgers van der Loeff and Lavaleye (1986) and Rutgers van der Loeff (1991) (*area* 11; Figure 2). Oxygen invariably penetrates to a depth of at least several decimetres and usually much deeper. At Discovery Station 10554 on the Madeira Abyssal Plain (*area* 23), oxygen levels were reduced to zero at a depth of less than 30 cm (Wilson *et al.*, 1985, 1986). This was attributed by Wilson *et al.* (1985) to the oxidation of organic matter in a turbidite layer. A relatively shallow depth (25–30 cm) of oxygen penetration was also observed by Rutgers van der Loeff (1991) (*area* 11) in a core from the Porcupine Abyssal Plain. However, even at such localities, meiobenthos living in the upper 5 cm of sediment in the northeast Atlantic should not normally experience a lack or shortage of oxygen.

3.3. Food Supply

Food availability often exerts a decisive influence on faunal densities. Except around hydrothermal vents, the food that sustains benthic communities originates mainly from surface water primary production

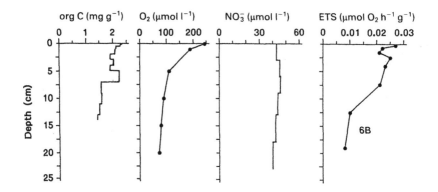

Figure 2 Vertical profiles in the sediment of organic carbon, O_2, NO_3^- and ETS activity at the DORA site (*area* 11; sample 6B) (after Rutgers van der Loeff and Lavaleye, 1986).

with some input from terrestrial sources. The most recent map of surface production in the Atlantic Ocean (the "Dahlem Map"; Berger, 1989: Figure 3) shows decreasing values from the northern margin of our region to about 35°N. Further south surface productivity increases in association with the coastal upwelling off northwest Africa. Export production (net particulate flux of carbon across the thermocline) shows a similar pattern (Berger *et al.*, 1988).

Only scattered data are available regarding the distribution of organic matter in northeast Atlantic sediments. Emery and Uchupi (1984) present a generalized map, based on literature sources, showing the organic carbon content of dried sediment for the whole of the Atlantic. In the northeastern part, values tend to decrease with increasing bathymetric depth. Seaward of the continental margin the organic carbon content generally lies between 0.25 and 0.50% with a southwest to northeast trending zone of higher values (0.5–1.0%) between about 25–35°N and 20–40°W. According to Sibuet (1984), the organic carbon content of sediments from deep basins in the Atlantic Ocean varies from 0.17 to 0.85%. Auffret (1985) gives values ranging from 0.2 to 0.7% at depths between 2000 and 4800 m in the Bay of Biscay. CHN analyses of Porcupine Seabight sediments after acid digestion yield organic carbon values of 0.5% dry weight with no distinct trend with depth (Rice *et al.*, 1991). Higher values (0.5–4.0%) have been recorded in late Pliocene and Pleistocene sediments from ODP (Ocean Drilling Project) site 658, situated under the upwelling area off Cap Blanc (Stein, 1991). As noted above, higher values are also associated with turbiditic sediments.

A more meaningful indication of food availability is provided by

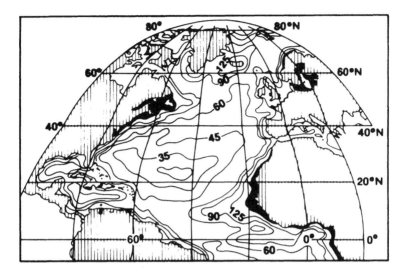

Figure 3 Surface primary production (g C m^{-2} year^{-1}) map of the North Atlantic (after Berger, 1989). Black areas near the coast indicate values >150 g C m^{-2} year^{-1}.

chloroplastic pigment equivalents (CPE) which reflect the amount of organic matter originating from primary production. Detailed information of CPE values from particular areas off northwest Africa and Europe are given by Thiel (1978, 1983), Pfannkuche *et al.* (1983) (*area* 18), Pfannkuche (1985) (*areas* 1–4). Off northwest Africa, particularly high CPE values in the sediments are associated with intense upwelling; although the patterns are complicated by local currents which influence the deposition of organic matter (Thiel, 1978, 1982; Pfannkuche *et al.*, 1983).

Most chloroplastic pigments probably originate from phytodetritus which represents an important mechanism for rapidly delivering organic matter originating from surface production to the ocean floor (Gooday and Turley, 1990). Large amounts of this material have been observed on the ocean floor in the Porcupine Seabight (*areas* 1–3), on the Porcupine Abyssal Plain (*area* 4) (Billett *et al.*, 1983; Lampitt, 1985; Rice *et al.*, 1986, 1991), the BIOTRANS area (*area* 12) (Thiel *et al.*, 1989–1990), the Rockall Trough and the Bay of Biscay (Gooday and Turley, 1990). In these northern areas, phytodetritus deposition seems to be associated with deep (>500 m) winter mixing (Robinson *et al.*, 1979), in addition to relatively high surface productivity. However, recent evidence indicates that phytodetritus is also deposited further south in areas where the mixed layer is <150 m in depth. Christiansen and Thiel (1992) observed abundant flocculent material in depressions and around mounds on the Madeira Abyssal Plain (34°N) and small amounts were present in multiple corer samples taken at the IOSDL southern site (31°N). Phytodetritus has also been photographed on the seafloor even further south, at 19°S (McCave, 1991). Pfannkuche (1993b) provided data on seasonal variation in CPE.

3.4. Deep Bottom Water Masses

In the North Atlantic, and other oceanic regions, some modern foraminiferal species assemblages show good local correlations with the water masses which bathe the ocean floor (Streeter, 1973; Schnitker, 1980; Weston and Murray, 1984). A similar correlation has not been established for meiobenthic taxa other than foraminifera, but this may reflect a general lack of species level distributional data for the metazoan meiobenthos. The only other examples of a correlation between water mass and animal distributions in the deep sea are provided by Patterson et al. (1982) for five North Atlantic species of the ophiuroid genus Ophiocten and by Tyler and Zibrowius (1992) for suspension feeding echinoderms southwest of Iceland. In the latter case the associations clearly are linked to the current activity rather than to water masses as such.

In the eastern North Atlantic, the deep waters consist mainly of North Atlantic Deep Water (NADW), a composite water mass with an upper, low salinity (<34.94‰) layer, a middle layer of Norwegian Overflow Water (salinity 34.96–35.03‰), and a lower layer which consists of Norwegian Overflow Water diluted by mixing with Denmark Strait Overflow Water (Worthington, 1976; Weston and Murray, 1984; Gage and Tyler, 1991). The upper NADW layer, which unlike the lower layer originates from the Labrador Sea, is often considered as a separate entity, the North East Atlantic Deep Water (NEADW). At depths less than about 2000 m, the dominant water mass in the northeast Atlantic is Mediterranean Water which flows through the Straits of Gibraltar, around the European continental margin, through the Rockall Trough and into the Norwegian Sea, becoming progressively diluted as it proceeds northwards. Mediterranean Water is characterized by its high density and salinity and by its variable physical properties (Weston, 1985).

Below about 4500 m, the deep basins off the African coast are occupied by a distinct water mass which originates from the Antarctic. This Antarctic Bottom Water (AABW) is restricted to areas south of about 36°N, where its passage is blocked by ridges associated with the Azores

Fracture Zone. AABW is characterized by low temperatures and salinities and by relatively high dissolved oxygen and nutrient concentrations.

3.5. Near-bottom Currents

In the northwest Atlantic, unusually high current speeds (20–25 cm s^{-1}) occur during periodic "benthic storms" which originate when eddy kinetic energy derived from near-surface eddies impinges on the ocean floor (Richardson *et al.*, 1981; Hollister and McCave, 1984; Gross *et al.*, 1988). These events erode and redistribute bottom sediments and strongly influence the character of meiobenthic communities (e.g. Thistle *et al.*, 1985). Recently there has been some evidence for benthic storms in the northeast Atlantic; currents capable of eroding and depositing sediments are associated with the sediment drifts which are extensively developed to the west of the Rockall Trough and elsewhere (Emery and Uchupi, 1984). At BIOTRANS (*area* 12), the deep mean flow is 3–5 cm s^{-1} but benthic storms occur with durations between 3 and 27 days with maximum velocities of 27 cm s^{-1} (Klein and Mittelstaedt, 1992). Maximum current velocities here may approach 40 cm s^{-1} (Dickson and Kidd, 1987) and must have an impact on the benthic fauna similar to that described by Thistle *et al.* (1985). Away from such areas, the near-bottom currents are fairly gentle and probably have little impact on the meiobenthos beyond the transportation of nutrients and larvae. A residual boundary current flows northward along the continental margin of northwestern Europe and has superimposed on it diurnal or semidiurnal tidal currents (Dickson *et al.*, 1986). Rice *et al.* (1991) summarize information on currents flowing near the floor of the Porcupine Seabight. At a depth of 4025 m, near the mouth of the Seabight, current speeds varied semidiurnally and reached a maximum on the western side of the Porcupine Bank (Norris and McDonald, 1986). Measurements made during the BIOGAS programme indicate current speeds of generally less than 10 cm s^{-1} at abyssal depths (Vangriesheim, 1985). Current speeds in more central oceanic regions are presumably lower than along the continental margins.

4. HORIZONTAL SPATIAL PATTERNS

When comparing quantitative meiobenthos data at an ocean-wide scale, one must take account of the variability in population density on different scales: from small-scale (within one corer) and medium-scale (between corers from nearby locations) to large-scale (ocean-wide scale; e.g. Thiel, 1983). In this section both bathymetric and latitudinal trends in meioben-thic density, and composition at different taxonomic levels, will be discussed. The lack of standardization in sampling (Bett *et al.*, 1994) and in processing techniques (e.g. different mesh sizes) complicates the comparison of data from different meiobenthic investigations. Data evaluation is therefore a challenge and observations of differences in faunal densities and changes in faunal composition have to be made with some caution.

4.1. Bathymetric Trends

4.1.1. *Faunal Densities*

There is a general tendency (see reviews Thiel, 1983 and Tietjen, 1992) for metazoan meiobenthic densities to decrease with increasing bathymet-ric depth within limited geographical areas (i.e. under constant climatolo-gical and ecophysiological life conditions). This tendency does not hold for the total northeast Atlantic ocean, as illustrated in Figure 4, in which the mean densities per *area* are plotted as a function of mean water depth. In general the lowest densities are recorded at abyssal depths where mean densities for each *area* are never higher than 600 individuals (ind)/10 cm^2, and the highest densities are found in the shallowest *areas* (< 1000 m: mean per *area* up to 2000 ind/10 cm^2), but occasionally densities of less than 500 ind/10 cm^2 are recorded on the upper slope and in the shelf break area.

Densities do not always decrease in relation to depth within smaller geographical regions (Figure 5) which may show different and sometimes conflicting patterns of density gradients: for example off Mauretania, off Morocco, off Gibraltar, in the Bay of Biscay, and on the Iberian, Madeiran and Porcupine abyssal plains.

In the southern part of the northeast Atlantic, off the coast of Mauretania, densities were relatively high at the deepest sites (*areas* 27, 28 and 29: means 500–1000 ind/10 cm^2), higher than on the upper slope and in the shelf break area (*area* 26: 500 ind/10 cm^2) (Thiel, 1975; Dinet, unpublished). The lack of any depth-associated gradient in this area, and the relatively high densities on the lower part of the slope, are due to increased primary production resulting from intense upwelling activity in the water column, and providing high food input to these deeper *areas*. Low densities on the upper slope can be explained by hydrodynamic processes, which cause erosion and prevent organic matter from reaching

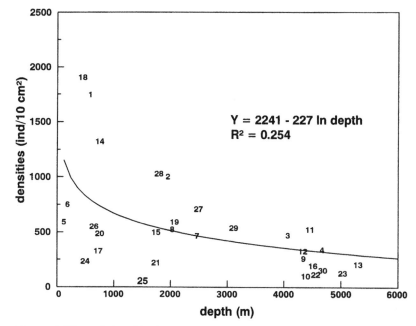

Figure 4 The relationship between mean metazoan density and mean water depth of each *area* (30 *areas* as defined in Table 1) in the northeast Atlantic.

the seafloor (Thiel, 1983). At the oligotrophic abyssal site (*area* 30) situated on the same latitude, and intensively studied by the EUMELI project (Dinet, unpublished), much lower densities are found in comparison to the slope (200 ind/10 cm^2).

Along the NW African coast, off Morocco, densities are in general low. Some higher values were recorded for the upper slope (*area* 20: 500 ind/10 cm^2), but deeper down the meiobenthos is very poorly represented (*areas* 21 and 22: <250 ind/10 cm^2). From 1000 m depth, the densities are similar to those of *area* 23 on the Madeira Abyssal Plain, which has been intensively investigated by IOSDL (Gooday, unpublished).

Two relatively shallow (<1750 m) sites located on the Great Meteor seamount, near the western border of the Madeira Abyssal Plain, also have low meiobenthic densities (*areas* 24 and 25: <250 ind/10 cm^2). Beside their central oceanic location, a higher predation pressure and current activity are possible causes of the low abundances (Thiel, 1983).

Off Gibraltar, Thiel (1975) found much lower densities on the upper slope (*area* 17: 300 ind/10 cm^2) than Pfannkuche *et al.* (1983) on a nearby transect off the African coast (*areas* 18 and 19: 600–2000 ind/10 cm^2). In the latter study, meiobenthic abundances showed a significant correlation with CPE concentrations, which decreased with water depth. Thiel (1983)

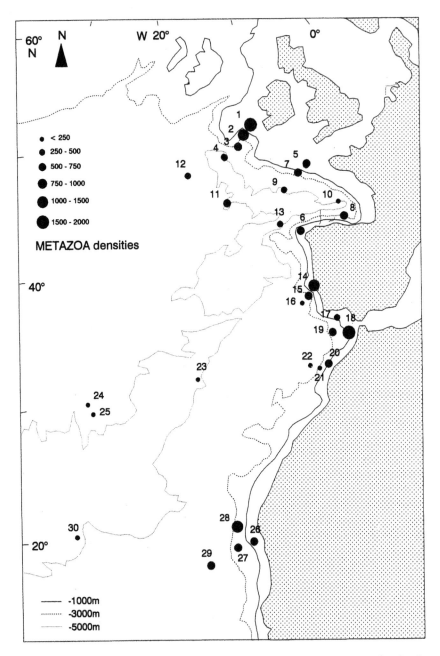

Figure 5 Map of the northeast Atlantic showing the mean metazoan density for each *area* (as defined in Table 1) (ind/10 cm^2).

explained the lower densities found during the *Meteor 21* cruise by the outflow of warm Mediterranean water, which accelerates along the Atlantic slope due to its higher density, preventing the sedimentation of detritus.

On a transect near the Portuguese coast (Thiel, 1975), meiobenthic densities decrease from 1400 to 100 ind/10 cm^2 in relation to bathymetric depth (*areas* 14, 15 and 16).

In the Bay of Biscay, densities decrease more-or-less with depth, although the upper slope and the shelf break sites are characterized by relatively low numbers (*areas* 5, 6, 7 and 8: 450–750 ind/10 cm^2) (Vanreusel *et al.*, 1992 and unpublished; Dinet and Vivier, 1977; Thiel, 1975). Vanreusel *et al.* (1992) explained these low numbers in terms of the low chlorophyll-a content and/or the coarseness of the sediments. In adjacent abyssal regions (*areas* 9 and 10), Dinet and Vivier (1977) observed a larger variability, which they associated with sedimentological heterogeneity and hydrodynamical disturbance. In these areas, the sediments seem to be unusually coarse and unstable, due to the presence of submarine canyons which elevate current activity (Dinet and Vivier, 1977).

Finally, an obvious decrease in the meiobenthic densities, related to diminishing CPE content with depth, is found in the Porcupine Seabight (from 2000 to 450 ind/10 cm^2) (Pfannkuche, 1985).

Tietjen (1992) described the distribution of metazoan meiofauna of the global Atlantic Ocean by the following logarithmic function of depth: $Y = 6238 - 710.5 \ln$ depth ($R = -0.89$). As shown in Figure 4 this relationship is not found for data of the northeast Atlantic. This function is only approached (Figure 6; $Y = 5837 - 666 \ln$ depth, $R^2 = 0896$) when mainly shallow *areas* which were characterized by much lower densities than expected (as discussed above) because of hydrodynamical processes (*areas* 17, 20, 21, 26), their off-shore location (*areas* 24, 25) or sedimentological characteristics (*areas* 5, 6) are omitted. Tietjen (1992) also considered only four sites shallower than 1000 m in describing this strong functional relationship.

Nematode densities range from 5 ind/10 cm^2 in the Gulf of Biscay (Dinet and Vivier, 1977) to 2382 ind/10 cm^2 in the Porcupine Seabight (Pfannkuche, 1985). Nematodes represent 80–99% of the total metazoan abundances and are therefore mainly responsible for the depth-related tendencies found for the metazoans (see Figures 4, 6). This is illustrated by Figure 7 which shows a weak negative relationship between nematode densities and water depth. However, excluding the two most shallow *areas* 5 and 6 at the shelf break of the Bay of Biscay strengthens the relationship ($R^2 = 0.870$, $Y = 5400 - 613 \ln$ depth). The exclusion of these two *areas* can be justified by their exceptional coarse sediment and their shallow location at the shelf break.

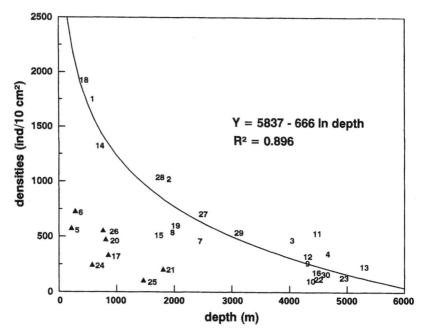

Figure 6 The relationship between mean metazoan densities and mean water depth for each *area* in the northeast Atlantic. The *areas* 5, 6, 17, 20, 21, 24, 25, 26 are not included in the regression.

Considering smaller geographic regions, nematode densities also decrease with water depth in the Porcupine Seabight, near the coast of Mauretania and to a lesser extent in the Bay of Biscay. In the latter case, densities are low (< 1000 ind/10 cm^2) at the shelf break and on the upper slope (Figure 8).

Copepods represent the second major metazoan meiobenthic taxon after the nematodes. Their densities vary from 166 individuals/10 cm^2, the highest abundance recorded by Vanreusel and Vincx (unpublished) in the Bay of Biscay, to zero as observed in a few cases by Dinet (unpublished) in some samples from the EUMELI site (*areas* 28, 29, 30). While the general distribution pattern exhibits a decrease with increasing depths (Figures 9 and 10), high densities of copepods (50–60 ind/10 cm^2), are recorded in the northwestern abyssal zone. However, low abundances are encountered in all zones and at all depths (Figure 9), probably because of the patchy distribution of copepods (Dinet *et al.*, 1985).

The decrease in copepod densities with water depth within smaller geographic areas is pronounced in the Bay of Biscay (Figure 10). In the Porcupine area densities are higher on the lower (*area* 2) than on the

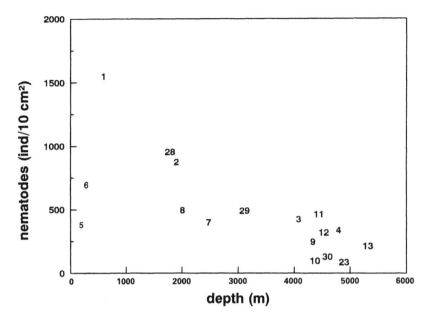

Figure 7 The relationship between mean nematode density and mean water depth for each *area* (cf. Table 1) in the northeast Atlantic.

upper slope (*area* 1). In the upwelling near Mauretania, the copepods show low densities indicating another response than that of the nematodes, as indicated by their overall low densities in this area.

The densities of all other numerically important meiobenthic taxa are given in Table 2.

Polychaetes were found at all sites where a detailed meiobenthos investigation was carried out. Densities range from 1 to 45 ind/10 cm^2 and a decreasing trend in numbers is apparent with increasing water depth (below 2000 m water depth, densities are always lower than 20 ind/ 10 cm^2). Turbellaria are recorded from the southern *areas* 28, 29 and 30 with densities ranging from 1 to 45 ind/10 cm^2 (Dinet, unpublished) and from the Bay of Biscay (Vanreusel *et al.*, 1992) (*areas* 5 and 6) with densities up to 188 ind/10 cm^2. The soft bodied nature of the turbellaria presents practical problems so that a significant bias in the recorded density values is likely between studies. Oligochaetes and hydrozoans were only recorded from the shelf break area in the Bay of Biscay (*area* 5). Ostracods are common in deep-sea sediments and show a decreasing trend with increasing water depth. Gastrotrichs are only relatively abundant in the Porcupine Seabight and in the lower stations of the Bay of Biscay. Kinorhynchs are frequently found, albeit in low numbers.

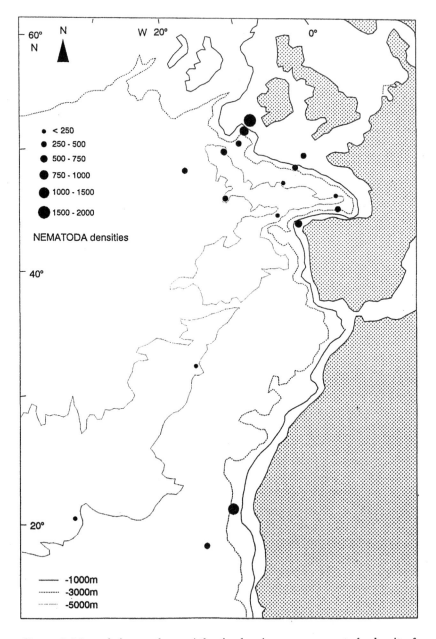

Figure 8 Map of the northeast Atlantic showing mean nematode density for each *area* (as defined in Table 1) (ind/10 cm^2).

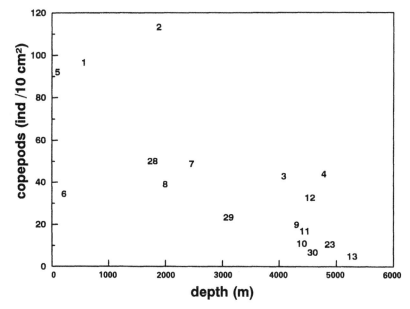

Figure 9 The relationship between mean copepod density and mean water depth for each *area* (cf. Table 1) in the northeast Atlantic.

Tardigrades, amphipods, loricifers, halacarids, isopods, bivalves and tanaids occur sporadically and always in very low numbers.

Protozoans, in particular foraminifera, are very successful in the deep sea. They often constitute at least 50% of deep-sea meiobenthic assemblages (Gooday, 1986a; Gooday and Lambshead, 1989; Pfann-kuche, 1993b). Abundance data for specimens stained with Rose Bengal are available for seven *areas* in the northeast Atlantic (Table 2). In relation to water depth, the highest densities are found at the shallowest sites (1000–2000 ind/10 cm^2), while the lowest densities are found in abyssal areas, although relatively high numbers have been recorded from 4500 m depth (1000 ind/10 cm^2) (Figure 11). Data on the local distribution of foraminifera are only available from the Porcupine Seabight (Figure 12). At the deepest sites (<4500 m) densities are somewhat lower (500–1000 ind/10 cm^2) than on the slope (1000–3000 ind/10 cm^2).

4.1.2. *Faunal Composition*

To compare the composition of the metazoan meiobenthos from different deep-sea areas in the northeast Atlantic, we have concentrated on the most abundant taxa, for example nematodes, harpacticoid copepods

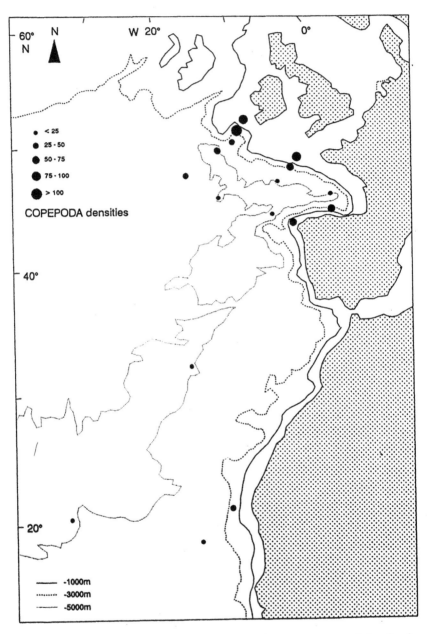

Figure 10 Map of the northeast Atlantic showing mean copepod density for each *area* (ind/10 cm^2).

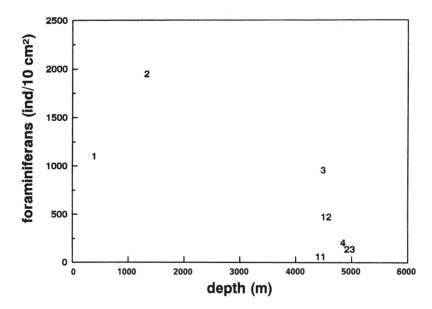

Figure 11 The relationship between mean foraminifera density and mean water depth for each *area* (cf. Table 1) in the northeast Atlantic.

(including nauplii) and polychaetes, and combined all the other taxa, which appear only occasionally, into one category ("others"). Nematodes clearly dominate the metazoan meiobenthic community: on average, more than 80% (maximum 99%) of the metazoan meiobenthos are nematodes. Copepods and nauplii together constitute the second most important group, representing up to 10% of the metazoans. Polychaetes and other taxa (mainly ostracods, kinorhynchs, turbellaria, gastropods and bivalves) appear in almost all water depths, but are most abundant on the shelf and the continental slope.

In contrast to the comparatively obvious influence of depth-related factors in determining the densities of meiobenthic taxa within the study area, global trends in the percentage composition of the fauna are not so apparent. Nevertheless, the evaluation of all meiobenthic data available from the northeast Atlantic shows changes in taxonomic composition with increasing water depth. Box plots illustrate the trends for different taxa (Figure 13 a–d). In these figures all 17 *areas* from which composition data are available are arranged in 11 depth classes.

The relative abundance of nematodes generally increases with depth, and correspondingly the percentage composition of all other taxa decreases with depth. ANOVA statistics (based on arcsine transformed percentage data) confirm significant differences ($p < 0.05$) between mean

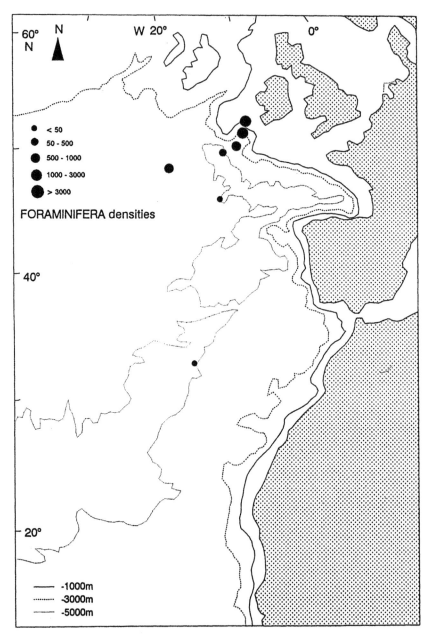

Figure 12 Map of the northeast Atlantic showing mean foraminifera density for each *area* (ind/10 cm^2).

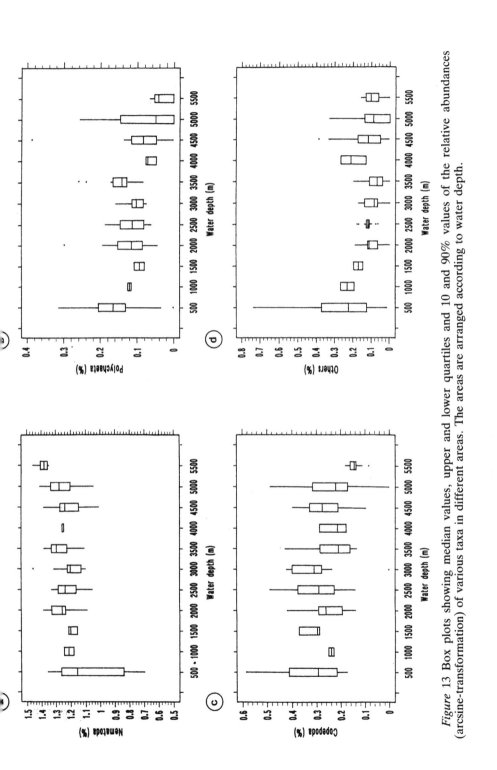

Figure 13 Box plots showing median values, upper and lower quartiles and 10 and 90% values of the relative abundances (arcsine-transformation) of various taxa in different areas. The areas are arranged according to water depth.

nematode percentages on the shelf (75.1% ± 19.2%) and at the deepest stations below 5000 m (96.5% ± 1.2%). Significant decreases ($p < 0.05$) in relative abundances, from shallow to deep water, were found for polychaetes (3.2% ± 2.7% to 0.2% ± 0.2%), copepods (11.2% ± 8.5% to 2.2% ± 0.7%), and "others" (10.6% ± 13.6% to 1.1% ± 0.8%). The highest coefficients of variation, representing variability within *areas*, occurred at shallow stations (128.3% for "others"), and the lowest variation was found for nematodes from the deepest stations. Also, within smaller geographical areas, most depth transects investigated in the northeast Atlantic are characterized by an increasing relative abundance of nematodes with depth (Thiel, 1972b; Dinet and Vivier, 1977; Pfannkuche, 1985).

Carney *et al.* (1983) described three different types of depth-associated factors controlling the zonation of the benthos over continental margins. The first includes gradients of physiologically important parameters such as temperature, salinity, oxygen concentration and hydrostatic pressure. The greatest vertical changes in these factors occur within the top 1000 m in the oceans. Therefore, their effect on zonation should be largely restricted to those depths.

The second group includes factors that change qualitatively with bathymetric depth, such as the transition from coarser to finer sediments. Sediment granulometry plays an important role in faunal zonation on continental shelves (Gray, 1974, and references cited therein). In most deep-sea areas, however, granulometric parameters are more uniform and do not appear to affect the large-scale distribution of the meiobenthos (Thiel, 1983).

The third type of factor controlling faunal zonation consists of those resources that change in availability with depth. A decrease in available food (quantitative and qualitative input of detritus to the benthos) or an increase in available space (generally lower abundances in deeper regions) with water depth are examples of the third type.

While availability of interstitial space probably would not affect smaller benthic size groups (such as the meiobenthos) significantly, the decisive influence of the quantitative and qualitative food supply on benthos is clear.

For the smallest benthic organisms such as protozoans (mainly foraminifera) it is clear that they are well adapted, by their short generation times and opportunistic feeding habits, to respond quickly to the variable food input at abyssal depths (Linke, 1992; Pfannkuche, 1993b; Altenbach, 1993). For comparable reasons small nematode species possibly might be well able to cope with the temporary availability of food. Besides, since it is generally accepted that a major part of the carbon flow in the deep sea is channelled through the bacteria (Lochte, 1992), those organisms that feed on bacteria will be favoured. These organisms include most of the dominant nematode species.

4.2. Latitudinal Trends

4.2.1. *Faunal Densities*

Large-scale latitudinal trends are also important in characterizing the meiobenthos. In Figure 14 all *areas* are arranged according to their latitude (from south (S) to north (N)), within three different depth zones; the northern abyssal sites (> 3000 m) are ordered from west (W) to east (E). These arrangements illustrate weak geographic gradients of the densities of the metazoan meiobenthos which, to some extent, are related to the surface primary productivity.

Among the deepest stations (> 3000 m), densities were higher in the northwest (*areas* 11, 12, 4 and 3—the BIOTRANS *area*, the DORA *area* and the Porcupine Abyssal Plain) with 300–500 ind/10 cm^2, than in the south (*areas* 30, 22, 23 and 16—the EUMELI *area*, off the NW African coast, the Madeira Abyssal Plain and off the Portuguese coast) with less than 250 ind/10 cm^2. An exception to this increasing south to north trend is the most southerly location, *area* 29 in the upwelling zone near the coast of Mauretania, which is characterized by high abundances. According to the surface primary productivity map for the Atlantic Ocean (Figure 3), the upwelling area off Mauretania has a primary productivity of more than 150 g C m^{-2} year^{-1}, the northern sites are situated between the 60 and 125 isolines, while at the other southern sites surface productivity varies between 45 and 90 g C m^{-2} year^{-1}.

From the northwestern *areas* 3, 4, 11 and 12, towards the Iberian Sea and the Gulf of Biscay in the northeast (*areas* 9, 10 and 13), densities also decrease, although not significantly. Sedimentological and hydrodynamical processes may explain the low abundances in the east (see above). The observed trend, however, is opposite to the increase in surface productivity. Furthermore, at the Porcupine Abyssal Plain (*area* 4), the DORA (*area* 11), and the BIOTRANS sites (*area* 12), large amounts of phytodetritus reach the seafloor within a relatively short period after the spring plankton bloom (Thiel *et al.*, 1989–1990; Rice *et al.*, 1991). This is likely to represent an important food input for the meiobenthos (Billett *et al.*, 1983; Gooday, 1988).

A similar south to north trend is found for the 1000–3000 m *areas* (Figure 14): densities were lowest on the Great Meteor seamount and off Morocco (*areas* 21 and 25), where the surface productivity varies between 45 and 90 g C m^{-2} year^{-1}. They were somewhat higher off Gibraltar, the

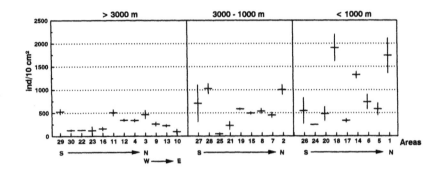

Figure 14 The mean metazoan densities (± SE) for each *area* in the northeast Atlantic. The areas are grouped into depth zones and arranged by latitude (N, North; S, South) and/or longitude (W, West; E, East).

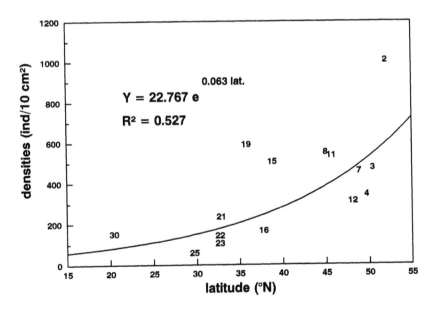

Figure 15 The relationship between mean metazoan density and mean latitude for all *areas* in the northeast Atlantic deeper than 1000 m, except *areas* 27, 28 and 29 in the upwelling near Mauretania, and the northeastern *areas* 13, 9 and 10 off the Bay of Biscay.

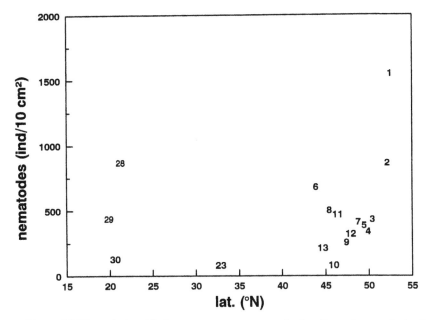

Figure 16 The relationship between mean nematode density and mean latitude of each *area* in the northeast Atlantic.

Portuguese coast and the Bay of Biscay (*areas* 7, 8, 15 and 19), situated between the isolines of 60 and 125 g C m^{-2} year^{-1} and highest in the Porcupine Seabight between the isolines of 90 and 125 g C m^{-2} year^{-1} (*area* 2). Off Mauretania (*areas* 27 and 28) meiobenthos was again exceptionally abundant due to upwelling (productivity > 150 g C m^{-2} year^{-1}).

On the upper slope and around the shelf break (<1000 m) the same geographic tendency is no longer evident. Meiobenthos is most abundant in the Porcupine Seabight, off the North African coast and to a lesser extent near the Portuguese coast (*areas* 1, 14 and 18). Densities are lowest on the Great Meteor seamount (*area* 24) and the remaining sites having intermediate abundances (*areas* 26, 20, 17, 6 and 5).

A plot of the mean densities of all *areas* deeper than 1000 m (Figure 15), with the exception of those in the southern upwelling zone near Mauretania and the northwestern sites near the Bay of Biscay, as a function of their latitude, summarizes these weak south–north gradients.

Nematodes show a similar relationship with latitude. Again densities are much higher in the northwest in comparison to the south, with the exception of two upwelling sites near Mauretania (*areas* 28 and 29) (Figure 16). The same pattern applies to mean copepod densities,

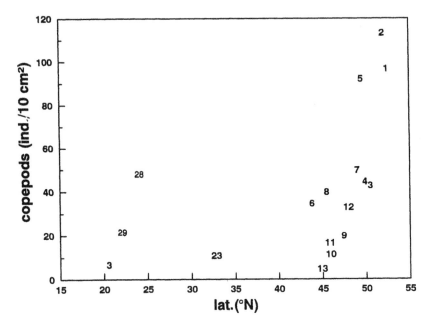

Figure 17 The relationship between mean copepod density and mean latitude for each *area* in the northeast Atlantic.

although abundances are much lower (20 ind/10 cm^2) over the total gradient (Figure 17). These latitudinal trends are presumably related to the higher productivity of the northern areas.

Figure 18 illustrates latitudinal variation for the foraminifera, although note that relatively few data are available from the south. Densities reach 1000–2000 ind/10 cm^2 at bathyal depths in the Porcupine Seabight (*areas* 1 and 2) and are almost as high (956 ind/10 cm^2) on the Porcupine Abyssal Plain near the mouth of the Porcupine Seabight (*area* 3). Values are variable but lower in *area* 4 on the Porcupine Abyssal Plain (102–283 ind/10 cm^2) (Gooday and Lambshead, 1989, not included in Table 2), and at the BIOTRANS site (*area* 12; 196–864 ind/10 cm^2). The lowest values (<50/10 cm^2) were recorded from *area* 11 (DORA) and *area* 23 (Madeira Abyssal Plain).

4.2.2. *Faunal Composition*

Quantitative and qualitative aspects of food availability are important in controlling the composition of the benthos, enabling a taxonomically richer community to develop in areas with a higher input. Changes in

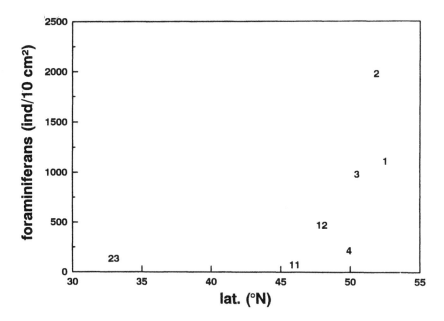

Figure 18 The relationship between mean foraminifera density and mean latitude of each *area* in the northeast Atlantic.

taxonomic composition attributable to this factor should be detectable in the northeast Atlantic. The greater depth of winter mixing in higher latitudes facilitates higher productivity in those regions, whereas at the southern edge of our area of interest, there is increased surface productivity associated with the coastal upwelling off NW Africa. In both cases this may affect the composition of the meiobenthos.

Figure 19a–d shows median percentages of different taxa with increasing latitude (note that some *areas* have been combined according to their latitude, namely *areas* 8 and 11 at 46°N, *areas* 7 and 13 at 49°N, and *areas* 3 and 4 at 51°N).

The percentage of copepods and particularly polychaetes is low around 45°N and significantly (ANOVA $p < 0.05$) higher further to the south and to the north. The relative abundance of nematodes shows a tendency for higher values at lower latitudes and lower values at higher latitudes. The reverse is true for the group of other taxa, where slightly increased percentages in higher latitudes were observed. However, ANOVA indicated no significant differences in the relative abundances of nematodes or "others" with latitude. The results for 50°N represent shallow water stations on the shelf break area in the Bay of Biscay exclusively

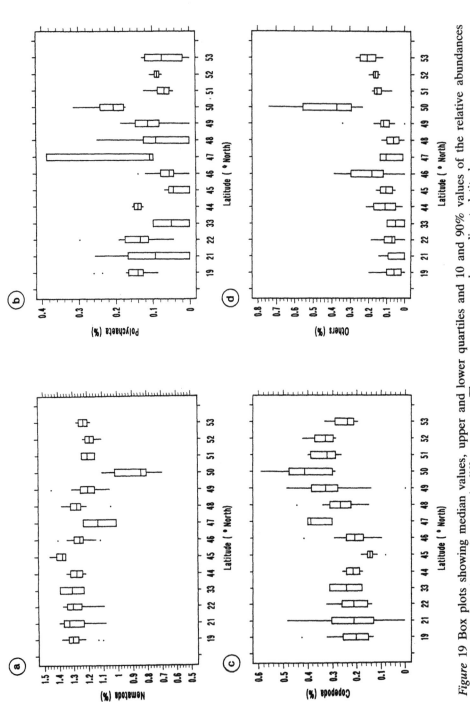

Figure 19 Box plots showing median values, upper and lower quartiles and 10 and 90% values of the relative abundances (arcsine-transformation) of various taxa in different areas. The areas are arranged according to latitude.

(*area* 5) (Vanreusel and Vincx, unpublished) and should be regarded separately.

4.3. Multivariate Analysis

The various depth and latitude (primary production) related densities trends for different taxa in the northeast Atlantic, as illustrated above, are confirmed overall by a multivariate analysis. This analysis has been performed for the 17 *areas* from which taxon composition data are available: *areas* 1–13, *area* 23, *areas* 28–30; the replicate density data (Table 2) were summarized to geometric mean values before analysis (Figures 20 and 21).

The abundance of each taxon examined shows a significant negative rank correlation with depth (nematodes, $p < 0.002$; harpacticoids, $p < 0.01$; polychaetes, $p < 0.002$; combined other taxa, $p < 0.002$). When variations in the abundance of these taxa are analysed simultaneously in a multivariate ordination (non-metric multidimensional scaling, using root–root transformation of abundances and the Bray–Curtis similarity measure) the strongest trend again is related to depth; rank correlation of the multidimensional scaling (MDS) x-coordinate with depth is highly significant ($p < 0.002$).

Water depth is itself unlikely to be the causal factor, but rather a suite of other factors linked to water depth is more likely to have produced the observed distributions. Over the large geographical area covered in this analysis the supply of organic matter to the seafloor is likely to be a major factor controlling meiobenthic abundances. In addition to water depth, surface primary production and distance off-shore also play important roles in determining the supply of organic matter to the seafloor. Surface primary production, as estimated from the "Dahlem map" (Berger, 1989), exhibits a significant ($p < 0.05$) rank correlation with the y-coordinate of the MDS ordination. The distance of sampling stations from the nearest continental land mass is significantly correlated with both the x- and y-coordinates of the ordination. In the present study area, however, distance off-shore can be expected to correlate with both depth and surface primary production.

Multivariate ordination (MDS, arcsine transformation, Bray–Curtis similarity) of the percentage composition data yields few obvious trends (Figures 22 and 23). The distribution of sampling stations in ordination space shows no correlation with depth, surface primary production, or distance off-shore.

Although the multivariate analysis of percentage composition did not reveal any general trend in the data, there is nevertheless a quite

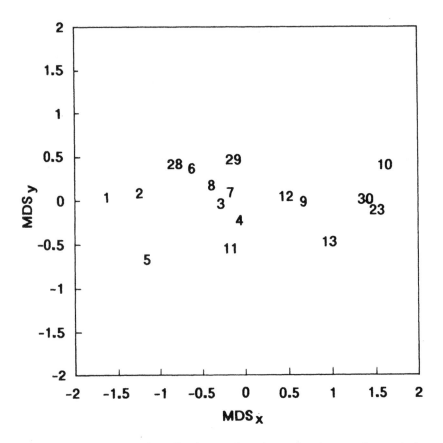

Figure 20 Ordination (MDS) of *areas* based on abundances of nematodes, copepods, polychaetes and combined other taxa, using root–root transformation.

appreciable variation in the composition of the meiobenthos among sampling stations. This variation is clearly reflected in the relative dominance of the fauna by nematodes or, conversely, the relative importance of all other taxonomic groups. Stations located to the left on the ordination have comparatively low nematode dominance and relatively high proportions of other taxonomic groups (i.e. harpacticoids and polychaetes), as examplified by *area* 5 (nematodes 62.4%, copepods 16.4%, polychaetes 4.9%, other taxa 16.3%). Stations to the right of the ordination are strongly dominated by nematodes with relatively low contributions from other taxa, *area* 13 for example (nematodes 96.8%, copepods 2.0%, polychaetes 0.2%, other taxa 1.0%). The most obvious feature of the ordination is the clear distinction of *area* 5 from all other

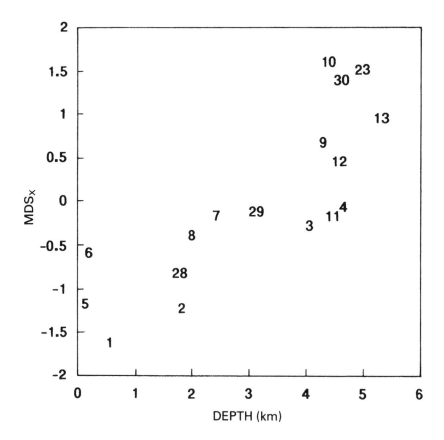

Figure 21 Plot of MDS (from Figure 20) against water depth.

stations. This may be explained by reference to sediment characteristics. *Area* 5 has medium to coarse sandy sediments while the other stations have rather finer sediments. Sediment grain size is generally regarded as one of the most important factors controlling meiobenthic composition (Coull, 1988). Coarser sediments generally support a more diverse fauna with reduced nematode dominance, as at *area* 5. *Area* 5 is also, by far, the shallowest of all the deep-sea *areas* considered.

With the exception of the distinctiveness of *area* 5 there appear to be no other obvious trends in the data. Presumably some of the remaining variation is attributable to local environmental conditions and some to biases in the data (e.g. limited replication, variations in sample collection and processing techniques).

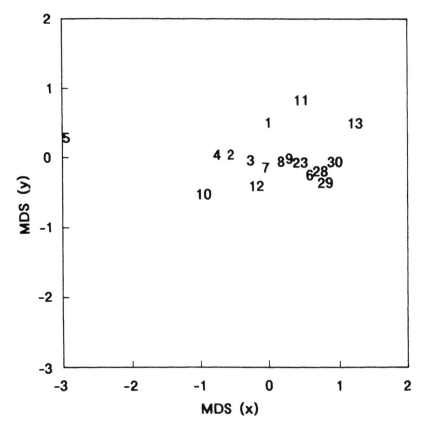

Figure 22 Ordination (MDS) of areas based on percentage composition of nematodes, copepods, polychaetes and combined other taxa, using arcsine transformation.

4.4. Comparison with Other Regions

Data on meiobenthic distributions are available for a number of geographical regions other than the northeast Atlantic: the northwest Atlantic (Wigley and McIntyre, 1964; Tietjen, 1971; Coull *et al.*, 1977; Sibuet *et al.*, 1984; Thistle *et al.*, 1985), the Pacific Ocean (Thiel, 1975; Shirayama, 1984a, b; Snider *et al.*, 1984), the Mediterranean and Red Sea (Dinet *et al.*, 1973; Thiel, 1975; Dinet, 1976; Vivier, 1978; Thiel, 1979a, 1983; Thiel *et al.*, 1987; Soetaert *et al.*, 1991a, b; Pfannkuche, 1993a), from the Norwegian Sea (Dinet, 1979), from the Greenland, Norwegian Sea and Indian Ocean (Thiel, 1975; Jensen, 1988), the Arctic Ocean (Pfannkuche and Thiel, 1987), and the southeast Atlantic (Soltwedel, 1993).

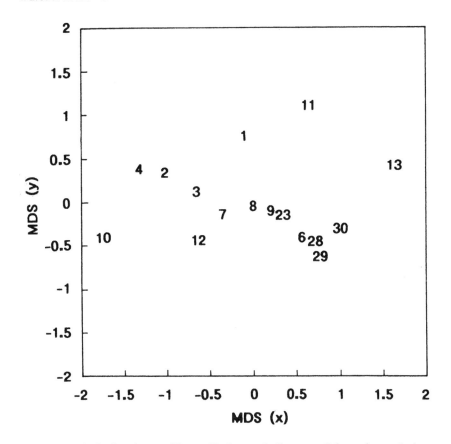

Figure 23 Ordination, as Figure 22, but excluding *area* 5 from the analysis.

Thiel (1983) reviewed the Atlantic data and showed that densities in general are higher in the east than in the west Atlantic, a pattern which he associated with the higher primary productivity in the east (Figure 3). In the west Atlantic, densities are always lower than 1000 ind/10 cm^2 along the upper slope, and decrease to less than 100 ind/10 cm^2 on the abyssal plain (Thiel, 1983). Tietjen *et al.* (1989) recorded densities of $114 \pm 26/10$ cm^2 on the Hatteras Abyssal Plain, similar to other low values reported for the west Atlantic.

The cold waters off northern Europe (north Atlantic and the Norwegian Sea) are characterized by higher densities than the northeast Atlantic, and also by a larger variability within the different depth zones used here. On the upper slope (<1000 m) densities vary from 200 to 3500 ind/10 cm^2; below 3000 m, densities still reach values from 10 to 800 ind/10 cm^2. The comparatively high numbers are related to the higher planktonic mortality and the low temperatures which reduce organic degradation rates stimulate and hence the long term availability of food (Thiel, 1983). Pfannkuche and Thiel (1987) reported even higher abundances on the upper slope (from 1100 to 4300 ind/10 cm^2) along a high Arctic depth transect. On the lower slope (>3000 m) densities of 230–350 ind/10 cm^2 were recorded. The authors invoke hydrodynamical processes which cause the downslope transport of under-ice primary production and sustain the food supply.

The low primary productivity in the Mediterranean results in lower densities compared to the northeast Atlantic. Even on the upper slope densities never exceed 1000 ind/10 cm^2 (Thiel, 1983). The results of a study by Soetaert *et al.* (1991a, b) along an upper slope depth transect near Corsica, illustrate this tendency. Only canyons, which have a higher food supply due to specific hydrodynamical processes, are characterized by relatively high meiobenthic abundances (Soetaert *et al.*, 1991a, b).

In the Pacific Ocean meiobenthic densities also are generally low. Alongi and Pichon (1988) found that densities never exceed 200 ind/10 cm^2, and are only exceptionally higher than 100 ind/10 cm^2 along the upper slope (<1000 m) in the western Coral Sea (SW Pacific). At abyssal depths in the east central Pacific (Renaud-Mornant and Gourbault, 1990) and the south east Pacific (Schriever *et al.*, 1991) densities are always lower than 200 ind/10 cm^2. These areas are all considered to be oligotrophic zones. Only Shirayama (1984a) found higher densities at bathyal (up to 1300 ind/10 cm^2), abyssal (up to 500 ind/10 cm^2) and even hadal depths (up to 430 ind/10 cm^2) in the western Pacific.

Changes in faunal composition at an ocean-wide scale are to be expected between areas of different productivity, such as from shelf to abyss, upwelling to non-upwelling, and low to higher latitudes. Table 3 compares the results of meiobenthos investigations on a global scale. Only deep stations (below 2000 m) have been included to reduce variations caused by depth-related factors, but differences in the relative abundances of certain taxa between different regions are still noticeable. These variations may be attributed to different sediment granulometry, different hydrographic conditions and/or the general availability of food in regions with different surface primary production.

4.5. Patterns at Different Taxonomic Levels

Meiobenthic taxonomy is always difficult and time consuming; in the deep sea this problem is compounded by high species diversity and low species dominance. Most deep-sea meiobenthic investigations have, therefore,

Author	Area	Provinces	Latitude	Depth (m)	Nematoda (%)	Copepoda (%)	Others (%)
Pfannkuche and Thiel, 1987	Arctic Ocean Nansen Basin	Arctic/subarctic regimes	82°	2500–3920	90.2	6.6	3.2
Dinet, 1979	Norwegian Sea Norwegian Basin		64–74°N	2465–3709	96.0	3.3	0.9
This chapter[a]	West European Basin	Coastal boundary regimes	45–52°N	2000–5340	89.8	7.1	2.1
Tietjen, 1971; Coull et al., 1977	Northwestern Atlantic, NW Atlantic Basin		34°N	2000–4000	81.8	5.5	9.8
This chapter[b]	Madeiran Abyssal Plain	Subtropic gyral province	33°N	4856–5120	92.8	6.3	1.0
This chapter[c]	Cape Verde Ridge	Provinces with pronounced seasonality	19–22°N	2041–4652	93.8	4.4	2.1
Romano and Dinet, 1981	NW Indian Ocean, Arabian Sea		14–23°N	2390–4727	88.0	3.3	10.1
Soltwedel (unpubl.)	Central Atlantic, Middle oceanic ridge		1–6°S	2473–5213	84.8	10.4	3.5
Dinet, 1973	SE Atlantic, Walvis ridge		19–22°S	2800–5170	94.9	4.1	1.4
Shirayama, 1984a	Western Pacific, NW Pacific Basin	Subtropical gyral province	24–33°N	2430–8260	86.2	4.3	9.2
Shirayama, 1984a	Western Pacific, Eastern Carolina Basin		0–11°N	2090–5730	86.2	5.2	8.4
Woods and Tietjen, 1985	Venezuela Basin	Subtropical basin	12–15°N	3517–5054	85.2	9.0	5.8
Alongi, 1992	Western Pacific, Coral and Solomon Sea	n.d.	12–14°S	2395–4350	43.7	33.3	19.5

n.d., not defined.
[a]Dinet and Vivier, 1977; Pfannkuche, 1985; Rutgers van der Loeff and Lavaleye, 1986; Thiel, 1972; Gooday (unpubl.) + Vanreusel and Vincx (unpubl.).
[b]Gooday and Soltwedel (unpubl.).
[c]Dinet (unpubl.).

dealt with only taxonomic levels higher than species (Thiel, 1983, and references therein). Species level ecological data are only available for the nematodes, copepods and foraminiferans from a limited number of areas.

4.5.1. *Nematodes*

In the deep northeast Atlantic, only two studies (*areas* 6–10), from the Bay of Biscay, have dealt with the taxonomic composition of nematode communities. One study is of two stations in the shelf break area along the Spanish coast (Vanreusel *et al.*, 1992), and the other of several abyssal stations (Dinet and Vivier, 1979). Rutgers van der Loeff and Lavaleye (1986) examined the taxonomic composition of the nematode communities at the abyssal DORA *area*, to study the composition of different feeding types.

Thirty-five families are recorded from the Gulf of Biscay, of which 11 are common to both the shelf break and the abyssal locations. Dominant (>10%) families from the abyssal plain in the Gulf of Biscay are the Monhysteridae and Chromadoridae, followed by the Oxystominidae, Desmoscolecidae, Microlaimidae and Axonolaimidea (>5%). At the shelf break stations along the Spanish coast, the Xyalidae and, again, Chromadoridae are the dominant families (>10%), while the Oxystominidae, Desmoscolecidae and Microlaimidae are subdominant (>5%). Other important families at these stations included the Comesomatidae, Cyatholaimidae and Selachinematidae. At the genus level the similarity between abyssal and shelf break samples in the Gulf of Biscay is less striking. Only *Halalaimus*, *Microlaimus*, *Acantholaimus*, *Leptolaimus*, *Diplopeltula* and *Sabatieria* are common dominant to subdominant genera (with a relative abundance >1%) for both depth zones. The dominant genus at the abyssal sites, *Theristus* (represented by 42 different species), is replaced by two closely related genera, *Daptonema* and *Trichotheristus* at the shelf break sites.

Dinet and Vivier (1979) investigated species composition at the abyssal sites, noting that the majority were new to science; a *Syringolaimus* species was dominant, followed by three *Microlaimus* spp., three species of Spilipherinae, and eight *Theristus* spp.

In the Bay of Biscay, diversity is usually high, both in terms of species numbers and evenness. At the abyssal sites (six stations) 317 species in 109 genera were recorded, with between 50 and 20 species per station, the rank 1 species dominance did not exceed 11%. At the shelf break a total of 79 genera was found at the two stations with, respectively, 62 and 46 genera per station. Here again the high diversity is due to the numbers of genera present (and thus even more for species) and the low rank 1 generic dominance (maximum 10–18%).

Rutgers van der Loeff and Lavaleye (1986) recorded a lower density at the DORA site and suggest that, referring to Dinet and Vivier (1979), ". . . this may be explained by their more accurate way of identifying the different nematode species . . .". Nevertheless the study area still shows the high nematode diversity thought to be characteristic at the deep-sea benthos.

One of the oldest, but still frequently used, classifications of nematodes into functional groups is that of Wieser (1953) who defined four feeding types (i.e. selective and non-selective deposit feeders, epistrate feeders and predators/omnivores) based on the buccal morphology. At the DORA site (Rutgers van der Loeff and Lavaleye, 1986) non-selective deposit feeders are dominant (50%). Selective deposit feeders are represented by 25.6%, epistratum feeders and predators/omnivores by, respectively, 14.7 and 9.7%. In the Bay of Biscay, all dominant genera except *Microlaimus* are defined by Wieser (1953) as deposit feeders.

Depth-related factors are assumed to be the main controlling influence for the nematode communities in the Bay of Biscay. Dinet and Vivier (1979) found that 135 species were present only at shallower stations than 3000 m, while 61 species were restricted to the zone deeper than 4000 m. Diversity decreased from 2000 to 4700 m depth. According to Tietjen (1976, 1984) species diversity is directly related to sediment heterogeneity, predictability of the environment and/or the proximity to areas of high surface productivity. More recently, he confirmed the presumed relation with sediment heterogeneity (Tietjen, 1989), and also suggested that greater fluxes of sedimenting material and bioturbation rates can enhance species diversity in some areas of the deep sea (Tietjen, 1992).

Elsewhere, published studies of deep-sea nematodes carried out at lower taxonomic levels (family/genus/species) are mainly concentrated in the west Atlantic: off N Carolina (Tietjen, 1971, 1976), in the Venezuela Basin (Tietjen, 1984), near the Scotian Rise (Thistle and Sherman, 1985) and from the Puerto Rico Trench and the Hatteras Abyssal Plain (Tietjen, 1989). Jensen (1988) investigated some structural aspects of the nematode communities from the Norwegian Sea.

High relative abundances of the Xyalidae or Monhysteridae, Oxystominidae and Desmoscolecidae are general in the west Atlantic (Tietjen, 1976, 1984, 1989; Thistle and Sherman, 1985), and similar to the abyssal region of the Bay of Biscay. The Chromadoridae are dominant at the abyssal sites in the northeast Atlantic and in some areas of the west Atlantic (Thistle and Sherman, 1985; Tietjen, 1989) but not at the deeper sites (> 800 m) off N Carolina. The dominance of the Linhomoeidae off N Carolina is in marked contrast to the results from the Bay of Biscay. The high numbers of the Cyatholaimidae and Comesomatidae off N Carolina, especially along the lower slope, are similar to that reported in the shelf break region of the Bay of Biscay. At the genus level the affinity between the west and east Atlantic is more pronounced. *Theristus, Halalaimus, Microlaimus, Desmoscolex* and *Acantholaimus* are the dominant genera in the east as well as in the west. This is illustrated by Table 4 which also indicates that the deepest sites (>1500 m) off N Carolina are exceptional: only *Microlaimus, Theristus* and *Desmoscolex* are well represented here. The genera *Porocoma* and *Aegialoalaimus*, which are dominant off N Carolina, were not recorded by Dinet and Vivier (1979). The high abundance of *Desmodora, Sabatieria,* and *Pselionema* on the lower slope off N Carolina is more in accord with the situation at the shelf break in the east, as reported by Vanreusel *et al.* (1992). However, *Daptonema,* dominant in the east, seems to be replaced by *Monhystera* in the west.

The dominance of deposit feeders at the DORA site in the northeast Atlantic (Rutgers van der Loeff and Lavaleye, 1986) is of the same order

Table 4   Dominant nematode genera in the northeast Atlantic (Bay of Biscay) and their presence in the west Atlantic

Genera	Dinet and Vivier, 1979	Vanreusel et al., 1992 <1500 m	Tietjen 1971, 1976 <1500 m	Thistle and Sherman, 1985	Tietjen, 1989
Theristus	**		**	**	**
Spiliphera	**				
Halalaimus	**	**	**	**	**
Microlaimus	**	*	**	**	**
Acantholaimus	**	*		**	**
Desmoscolex	**		**	*	*
Syringolaimus	**			*	*
Leptolaimus	*	*	**		**
Campylaimus	*				
Diplopeltula	*	**			*
Thalassoalaimus	*				*
Sphaerolaimus	*		**	**	**
Longicyatholaimus	*			**	
Tricoma	*		**		**
Cervonema	*				
Sabatieria	*	**	**	**	*
Daptonema		**			
Richtersia		**			
Prochromadorella		**			
Actinonema		**			**
Pselionema		**	*		

         ⌞————— East Atlantic —————⌟   ⌞———————— West Atlantic ————————⌟

** >3%, * >1%.

of magnitude as in the west Atlantic (50–80%) (Tietjen, 1976, 1984; Thistle and Sherman, 1985) and in the Norwegian Sea (Jensen, 1988). However, Tietjen (1971) found changes in the distribution of the feeding types (especially the epistrate feeders) which were highly correlated with changes in sediment composition. Tietjen (1984) suggested that poorly sorted sediments, which are characterized by a greater diversity of microhabitats and thus also a greater variety of food sources, may enhance the abundance of epistrate feeders in some deep-sea areas.

Estimates of nematode species diversity and evenness in the northeast Atlantic are within the range of those for the west Atlantic (Tietjen, 1984, 1989) and the Norwegian Sea (Jensen, 1988). The high diversity is generally maintained by a high number of species and an even distribution of specimens among the species. Only Tietjen (1976) described somewhat lower values for the deep sea off N Carolina.

4.5.2. Copepods

Also in the case of copepods, no complete diversity analysis has been made of northeast Atlantic deep-sea communities. Nevertheless, some taxonomic studies suggest that these assemblages are highly diverse and comprise many undescribed species. Bodin (1968) recorded 29 species, including 25 new to science, from a collection of 47 individuals collected in the deep Bay of Biscay. In the same area, 11 new species were found by Dinet (1977, 1981) in a study of a single genus *Pontostratiotes*, which predominates the large-size fraction of meiobenthos. Although most harpacticoid families are represented in the area, the assemblages are dominated by Cletodidae, Diosaccidae, Ectinosomatidae, Tisbidae and particularly the Cerviniidae which can be considered as a typical deep-sea taxon.

A bathymetric zonation, mainly characterized by a change in species composition occurring between 2000 and 3000 m depth, was identified by Dinet (1985) using the genus *Pontostratiotes*.

4.5.3. Foraminifera

Deep-sea foraminiferal populations are also highly diverse. Gooday (1986a) and Gooday and Lambshead (1989) found 78–105 species in the top 1 cm of sediment samples (3.46 cm^2) from the bathyal Porcupine Seabight (*area* 2, 1340 m). Rather lower numbers of species were present in similar samples from *area* 1 (395 m; 55–64 species) and *area* 3 (4494 m; 53–76 species) in the Seabight (Cartwright, 1988) and from the abyssal BIOTRANS site (*area* 12; 62 species) (Gooday, unpublished). Diversity is also high in abyssal plain samples; the top 1 cm of complete multiple cores (25 cm^2) from the Porcupine Abyssal Plain (*area* 4) and the Madeira Abyssal Plain (*area* 23) yielded 110–143 species and 114 species respectively.

Calcareous taxa of the kind normally studied by geologists make up 20–45% of meiobenthic foraminiferal assemblages in bathyal samples but a much lower proportion (often <10%) at abyssal sites (Gooday, 1986a, 1990; Gooday and Lambshead, 1989; Gooday, unpublished). Much more abundant in abyssal assemblages are small, delicate, soft-bodied taxa such as saccamminids and allogromiids (Gooday, 1986a, b, 1990). Hormosinacean genera such as *Reophax* and *Leptohalysis* may also be common.

Corliss and Chen (1988) observed that different morphotypes of deep-sea benthic foraminifera have distinct distribution patterns with water depth. Moreover, the relative abundances of epi- and infaunal species appeared to be related to the organic carbon content of surface sediments. The authors concluded that bathymetric groupings based on morphology and microhabitat preferences are related to the flux of organic carbon to the seafloor.

4.6. Small-scale Spatial Patterns

It is generally accepted that metazoan meiobenthos densities are much more variable on a small scale in the deep sea than in shallow waters. For instance, Pfannkuche (1993b) found no seasonal variations in abundance or biomass since the variability between samples from a single cruise in the BIOTRANS area was greater than the seasonal changes. However, there is little information on the scale of this spatial variation and only a few studies consider possible causes. Biogenic structures such as burrows, depressions and mounds produced by the larger macrobenthos clearly influence the distribution of the meiobenthos on small scales, for example within box corers (Thistle and Eckman, 1988). Particulate organic matter is trapped occasionally by such structures. Opportunistic taxa, such as nematodes (Aller and Aller, 1986), may become unusually abundant early in the colonization of these patches, increasing variability in faunal composition within a sampling station (Grassle and Mosse-Pocteous, 1987). Thistle and Sherman (1985) found evidence for the association of nematode species with pebbles which they assumed to provide a refuge for meiobenthos from some type of predators. At the species level, there is no evidence of an association between small-scale variations in nematode abundance and variance in the abundance of bacteria or several classes of biogenic structures. However, individual nematode species may respond to organic matter inputs. Thiel *et al.* (1989–1990) observed the association of one species of the genus *Monhystera* with aggregated phytodetritus deposits at the BIOTRANS site. Numerous exuvia of ectinosomatid harpacticoids were also observed in phytodetritus from this site. One of several possibilities is that these exuviae are derived from an annual reproductive event related to the seasonal phytodetrital input. The small-scale patchiness of the food supply is often presumed to be the main source of variation in densities (Grassle and Mosse-Porteous, 1987). It was hypothesized by Thistle (1983) that, as the rates of physical and biological disturbance decrease with depth, the habitat heterogeneity created by organisms becomes increasingly available to other organisms for habitat partitioning or as prey refuges, allowing larger numbers of species to be accommodated locally. However, for copepods, no increase in diversity was measured in sites with a greater degree of association (Thistle, 1983).

5. VERTICAL SPATIAL PATTERNS

5.1. Suprabenthic Microhabitats

Deep-sea foraminifera commonly live attached to surfaces elevated above the sediment water interface. In the northeast Atlantic, calcareous species (e.g. *Cibicides weullerstorfi, Planulina ariminesis*) with plano-convex tests have been found on substrata such as stones, sponge skeletons, hydroid stalks, ascidians, polychaete tubes and the tests of other foraminifera (Weston, 1985; Lutze and Thiel, 1989). Agglutinating species, mainly trochamminaceans (e.g. *Tritaxis conica*), also commonly occur on hard substrata in samples from the Porcupine Seabight and off NW Africa (Gooday, unpublished observations).

Phytodetritus represents another elevated habitat. Gelatinous aggregates of this material harbour large foraminiferal populations dominated by species such as *Epistominella exigua* and *Tinogullmia riemanni* (Gooday, 1988, 1990; Gooday and Lambshead, 1989). Nematodes, many belonging to the genus *Monhystera*, also occur within phytodetritus (Thiel *et al.*, 1989–1990). In one case, 27 nematodes occurred in a single aggregate (0.3 ml volume) from the BIOTRANS site.

5.2. Vertical Distribution Within the Sediment

Few studies on the meiofauna of the northeast Atlantic have quantified the vertical profiles of meiofauna abundances within the sediment. The data used were supplied by Dinet and Vivier, 1977; Gooday, unpublished; Lambshead and Ferrero, unpublished; Pfannkuche, 1985; Rutgers van de Loeff and Lavaleye, 1986; Soltwedel, unpublished; Thiel, 1972a; Vanreusel *et al.*, 1992 and Vanreusel and Vincx, unpublished. In order to make meaningful comparisons between data-sets, we have restricted our analysis to data describing the top 5 cm of sediment divided into 1-cm sections.

Data for total metazoan meiofauna, nematodes, foraminiferans and harpacticoid copepods and nauplii are considered. Figures 24–26 show the data used for the following results. The number of observations available for each site varied greatly from $n = 1$ to $n = 12$. Therefore, mean values were calculated for all sites where $n > 1$ and these "summary means" were used for the production of individual vertical profile figures and the subsequent statistical analyses.

The following analysis is based on values for percentage total abundance at each sediment depth level calculated for each site. Vertical profiles where the organisms are concentrated in the surface sediment levels will be referred to as "superficial" and those where the organisms extend to the lower sediment levels in greater numbers will be referred to as "deep".

To gain a generalized picture of differences between the four faunistic groups, average vertical profiles for the entire study area were calculated using the summary means (Table 5, Figure 24). Percentage total

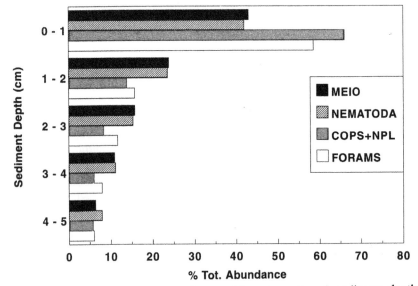

Figure 24 Percentage total abundance of four taxa at each sediment depth: average vertical profile for the entire study area (northeast Atlantic).

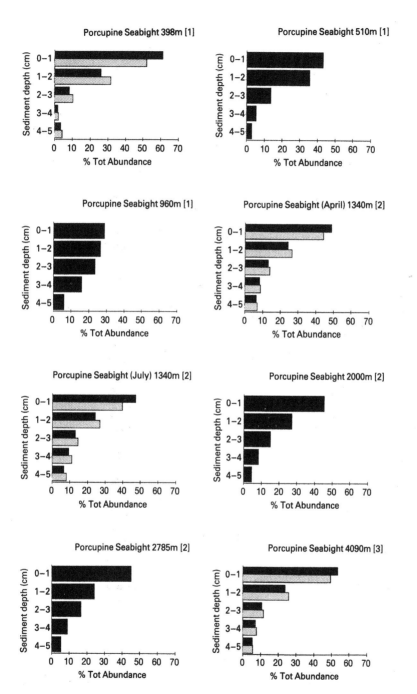

Figure 25 Individual vertical profiles percentage total abundance of all meiofauna (black) and nematodes (grey) for study areas 1 and 2 and part of 3 (cf. Table 1); number of observations in brackets.

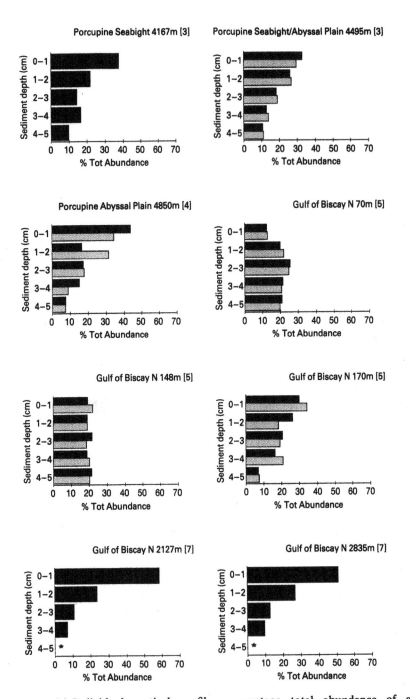

Figure 26 Individual vertical profiles percentage total abundance of all meiofauna (black) and nematodes (grey) for study areas 3 (part), 4, 5, 7 (cf. Table 1); number of observations in brackets.

Table 5 Average values and statistics for percentage total abundance of metazoan meiofauna, nematodes, copepods and nauplii and foraminifera at five sediment depth levels. Averages calculated from sample means.

Sample means	N	Average	Variance	SD	SE	Minimum	Maximum	Range
0–1 cm								
Meiofauna	27	43.0444	146.188	12.0908	2.32688	12.4500	61.9000	49.4500
Nematodes	20	41.9975	178.013	13.3421	2.98339	12.8200	63.2500	50.4300
Copepods and nauplii	11	65.9373	476.302	21.8244	6.58029	33.6787	98.3871	64.7084
Foraminifera	10	58.5846	314.117	17.7234	5.60462	22.1959	80.7829	58.5870
1–2 cm								
Meiofauna	27	23.9130	16.8529	4.10523	0.79005	16.5400	35.3400	18.8000
Nematodes	20	23.6000	24.8725	4.98724	1.11518	13.4600	31.8500	18.3900
Copepods and nauplii	11	13.8572	59.5248	7.71523	2.32623	1.61290	25.0000	23.3871
Foraminifera	10	15.7483	31.9869	5.65570	1.78849	9.48991	23.8536	14.3637
2–3 cm								
Meiofauna	27	15.7548	22.6977	4.76421	0.91687	5.40000	25.4800	20.0800
Nematodes	20	15.3310	24.2667	4.92612	1.10151	4.63000	24.7400	20.1100
Copepods and nauplii	11	8.33085	41.3939	6.43377	1.93986	0.00000	17.0483	17.0483
Foraminifera	10	11.6756	40.8625	6.39235	2.02144	3.79597	25.7266	21.9306
3–4 cm								
Meiofauna	27	10.9044	25.9985	5.09887	0.98128	1.59000	21.3600	19.7700
Nematodes	20	11.1475	32.7532	5.72304	1.27971	1.64000	20.8200	19.1800
Copepods and a nauplii	11	6.10382	54.4696	7.38036	2.22526	0.00000	20.2073	20.2073
Foraminifera	10	7.96185	19.1303	4.37381	1.38312	2.90698	15.5651	12.6581
4–5 cm								
Meiofauna	27	6.38370	27.5955	5.25314	1.01097	0.00000	21.3700	21.3700
Nematodes	20	7.92250	22.2688	4.71898	1.05520	1.45000	20.1900	18.7400
Copepods and nauplii	11	5.77072	64.1662	8.01038	2.41522	0.00000	25.0000	25.0000
Foraminifera	10	6.02957	10.6683	3.26624	1.03288	2.60973	12.6588	10.0491

Table 6 Mann–Whitney comparisons of percentage total abundance of metazoan meiofauna (M), nematodes (N), copepods and nauplii (C) and foraminifera (F) at five sediment depth levels. N/S = not significant;* = $p < 0.05$; ** = $p < 0.01$; *** = $p < 0.001$.

		M	N	C
0–1 cm				
	N	N/S		
	C	**	**	
	F	**	**	N/S
1–2 cm				
	N	N/S		
	C	***	**	
	F	***	**	N/S
2-3 cm				
	N	N/S		
	C	**	**	
	F	*	N/S	N/S
3–4 cm				
	N	N/S		
	C	*	N/S	
	F	N/S	N/S	N/S
4–5 cm				
	N	N/S		
	C	N/S	N/S	
	F	N/S	N/S	N/S

abundances of the four faunistic groups at each sediment depth level were compared using a Mann–Whitney U-test (Table 6).

Metazoan meiofauna and nematodes showed no significant differences in percentage total abundance at any sediment depth. Average percentage abundance in the surface centimetre was just over 40% decreasing gradually with depth. Both copepods and foraminifera were significantly more abundant than metazoan meiofauna and nematodes in the surface centimetre (average percentage abundance 66% copepods, 57% foraminifera) of sediment and correspondingly less abundant in many of the lower sediment layers. There were, however, no significant differences in the deepest (4–5 cm) sediment level or between copepods and foraminifera. This may have been due both to low numbers of individuals and in the case of the test between copepods and foraminifera due to the smaller data-set available for analysis.

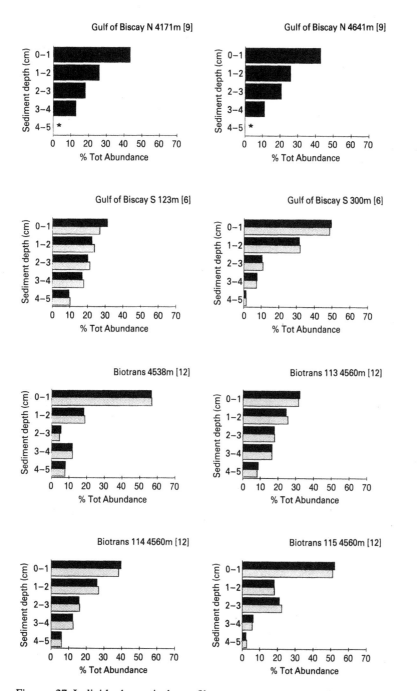

Figure 27 Individual vertical profiles percentage total abundance of all meiofauna (black) and nematodes (grey) for study areas 6, 9, 12 (cf. Table 1); number of observations in brackets.

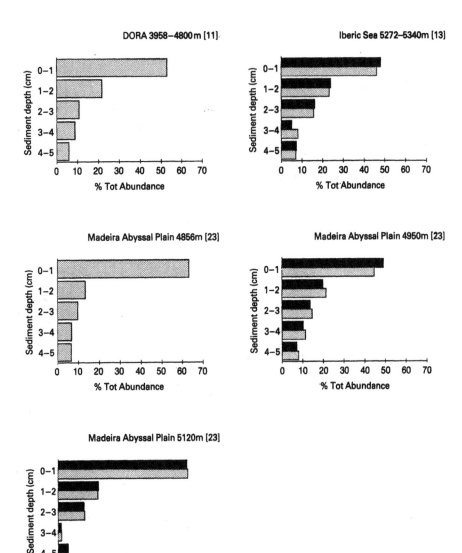

Figure 28 Individual vertical profiles percentage total abundance of all meiofauna (black) and nematodes (grey) for study areas 11, 13, 23 (cf. Table 1); number of observations in brackets.

Vertical profiles for 25 sites at 11 of the designated study *areas* (see Table 1) are shown in Figures 25–28. In general, abundances of organisms were greatest in the top centimetre of sediment and decreased gradually with depth. Twelve of the 27 profiles had between 40 and 50% of the metazoan meiofauna to be found in the top centimetre. A further six profiles each were found to have 30–40% and 50–60% of total metazoan meiofauna in the top centimetre of sediment. In all but three of the profiles, less than 10% of the total abundance was found at the 4–5 cm sediment depth. Dinet and Vivier (1977) measured abundances to a depth of 4 cm but still found less than 10% of total abundance at the lowest sediment depth at two of the four sites studied.

Two geographical areas, the Porcupine Seabight/Abyssal Plain (*areas* 1–4) and Bay of Biscay (*areas* 5, 6, 7 and 9), have received most study. In the Porcupine Seabight, profiles showed a tendency to become deeper with increasing depth to 960 m where only 29% of the individuals were found in the surface centimetre. However, at 1340 m, the profiles were more superficial with 40–50% of individuals in the surface centimetre and this profile remained quite consistent at all sites down to a depth of 2785 m. At the deeper sites, profiles were more variable showing both deeper and more superficial distributions. Sites in the Bay of Biscay were taken from two different localities in the north (*areas* 5, 7 and 9) and south (*area* 6) of the Bay. In the north, the two shallowest stations at 70 and 148 m showed remarkably deep vertical profiles both having subsurface maxima. At 170 m the profile remained rather deep but was more typical in shape having a surface maximum of 30%. The trend for increasingly superficial profiles continued at the 2127 and 2835 m sites where >50% of individuals were found in the surface centimetre but at the deepest sites (4171 and 4641 m), the profiles were slightly deeper.

Profiles from the four BIOTRANS sites (*area* 12) exhibited considerable variation with both superficial, intermediate and deeper vertical distributions recorded and values for the top centimetre ranging from 32 to 56%.

At the deepest sites sampled, the Iberian Sea (*area* 13) and Madeira Abyssal Plain (*area* 23), profiles were rather superficial with >45% of individuals in the surface centimetre and particularly on the Madeira Abyssal Plain at 5120 m where 62% of the individuals were found in the surface centimetre.

5.2.1. *Nematodes*

Vertical profiles for 20 sites at 10 of the designated study areas (see Table 1) are shown in Figures 25–28. Profiles from nine of the *areas* and 17 of

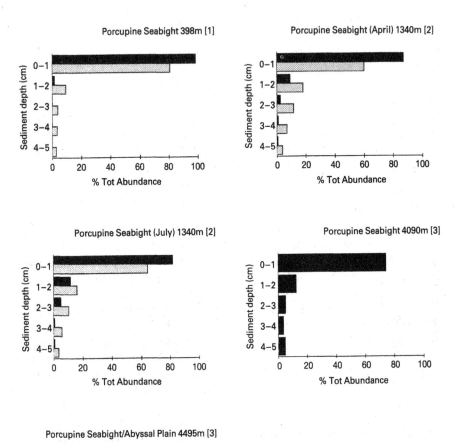

Figure 29 Individual vertical profiles percentage total abundance of copepods (black) and foraminifera (grey) for study areas 1, 2, 3 (cf. Table 1); number of observations in brackets.

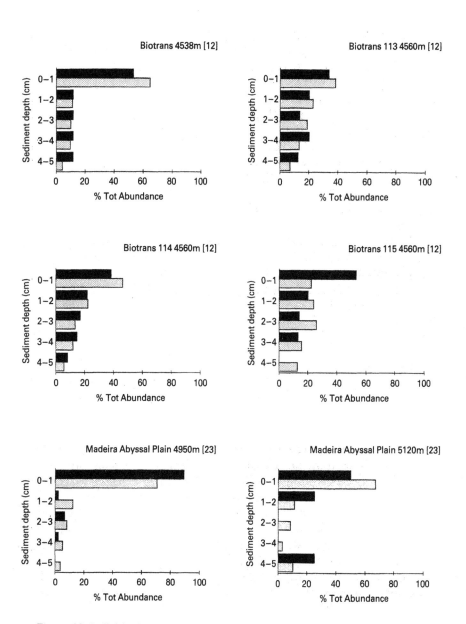

Figure 30 Individual vertical profiles percentage total abundance of copepods (black) and foraminifera (grey) for study areas 12 and 23 (cf. Table 1); number of observations in brackets.

the sites were common to those recorded for metazoan meiofauna. Owing to the numerical abundance of nematodes compared with the other meiofaunal taxa, these profiles were similar to those for metazoan meiofauna but generally showed a lower percentage abundance in the surface centimetre of sediment and subsequent higher values in the deeper sediment layers due to the more superficial distribution of copepods and foraminifera. Five of the 20 profiles had between 40 and 50% of the nematodes in the top centimetre. A further five profiles were found to have 30–40% and four profiles to have 50–60% of total nematodes in the top centimetre of sediment.

Additional profiles from the Porcupine Abyssal Plain (4850 m) and Madeira Abyssal Plain (4856 m) were not markedly different from those at previously described sites. The profile from the Madeira Abyssal Plain was very superficial with >60% of individuals occurring in the surface centimetre of sediment and was similar to the profile at 5120 m.

The profile from the DORA station (3958-4800 m) situated between the BIOTRANS and Iberian Sea stations showed a superficial distribution with >50% of individuals occurring in the top centimetre of sediment and within the range of profiles described for the BIOTRANS and Iberic Sea stations.

5.2.2. *Copepods and Nauplii*

Vertical profiles for 11 sites at five of the designated study *areas* (see Table 1) are shown in Figures 29 and 30. Five of the profiles showed more than 70% of individuals to be in the surface centimetre of sediment and only two profiles showed less than 50% of individuals in the surface centimetre. Sites from the Porcupine Seabight/Abyssal Plain area exhibited very superficial profiles. At 398 m, 98% of the copepods and nauplii occurred in the surface centimetre of sediment with none recorded below 2 cm. The percentage abundance in the surface centimetre reduced with increasing depth but was still >65% at the deepest site (4495 m). Percentage abundances were very low below 3 cm sediment depth and no individuals were found below this level at 4495 m.

Profiles from sites at the BIOTRANS (4538–4560 m) station were, like those for metazoan meiofauna and nematodes, rather variable showing both rather superficial and more intermediate distributions. Penetration of copepods and nauplii was generally greater than in the Porcupine Seabight and Abyssal Plain.

On the Madeira Abyssal Plain, one site at 4950 m showed a very superficial vertical profile with 89.1% of individuals in the surface centimetre of sediment. The other site showed a less superficial profile but abundances/10 cm^2 were very low and it was likely that the observed profile was affected by mathematical limitations.

5.2.3. Foraminifera

Vertical profiles for 10 sites at five of the designated study areas (see Table 1) are shown in Figures 29 and 30. Seven of the profiles showed more than 60% of individuals to be in the surface centimetre of sediment. Sites from the Porcupine Seabight/Abyssal Plain showed very superficial vertical profiles similar to those for copepods and nauplii but with generally greater penetration to the deeper sediment depths. At 398 m, >80% of individuals were found in the surface centimetre of sediment; >70% at 4495 m and >60% at 1340 m.

Sites at the BIOTRANS (4538–4560 m) station were again very variable with one site having a very superficial profile, two sites having a more intermediate profile and one site having a very deep profile with a subsurface maximum of 26% at the 2–3 cm sediment depth.

The two sites from the Madeira Abyssal Plain showed very superficial profiles with >65% of individuals occurring in the surface centimetre.

5.2.4. Multivariate analysis

To gain an overview of any pattern in the distribution of the individual vertical profiles described above, the data were subjected to multivariate analysis by non-metric MDS (Field et al., 1982). Percentage total abundance data were used with the five sediment depths being treated as analogous to species. The data were first subjected to an arcsin transformation. The resulting two-dimensional ordinations were superimposed with the station numbers (see Table 1) and a series of symbols, the size of which was scaled to represent the values of a series of environmental parameters available for each station. The environmental parameters used were depth, degrees north, degrees west, distance off-shore, primary production and organic flux. The values for primary production were taken from the "Dahlem Map" (Berger, 1989) and values for organic flux were calculated using these data and the depth data according to the equation of Pace et al. (1987). Many of the environmental parameters used were either directly or indirectly correlated, for example depth and degrees west over some or all of the study areas.

The data for environmental parameters were used in conjunction with the percentage total abundance data for all samples at each sediment depth to calculate Spearman's rank correlation coefficients.

The MDS ordinations for each faunistic group are shown in Figures

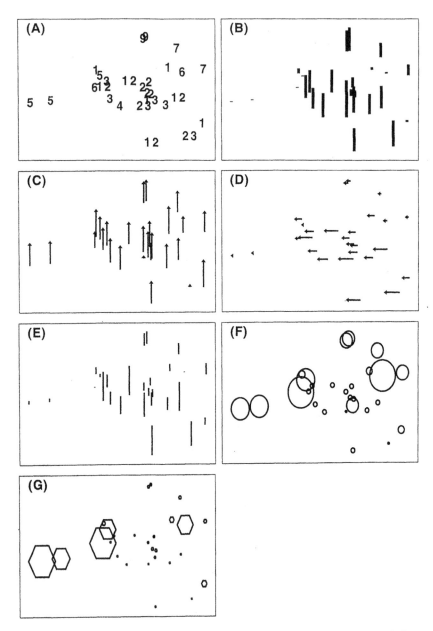

Figure 31 MDS ordination of root–root transformed metazoan meiofauna vertical profile data (percentage total abundance). Ordination superimposed with (A) station numbers and symbols scaled to show values of environmental parameters; (B) depth; (C) degrees north; (D) degrees west; (E) distance off-shore; (F) primary production; (G) organic flux (*area* numbers: 1-2-3-4-5-6-7-9-12-23).

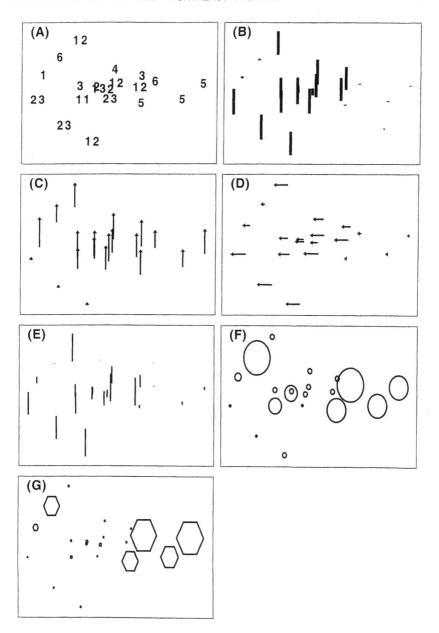

Figure 32 MDS ordination of root–root transformed nematode vertical profile data (percentage total abundance). Ordination superimposed with (A) station numbers and symbols scaled to show values of environmental parameters; (B) depth; (C) degrees north; (D) degrees west; (E) distance off-shore; (F) primary production; (G) organic flux (area numbers: 1-2-3-4-5-6-12-23).

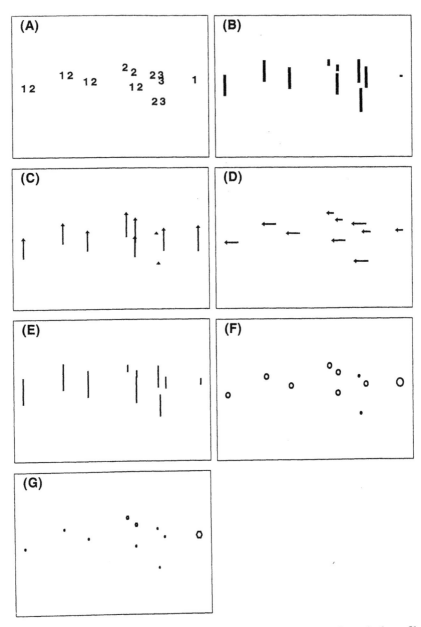

Figure 33 MDS ordination of root–root transformed copepod vertical profile data (percentage total abundance). Ordination superimposed with (A) station numbers and symbols scaled to show values of environmental parameters; (B) depth; (C) degrees north; (D) degrees west; (E) distance off-shore; (F) primary production; (G) organic flux (*area* numbers: 1-2-3-12-23).

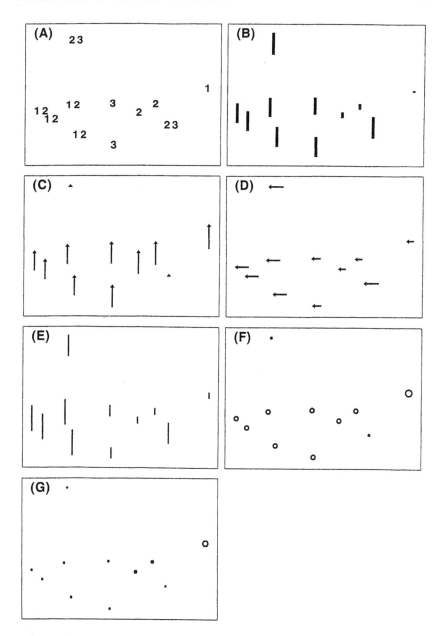

Figure 34 MDS ordination of root–root transformed foraminifera vertical profile data (percentage total abundance). Ordination superimposed with (A) station numbers and symbols scaled to show values of environmental parameters; (B) depth; (C) degrees north; (D) degrees west; (E) distance off-shore; (F) primary production; (G) organic flux (*area* numbers: 1-2-3).

31–34 (method explained in Clarke and Ainsworth, 1993). In general, a clear trend in the ordinations was seen from the most superficial profiles (sites at *areas* 1 and 23) to the deepest profiles at *area* 5 for metazoan meiofauna and nematodes and at *area* 12 for copepods and nauplii and foraminifera. Perhaps the most striking overall feature of the ordinations was found in those for the metazoan meiofauna and nematodes. Shallower sites having generally high organic flux and to a lesser extent, higher primary production and being closer to shore, appeared in the ordinations at both ends of the range of profiles. This indicated that similar environmental parameters could be associated with both superficial and deep vertical profiles. The ordinations superimposed with environmental characteristics did not show any strong overall trends. However, some local trends could be detected, especially in the metazoan meiofauna and nematode ordinations where more sites were available for analysis. In these ordinations, local trends were seen for increasingly superficial profiles in the Bay of Biscay shallower samples (*area* 5, 70–170 m and *area* 6, 123–300 m) as depth increased and organic flux decreased. Interestingly, sites at the deeper Bay of Biscay *areas* (7 and 9) showed a reversal of this trend. Another local gradient was seen in the increase in profile depth shown by metazoan meiofauna in sites at *area* 1 as depth increased from 398 to 960 m and organic flux decreased.

In all ordinations, the *area* 12 sites were fairly widely separated reflecting the high degree of variation in vertical profiles found at that station.

The ordinations for copepods and nauplii and foraminifera did not show any strong relationships with environmental parameters as the range of each parameter was rather small and fewer sites were available for analysis.

Spearman's rank correlation coefficients for each environmental parameter and faunistic group at the 0–1 cm sediment depths are shown in Table 7. A general observation was that when a significant positive correlation was found at the 0–1 cm sediment layer, significant negative correlations were found in the deeper sediment layers and vice versa. A degree of autocorrelation was thus apparent and results for deeper layers are therefore not shown for simplicity. A positive correlation at the 0–1 cm sediment depth indicated that profiles had become more superficial as the environmental parameter had increased and a negative correlation would indicate that profiles had become deeper as the environmental parameter had increased.

The results for metazoan meiofauna showed vertical profiles to be significantly positively correlated with depth and degrees west and significantly negatively correlated with primary production and organic flux. Nematodes showed a similar pattern but also showed a significantly negative correlation with degrees north. In both sets of results, the

Table 7 Spearman rank correlation coefficients for metazoan meiofauna, nematodes, copepods and nauplii and foraminifera at 0–1 cm sediment depth against depth (m), degrees north (N/S), degrees west (E/W), distance off-shore (DIST), primary production (PP) and organic flux (FLUX). N/S = not significant; * = $p < 0.05$; ** = $p < 0.01$; *** = $p < 0.001$.

	Spearman rank correlations					
	Depth	N/S	E/W	DIST	PP	FLUX
Metazoan meiofauna	0.2737 *	−0.0980 N/S	0.2832 *	−0.0051 N/S	−0.4733 ***	−0.2837 *
Nematodes	0.3189 **	−0.3335 **	0.2739 **	0.0432 N/S	−0.2155 *	−0.2126 *
Copepods and nauplii	−0.7373 ***	0.7214 ***	−0.6984 ***	−0.7768 ***	0.2372 N/S	0.7366 ***
Foraminiferans	−0.2390 N/S	0.1994 N/S	−0.1071 N/S	−0.4497 N/S	0.0000 N/S	0.2014 N/S

correlations, although significant, were not particularly strong, generally being below ±0.35.

Copepods and nauplii showed a complete reversal of the trends shown by metazoan meiofauna and nematodes being significantly negatively correlated with depth, degrees west and distance off-shore and significantly positively correlated with degrees north and organic flux. The correlation coefficients were generally much higher than for metazoan meiofauna and nematodes. However, there was no significant correlation with primary production. No significant correlations were found in the foraminifera data.

5.2.5. *Distributions below 5 cm*

Few data are available describing the abundance of meiofauna at sediment depths below 5 cm in the northeast Atlantic. Dinet and Vivier (1977) comment that although in their study, 90% of all meiofauna were found in the surface 4 cm of sediment, the deeper sediment layers are by no means azoic. Rutgers van der Loeff and Lavaleye (1986) reported that approximately 10% of total meiofauna was found below 6 cm depth and were found down to 20 cm. Lambshead and Ferrero (unpublished data) also recorded nematodes down to 20 cm sediment depth in samples from the Porcupine Abyssal Plain and in one core found that almost 25% of individuals occurred below the 5-cm horizon.

5.2.6. *Discussion*

The vertical distribution of meiofauna in the northeast Atlantic agrees with the generally observed fact that abundances are generally highest in the surface layers of sediment and decrease with increasing depth, such that approximately 90% of individuals may be found within the top 5 cm of sediment. This result, though unremarkable in itself, is significant in terms of the choice of core depth chosen to sample adequately the meiofaunal component of the benthos. Thus, for many studies it may be adequate to sample only to a depth of 5 cm, thus reducing the amount of sediment to be processed for each sample. This is a most important factor as the processing of meiofauna samples is an extremely lengthy operation and is often the limiting factor in the choice of sample replicates for a given site. In many cases, the increase in statistical robustness afforded by taking more replicates would more than outweigh any drawbacks associated with accepting a sampling efficiency of 90%.

In the intertidal and shallow waters, it is generally agreed that oxygen is the single most important factor in determining the vertical distribution of meiofaunal organisms. However, as has been noted earlier in this chapter, oxygen is not normally a limiting factor for meiofauna over the surface 5–10 cm and oxygen has previously been shown to control only the maximum penetration depth of the meiofauna (Shirayama, 1984b).

In this review, meiofaunal vertical profiles were found to be correlated with a number of different environmental parameters. However, some care is required when assuming the independence of the parameters chosen. A brief study of the "Dahlem" map of surface primary production shows that, over the study area as a whole, the physical and geographical parameters (depth, degrees north, degrees west and distance off-shore) could all be taken as being correlated to some extent with surface primary production and thus organic flux. Therefore, if any overall conclusion may be drawn it is that food supply to the sediment was identified as a significant factor in determining the vertical distribution of meiofauna within the sediment.

This conclusion agrees to some extent with the findings of other studies in the northeast Atlantic which have supplied data for this review. Pfannkuche (1985) found a strong correlation between CPE and vertical distribution patterns of meiofauna. Vanreusel and Vincx (unpublished) found on the basis of correlations with chlorophyll-a and bacteria (food) and nitrate concentration (sediment oxygenation) that food levels were significant in determining the vertical profiles of meiofauna but that lower oxygen levels in the deeper layers of fine sediments could limit this effect. These authors found a strong correlation between meiofauna vertical profiles and both bacterial densities and nitrate concentration at sites in the Porcupine Seabight. However, Dinet and Vivier (1977) suggested that sediment water content was significant in controlling the depth of penetration of meiofauna into the sediment and Rutgers van der Loeff and Lavaleye (1986) associated a small subsurface peak in nematode biomass with the presence of a boundary layer between soft sediments overlying a more compact clay.

These conclusions from studies in the northeast Atlantic agree with studies from elsewhere in the world's deep oceans where meiofauna vertical profiles have been correlated with food availability (e.g. Thiel, 1983; Shirayama, 1984b; Tietjen et al., 1989) and sediment water content (e.g. Snider et al., 1984; Jensen et al., 1992).

Therefore, it seems likely that the vertical profiles of meiofauna in the deep sea are not controlled by one factor but may be a combination of factors such as food supply and sediment characteristics. This may help to explain why in this study, although a rather general correlation was found between food supply and vertical profiles, individual localized gradients (as shown by the MDS analysis) may differ in response under the influence of other factors. Thus, differences in sediment characteristics could explain the appearance of relatively shallow, high food supply sites in the Bay of Biscay (areas 5 and 6) and the Porcupine Seabight (area 1) at the extremes of the vertical distributions found over the study area.

The basic tendency for biological systems to be variable is demonstrated by the wide range of vertical profiles found in the four sites at the BIOTRANS station (area 12). At this station, the sediment is subject to a seasonal input of phytodetritus from the overlying surface water. The phytodetritus forms a rather patchy resource on a small scale being concentrated in hollows and around biogenic structures but also displays patchiness on a larger scale occurring in different quantities at different parts of the BIOGAS station (Thiel et al., 1989–90). Thus, a basic correlation with food supply may still exist but food supply may be variable over medium scales in a stochastic fashion.

The difference in response found between metazoan meiofauna and nematodes and copepods and nauplii may be the result of a behavioural response such that copepods adopt a more epibenthic lifestyle in the presence of an increased food supply, whereas the metazoan meiofauna—principally nematodes—may be better adapted to exploit increased food resources deeper in the sediment. Tietjen et al. (1989) observed a similar discrepancy between the responses of nematodes and copepods to decreasing bacterial densities in the sediment. In their study on the Hatteras Abyssal Plain, nematode densities decreased with depth as bacterial densities decreased. However, in this case, copepod densities increased with depth. The authors ascribed this discrepancy to the high abundance of particularly large and robust burrowing copepods at the site which could more easily penetrate the deeper sediment depths.

Another factor which may control vertical profiles of meiofauna in the deep sea is the role of interspecific and macro-meiofauna interactions. Gooday (1986a) provides some data for foraminiferans showing that different species can have very different vertical profiles and discusses other possible interactions in detail. Pfannkuche and Thiel (1987), quoting Hansen (1978), commented that predation on the meiofauna by holothuroideans and sipunculids may have an impact on vertical distributions. Tietjen et al. (1989) also found that high rates of bioturbation increased food supply to the deeper sediment layers and that this was reflected by deeper vertical distributions of meiofauna. The importance of such interactions should not be overlooked although they may be technically difficult to study.

Finally, the results of the comparison of sampling gear given earlier in this review have important implications for future studies of vertical profiles in the deep sea. It appears that box corer samples must be viewed with some caution due to their demonstrated tendency to reduce meiofaunal numbers in the surface sediment layers. Therefore, until superceded, the multicorer should be regarded as the sampling equipment of choice for this type of study.

6. TEMPORAL PATTERNS

In boreal areas, the communities inhabiting shallow water sediments are closely coupled to seasonal variations in phytoplankton primary production (reviewed by Graf, 1992). For example the spring and autumn phytoplankton blooms in the Kiel Bight are followed by a rapid increase in heat production, oxygen consumption and electron transport system activity and by an increase in bacterial and foraminiferal biomass due to growth and reproduction (Graf et al., 1982, 1983; Altenbach, 1992). The evidence for a similar response by the metazoan meiobenthos is limited and equivocal (Graf et al., 1982). At depths of 30–35 m on the Alaskan shelf, the sedimentation of the spring phytoplankton bloom apparently did not influence the abundance of major metazoan meiobenthic taxa, either seasonally or interannually (Fleeger et al., 1989). However, reproduction in two abundant harpacticoid species did appear to be linked in some way to phytoplankton sedimentation events (Fleeger and Shirley, 1990). Deep-sea meiobenthic communities show similar dynamic characteristics. Although Gooday (1988) and Pfannkuche (1993b) illustrated a pronounced response of deep-sea foraminiferans to aggregated phytodetritus deposition following the phytoplankton blooms in particular areas of the northeast Atlantic (Billet et al., 1983), as yet there is no equivalent evidence for any metazoan taxon (Gooday, 1988; Graf, 1992; Pfannkuche, 1993b). It is mainly the smaller benthic size groups (bacteria, protozoans) that react strongly to POM pulses, emphasizing their particular importance to sediment–water interface flux rates.

At bathyal depths in the Norwegian Sea, certain benthic foraminifera display a very rapid physiological response to organic matter inputs (Linke, 1992) and also a feeding response which leads, within days, to a biomass increase (Altenbach, 1992). Abyssal foraminifera that colonize and feed on phytodetritus probably undergo rapid growth and reproduction, as do species living in the organically enriched, bathyal San Pedro Basin (Corliss and Silva, 1992). At the BIOTRANS site, Pfannkuche *et al.* (1990) and Pfannkuche (1992, 1993b) observed that the foraminiferal meiobenthos switched from being less abundant (around 40%) than the metazoan meiobenthos in March and May to being more abundant (60–70%) after the deposition of phytodetritus in July. The numerical abundance of the metazoan meiobenthos and their vertical distribution profiles remained stable following the 1982 phytodetritus sedimentation event in the Porcupine Seabight (*areas* 1–3). It appears that the foraminiferal meiobenthos are more dynamic than the metazoans, at least in terms of their response to organic matter inputs.

In areas with pronounced seasonality in the supply of organic matter to the seafloor, the timing of sampling is of great importance. Previous attempts to demonstrate temporal variability in faunal composition may have failed mainly because of inadequate sampling. Recently, de Bovée *et al.* (1990) found changes in taxonomic composition at deep stations (2105–2367 m) in the NW Mediterranean between summer and autumn. With increasing sediment pigment content, representing the quantitative food supply to the benthos from planktonic primary production (Thiel, 1978), nematode percentages decrease in autumn, while all other metazoan taxa increase in relative abundance. Even in open oceanic regions where seasonal detritus input occurs, there is evidence of temporal variability in faunal composition (Pfannkuche *et al.*, 1990).

7. SUMMARY AND CONCLUSIONS

The ecology of the meiobenthos in the deep northeast Atlantic is reviewed from published accounts and new data collected in a cooperative European research effort funded by the CEC MAST programme.

Data on the protozoan and metazoan meiobenthos from various sampling stations in the northeast Atlantic (15–53°N, continental margin to Mid-Atlantic Ridge) have been compiled. For the purposes of the review these data are arranged into 30 study *areas* based on geographical and bathymetric proximity (< 1000 m, 1000–3000 m, 3000–4500 m, >4500 m) of the original sampling stations.

A variety of techniques have been used for the collection and processing of deep-sea meiobenthos samples; in some cases this has produced a degree of bias in the results (see Bett *et al.*, 1994). Following Thiel (1993), we suggest that standard procedures should be adopted, for example the use of a multiple corer for sample collection, and a 31 μm sieve mesh as the lower limit for meiobenthos.

The environment of the deep northeast Atlantic is also reviewed in terms of those variables that may exert an important influence on the distribution of the benthos, for example sediment type, sediment oxygen concentrations, availability of organic matter, bottom water masses, and near-bottom currents.

The density of meiobenthos generally decreases with water depth. This appears to be true for the meiobenthos as a whole and for individual component taxa. This general trend, however, is complicated by a number of local/regional influences, for example upwelling areas, elevated near-bottom currents, and sedimentology. In contrast to density, the contribution of major taxa to the meiobenthos appears to be influenced only slightly by water depth. Among the metazoans, nematodes tend to be relatively most abundant in samples from the deepest stations, while other taxa reach their highest relative abundances at the shallowest stations.

Examination of latitudinal variation in density of meiobenthos suggests that there is, to some degree, a relationship between surface primary production and density. Corresponding variations in the composition of major taxa are minimal.

Multivariate analyses of taxon densities confirm the importance of water depth and surface primary production. Together, these two factors exert considerable control on the supply of organic matter to the seafloor and presumably the densities of meiobenthic taxa. Corresponding analyses of major taxon composition did not reveal any general trends in the data, though local factors (e.g. sediment type) do clearly influence taxon composition.

Comparisons of meiobenthos densities from the northeast Atlantic with other areas also suggest a link with surface primary production (and potentially water temperatures). By comparison to the northeast Atlantic, densities are lower in the western Atlantic, Mediterranean, and eastern Pacific, where surface primary production is generally lower, and densities are higher in the northern Atlantic, where production is higher.

There have been few studies of the meiobenthos at the lower taxonomic levels (family, genus, species) in the deep northeast Atlantic. The available data suggest that nematode, copepod and foraminiferal assemblages have a high diversity, both in terms of high species richness and low dominance. This is in accord with the emerging view of the deep-sea floor as a generally high diversity environment (e.g. Grassle and Maciolek, 1992). Family and genus level comparisons of the nematode communities from western and eastern Atlantic locations indicate many similarities. The occurrence of parallel nematode communities in fine shelf and coastal sediments (e.g. Heip *et al.*, 1985) may extend to the fine sediments of the deep sea.

Small-scale (centimetres) spatial variation in the environment of the deep-sea floor is thought to play an important role in many biological processes (e.g. Rice and Lambshead, 1993). Small-scale variation in the meiobenthos has, to date, received little attention in the deep northeast Atlantic. It is, however, clear that at those sites where significant phytodetritus deposition occurs (e.g. *areas* 1–4, 12), the patchy distribution of this organic material over the seafloor influences the distribution of the benthos, particularly the foraminiferal meiobenthos.

The vertical distribution of meiofauna in the northeast Atlantic agrees with the generally observed fact that abundances are generally highest in the surface layers of sediment and decrease with increasing depth. In general, copepods and foraminiferans were distributed somewhat more superficially than nematodes and total metazoa. Multivariate and correlation analysis showed that food supply and sediment characteristics were significant factors in determining the vertical distribution of meiofauna within the sediment but the role of oxygen concentrations, often of most importance in intertidal and shallow sediments, was not identified as significant.

As approximately 90% of all individuals were found in the top 5 cm of sediment, it may be adequate to sample only to this depth, thus reducing the amount of sediment to be processed for each sample. The differences identified in sample quality between various sampling devices could affect recorded vertical profile data and thus the multiple corer is suggested as the equipment of choice.

Temporal variation in deep-sea meiobenthos has been studied at only a few sites. It occurs most notably at those locations which experience a highly seasonal input of phytodetritus (e.g. *areas* 1–4, 12). The biological response to seasonal phytodetritus deposition is particularly marked in certain opportunist foraminiferal species, and may also be detected in small metazoans (e.g. nematodes).

Overall, the importance of the supply of organic matter to the seafloor in determining the distribution of deep-sea meiobenthos is clear. Its influence is apparent at all spatial scales: in global comparisons of oceanic regions, across regional gradients in surface primary production and water depth, in the local influences of upwelling, near-bottom currents and sediment type, and in the small-scale patchy distribution of organic matter on the seafloor. Corresponding temporal variations are less well known; in the case of organisms which exploit phytodetritus, these could range from hourly variation due to tidal redistribution, to the 1-year deposition cycle, and potentially to much longer geological time-scales (see Smart et al., 1994) reflecting changes in ocean climate.

Meiobenthic organisms provide an ideal tool with which to improve our understanding of the structure and function of the deep-sea benthic ecosystem. In practical terms this faunal component can be sampled with a multiple corer, an extremely reliable device that collects high-quality samples suited to a number of studies (biological, physical and chemical). The comparatively high abundance and diversity make the deep-sea meiobenthos ideally suited to quantitative studies and suggest that this group plays an important role in ecological processes. To date most studies have concentrated on larger-scale structural characteristics, and although complicated by the lack of a standard protocol, the relatively simple data on the densities of major taxa obtained show clear relationships with environmental variables. There is, nevertheless, a need for smaller-scale structural studies and process/function studies generally. Improved taxonomic resolution, work at the genus and species level, coupled with studies at smaller physical scales, will enable the factors controlling community structure to be examined in more detail. One area of current interest is the high diversity (and its maintenance) of deep-sea benthos, an understanding of which is likely to require a knowledge of ecology in "patchy" environments. Meiobenthos can be sampled at the physical scales (centimetres) on which seafloor patchiness is likely to operate. At both large and small scales, process studies addressing energy and nutrient fluxes are now a dominant theme in biological oceanography. Given that members of the deep-sea meiobenthos are likely to play a significant role in these processes, that they are especially amenable to quantitative study, and that they appear to be more responsive (particularly the foraminiferal component) to variations in the input of organic matter to the seafloor than members of larger benthic size categories, the meiobenthos should be a key target group for any benthic process study. While many of the issues mentioned above can be addressed by general sampling programmes or through directed sampling as "natural experiment" some problems can only be tackled using a formal experimental or manipulative approach. The practical and logistic problems of deep-sea experimentation, either in situ or ship-board, favour the use of small "observational units". This should prompt the use of meiobenthos as the most appropriate research tool.

ACKNOWLEDGEMENTS

The authors acknowledge the CEC-MAST programmes 0037-C(EDB) and MAS2-CT92-0033 for the major part of the financial support. Also the Belgian National Science Foundation (FKFO 32.0094.92) and the University of Gent, Belgium (supporting project Concerted Actions 92/98-08) are greatly acknowledged. We are also grateful to Dr K. Hostens for the preparations of several figures on computer.

REFERENCES

Aller, J.Y. and Aller, R.C. (1986). Evidence for localized enhancement of biological activity associated with tube and burrow structures in deep-sea sediments at the HEBBLE site, western North Atlantic. *Deep-Sea Research* **33**, 755–790.

Alongi, D. M. (1992). Bathymetric patterns of deep-sea benthic communities from bathyal to abyssal depths in the western South Pacific (Solomon and Coral Seas). *Deep-Sea Research* **39**, 549–566.

Alongi, D.M. and Pichon, M. (1988). Bathyal meiobenthos of the western Coral Sea: distribution and abundance in relation to microbial standing stocks and environmental factors. *Deep-Sea Research* **35**, 491–503.

Altenbach, A.V. (1992). Short term processes and patterns in the foraminiferal response to organic flux rates. *Marine Micropaleontology* **19** (1/2), 119–129.

Apostolescu, V., Montarges, M. and Wanneson, J. (1978). Morphologie, structure et sediments de l'Atlantique—Nord (Étude bibliographique). *Institute Francais du Petrole, Division Geologie, Report 22 328*, 54 pp.

Auffret, G. (1985). Environnement morphologique et sédimentologique. *In* "Peuplements profonds du Golfe de Gascogne" (L. Laubier and C. Monniot, eds), pp. 43–70. IFREMER, Paris.

Barnett, P.R.O., Watson, J. and Connelly, D. (1984). A multiple corer for taking virtually undisturbed samples from shelf, bathyal and abyssal sediments. *Oceanologica Acta* **7**, 399–408.

Berger, W.H. (1975). Deep-sea carbonates: dissolution profiles from foraminiferal preservation. *Cushman Foundation for Foraminiferal Research Special Publication* **13**, 82–86.

Berger, W.H. (1989). Global maps of ocean productivity. *In* "Productivity of the Ocean: Past and Present" (W.H. Berger, V.S. Smetacek and G. Wefer, eds), pp. 429–455. *Dahlem Workshop Reports, Life Sciences Research Report* **44**. John Wiley, Chichester.

Berger, W.H., Fischer, F., Lai, C. and Wu, G. (1988). OCEAN carbon flux: global maps of primary production and export. *In* "Biogeochemical Cycling and Fluxes Between the Deep Euphotic Zone and Other Oceanic Realms" (Agrian ed.), pp. 131–176. *NOAA National Underseas Research Program, Research Report* **88–1**.

Bett, B., Vanreusel, A., Vincx, M., Soltwedel, T., Pfannkuche, O., Lambshead, P.J.D., Gooday, A.J., Ferrero, T. and Dinet, A. (1994) Sampler bias in the quantitative study of deep-sea meiobenthos. *Marine Ecology Progress Series* **104**, 197–203.

Billett, D.S.M., Lampitt, R.S., Rice, A.L. and Mantoura, R.F.C. (1983). Seasonal sedimentation of phytoplankton to the deep-sea benthos. *Nature, London* **302**, 520–522.

Biscaye, P.E., Kolla, V., and Turekian, K.K. (1976). Distribution of calcium carbonate in surface sediments of the Atlantic Ocean. *Journal of Geophysical Research* **81**, 2595–2603.

Blomqvist, S. (1991). Quantitative sampling of soft-bottom sediments: problems and solutions. *Marine Ecology Progress Series* **72**, 295–304.

Bodin, P. (1968). Copépodes harpacticoides des étages bathyal et abyssal du Golfe de Gascogne. *Mémoires du Muséum National d'Histoire naturelle, Paris, Series A, Zoologie* **55** (1), 1–107.

Brady, H.B. (1884). Report on the Foraminifera dredged by *H.M.S. Challenger* during the years 1873–1876. *Reports of the Scientific results of the Voyage of the H.M.S. Challenger* **9** (Zoology), 1–814.

Carney, R.S., Haedrich, R.L. and Rowe, G.T. (1983). Zonation of Fauna in the Deep Sea. *In*: "Deep-Sea Biology" (G.T. Rowe, ed.) **8**, 371–399. John Wiley, London.

Cartwright, N.G. (1988). Biological and ecological studies on Benthic Foraminifera from the Bathyal and Abyssal Northeast Atlantic. PhD Thesis: University of Reading.

Christiansen, B. and Thiel, H. (1992). Deep-sea epibenthic megafauna of the northeast Atlantic and biomass at three mid-oceanic locations estimated from photographic transects. *In* "Deep-Sea Food Chains and the Global Carbon Cycle" (G.T. Rowe and V. Pariente, eds), pp. 125–138. Kluwer, Dordrecht.

Clarke, K.R. and Ainsworth, M. (1993). A method of linking multivariate community structure to environmental variables. *Marine Ecology Progress Series*, **92**, 205–219.

Corliss, B.H. and Chen, C. (1988) Morphotype patterns of Norwegian deep-sea benthic foraminifera and ecological implications. *Geology* **16**, 716–719.

Corliss, B.H. and Silva, K. (1992). Rapid growth of deep-sea benthic foraminifera from the San Pedro Basin, California. *Program & Abstracts, Fourth International Conference on Paleoceanography, 21–25 September 1992, Kiel, Germany*, pp. 88–89.

Coull, B.C. (1988). Ecology of marine meiofauna. *In* "Introduction to the Study of Meiofauna" (R.P. Higgins and H. Thiel, eds), pp. 18–38. Smithsonian Institution Press, Washington, DC.

Coull, B.C., Ellison, R.L., Fleger, J.W., Higgins, R.P., Hope, W.D., Hummon, W.D., Rieger, R.M., Sterrer, W.E., Thiel, H. and Tietjen, J.H. (1977). Quantitative estimates of the meiofauna from the deep-sea off North Carolina, USA. *Marine Biology* **39**, 233–240.

de Bovée, F., Guidi, L. and Soyer, J. (1990). Quantitative distribution of deep-sea meiobenthos in the North occidental Mediterranean basin (Gulf of Lions). *Continental Shelf Research* **10**, 1123–1145.

Decreamer, W. (1983). Five new species of deep-sea desmoscolecids (Nematoda-Desmoscolecida) from the Bay of Biscay. *Bulletin de l'Institut royal de Science naturelle de Belgique*, **55**, 1–26.

Desbruyères, D., Deming, J.W., Dinet, A. and Khripounoff, A. (1985). Réactions de l'écosystème profond aux perturbations: nouveaux résultats expérimentaux. *In* "Peuplements profonds du Golfe de Gascogne" (L. Laubier and C. Monniot, eds), pp. 121–142. IFREMER, Paris.

Dickson, R.R. and Kidd, R.B. (1987). Deep circulation in the Southern Rockall Trough—the oceanographic setting of site 610. *Initial Reports of the Deep Sea Drilling Program*, **9A**, 1061–1074.

Dickson, R.R., Goulds, W.J., Medler, K.J. and Gmitrowicz, E.M. (1986).

Seasonality in currents of the Rockall Channel. *Proceedings of the Royal Society of Edinburgh (B)* **88**, 103–125.

Dinet, A. (1973). Distribution quantitative du méiobenthos profond dans la région de la dorsale Walvis (Sud-Ouest Africain). *Marine Biology* **20**, 20–26.

Dinet, A. (1976). Ètude quantitative du méiobenthos dans le secteur nord de la Mer Egée. *Acta Adriatica* **18**, 83–88.

Dinet, A. (1977). Données quantitative sur le méiobenthos bathyal de la Mer de Norvège. Pages 13–14. *In* "Géochimie Organique des Sédiments Marine Profond. Orgon I, Mer de Norvège, Aout, 1974". CEPM-CNEXO, Comité Etudes Geo-Chim, Mar., (R. Pelet and A. Combaz, eds) CNRS, Paris.

Dinet, A. (1979). A quantitative survey of meiobenthos in the deep Norwegian Sea. *Ambio Special Report* **6**, 75–77.

Dinet, A. (1981). Description de deux nouvelles espèces abyssales de *Pontostratiotes* (Crustacea, Copepoda, Harpacticoida). *Bulletin de la Société zoologique de France* **106**, 201–211.

Dinet, A. (1985). Répartition du genre *Pontostratiotes* (Copepoda, Harpacticoida). *In* "Peuplements profonds du Golfe de Gascogne" (L. Laubier and C. Monniot, eds), pp. 419–424. IFREMER, Paris.

Dinet, A. and Khripounoff, A. (1980). Rapports quantitatifs entre le meiobenthos et la matière organique en zone abyssale. *In* "Colloques Internationaux du C.N.R.S. no 293. Biogéochimie de la matière organique à l'interface eau-sédiment marin" (R. Dumas, ed), pp. 319–324. CNRS, Paris.

Dinet, A. and Vivier, M.H. (1977). Le méiobenthos abyssal du Golfe de Gascogne. I. Considération sur les données quantitatives. *Cahiers de Biologie Marine* **18**, 85–97.

Dinet, A. and Vivier, M.H. (1979). Le méiobenthos abyssal du Golfe de Gascogne. II. Les peuplements de nématodes et leur diversité spécifique. *Cahiers de Biologie Marine* **20**, 109–123.

Dinet, A. and Vivier, M.H. (1981). Ajustement de la loi de Motomura aux peuplements nématologiques abyssaux du Golfe de Gasgogne. *Téthys* **10**, 141–148.

Dinet, A., Laubier, L., Soyer, J. and Vitiello, P. (1973). Resultats biologiques de la Campagne Polymede II. Le méiobenthos abyssal. *Rapport de la Commission Internationale pour l'Exploration de la Mediterranee* **21**, 701–704.

Dinet, A., Desbruyères, D. and Khripounoff, A. (1985). Abondance des peuplements macro- et méiobenthiques: répartition et stratége d'échantillonnage. *In* "Peuplements profonds du Golfe de Gascogne" (L. Laubier and C. Monniot, eds), pp. 121–142. IFREMER, Brest.

Embley, R.W. (1976). New evidence for debris flow deposits in the deep-sea. *Geology* **4**, 371–374.

Emery, K.O. and Uchupi, E. (1984). "The Geology of the Atlantic Ocean". Springer, New York.

Field, J.G, Clarke, K.R. and Warwick, R.M. (1982). A practical strategy for analysing multispecies distribution patterns. *Marine Ecology Progress Series* **8**, 37–52.

Fleeger, J.W. and Shirley, T.C. (1990). Meiofaunal responses to sedimentation from an Alaskan spring bloom. II. Harpacticoid population dynamics. *Marine Ecology Progress Series* **59**, 239–247.

Fleeger, J.W., Thistle, D. and Thiel, H. (1988). Sampling equipment. *In* "Introduction to the Study of Meiofauna" (R.P. Higgins and H. Thiel, eds), pp. 115–125. Smithsonian Institution Press, Washington, DC.

Fleeger, J.W, Shirley, T.C. and Ziemann, D.A. (1989). Meiofaunal responses to sedimentation from an Alaskan spring bloom. I. Major taxa. *Marine Ecology Progress Series* **51**, 131–145.

Gage, J.D. and Tyler, P.A. (1991). "Deep-Sea Biology. A Natural History of Organisms at the Deep-Sea Floor". Cambridge University Press, Cambridge.

Gooday, A.J. (1986a). Meiofaunal foraminiferans from the bathyal Porcupine Seabight: size structure, taxonomic composition, species diversity and vertical distribution in the sediment. *Deep-Sea Research* **33**, 1345–1372.

Gooday, A.J. (1986b). Foraminifera in meiofauna samples from the Bathyal Northeast Atlantic. *Sarsia* **71**, 275–287.

Gooday, A.J. (1988) A benthic foraminiferal response to the deposition of phytodetritus in the deep-sea. *Nature London* **332**, 70–73.

Gooday, A.J. (1990). Recent deep-sea agglutinated foraminifera: a brief review. *In* "Paleoecology Biostratigraphy, Paleoceanography and Taxonomy of Agglutinated Foraminifera" (C. Hemleben, W. Kaminski and D.B. Scott, eds), pp. 271–304. *NATO ASI Series*, Vol. *C237*. Kluwer, Dordrecht.

Gooday, A.J. and Lambshead, P.J.D. (1989). Influence of seasonally deposited phytodetritus on benthic foraminiferal populations in the bathyal northeast Atlantic: the species response. *Marine Ecology Progress Series* **58**, 53–67.

Gooday, A.J. and Turley, C.M. (1990). Responses by benthic organisms to inputs of organic material to the ocean floor: a review. *Philosophical Transactions of the Royal Society of London* A **331**, 119–138.

Graf, G. (1992). Benthic-pelagic coupling : a benthic view. *Oceanography and Marine Biology Annual Review* **30**, 149–190.

Graf, G., Bengtson, W., Diesner, U., Schulz, R. and Theede, H. (1982). Benthic response to sedimentation of a spring phytoplankton bloom: process and budget. *Marine Biology* **67**, 201–220.

Graf, G., Schulz, R., Peinert, R. and Meyer-Reil, L.-A. (1983). Benthic response to sedimentation events during autumn to spring at a shallow-water station in the Western Kiel Bight. *Marine Biology* **77**, 235–246.

Grassle, J.F. and Maciolek N.J. (1992). Deep-sea species richness: regional and local diversity estimates from quantitative bottom samples. *American Naturalist* **139**, 313–341.

Grassle, J.F. and Mosse-Porteous, L.S. (1987). Macrofaunal colonization of disturbed deep-sea environments and the structure of deep-sea benthic communities. *Deep-Sea Research* **34**, 1911–1950.

Gray, J.S. (1974). Animal-sediment relationships. *Oceanography and Marine Biology Annual Review* **12**, 223–261.

Gross, T.F., Williams, A.J. and Nowell, A.R.M. (1988). A deep-sea sediment transport storm. *Nature* (London), **331**, 518–520.

Hansen, S.D. (1978). Nahrung und Fressverhalten bei Sedimentfressern dargestellt am Beispiel von Sipunculiden und Holothurien. *Helgoländer wissenschaftlichen Meeresuntersuchungen* **31**, 191–221.

Heip, C., Vincx, M. and Vranken, G. (1985). The ecology of marine nematodes. *Oceanography and Marine Biology Annual Review* **23**, 399–489.

Hessler, R.R. and Jumars, P.A. (1974). Abyssal community analysis from replicate box cores in the central North Pacific. *Deep-Sea Research* **21**, 185–209.

Hill, P.R. (1987). Characteristics of sediments from Feni and Gardas drifts, sites 610 and 611, Deep Sea Drilling Project Leg 94. *Initial Reports Deep Sea Drilling Program* **94**, 1075–1082.

Hollister, C.D. and McCave, I.N. (1984). Sedimentation under deep-sea storms. *Nature London* **309**, 220–225.

Huggett, P.J. (1986). Mapping of hemipelagic versus turbiditic muds by feeding traces observed in deep-sea photographs. *In* "Geology and Geochemistry of Abyssal Plains" (P.P.E. Weaver and J. Thomson, eds), pp. 105–112. *Geological Society, Special Publication* **31**.

Jacobi, R.D. (1976). Sediment slides on the northwestern continental margin of Africa. *Marine Geology* **22**, 157–173.

Jensen, P. (1988). Nematode assemblages in the deep-sea benthos of the Norwegian Sea. *Deep-Sea Research* **35**, 253–266.

Jensen, P., Rumohr, J. and Graf, G. (1992). Sedimentological and biological differences across a deep-sea ridge exposed to advection and accumulation of fine-grained particles. *Oceanologica Acta* **15**, 287–296.

Kidd, R.B. and Huggett, Q.J. (1981). Rockdebris on abyssal plains in the Northeast Atlantic: a comparison of epibenthic sledge hauls and photographic surveys. *Oceanologica Acta* **4**, 99–104.

Kidd, R.B., Hunter, P.M. and Simm, R.W. (1986). Turbidity-current and debris-flow pathways to the Cape Verde Basin: status of long-range side-scan sonar (GLORIA). *In* "Geology and Geochemistry of Abyssal Plains" (P.P.E. Weaver and J. Thomson, eds), pp. 33–48. *Geological Society Special Publication* **31**.

Klein, H. and Mittelstaedt, E. (1992). Currents and dispersion in the abyssal Northeast Atlantic. Results from the NOAMP field program. *Deep-Sea Research* **39**, 1727–1745.

Kornicker, L.S. (1989). Bathyal and abyssal myodocopid ostracoda of the Bay of Biscay and vicinity. *Smithsonian Contributions to Zoology* **461**, 1–134.

Lambshead, P.J.D. and Gooday, A.J. (1990). The impact of seasonally deposited phytodetritus on epifaunal and shallow infaunal benthic foraminiferal populations in the bathyal north-east Atlantic: the assemblage response. *Deep-Sea Research* **37A**, 1263–1283.

Lampitt, R.S. (1985). Fast living on the ocean floor. *New Scientist* **1445**, 37–40.

Lampitt, R.S., Billett, D.S.M. and Rice, A.L. (1986). Biomass of the invertebrate megabenthos from 500 to 4100 m in the northeast Atlantic Ocean. *Marine Biology* **93**, 69–81.

Linke, P. (1992). Metabolic adaptations of deep-sea benthic foraminifera to seasonally varying food input. *Marine Ecology Progress Series* **81**, 51–63.

Lochte, K. (1992). Bacterial standing stock and consumption of organic carbon in the benthic boundary layer of the abyssal North Atlantic. *In* "Deep-sea Food Chains and the Global Carbon Cycle" (G.T. Rowe and V. Pariente, eds), pp. 1–10. Kluwer, Dordrecht.

Lutze, G.F. and Thiel, H. (1989). Epibenthic foraminifera of elevated microhabitats: *Cibicidoides wuellerstorfi* and *Planulina ariminensis*. *Journal of Foraminiferal Research* **19**, 153–158.

Mart, Y., Auffret, G.A., Auzende, J.M. and Pastouret, L. (1979). Geological observations from a submersible on the western continental slope of the American Massif. *Marine Geology* **31**, 61–68.

Masson, D.G., Kidd, R.B., Gardner, J.V., Huggett, Q.J. and Weaver, P.P.E. (1994). Saharan continental rise: facies distribution and sediment slides. *In* "Geological Evolution of Atlantic Continental Rises" (V.W. Poag and P.C. Degrocianski, eds), pp. 3–10. Van Nostrand Reinhold, New York.

McCave, N. (1991). Deep flashes reveal fluffing beds. *BOFS News and Views* **5**, 5–6.

Murray, J.W. (1991). "Ecology and Palaeoecology of Benthic Foraminifera". Longman, Harlow.

Norris, S. and McDonald, N. (1986). Current meter observations near the Porcupine Bank 1981–1983. Ministry of Agriculture, Fisheries and Food. Directorate of Fisheries Research Lowestoft, Fisheries Data Report **8**, 103 pp.

Pace, M.L., Knauer, G.A., Karl, D.M. and Martin, J.H. (1987). Primary production, new production and vertical flux in the eastern Pacific Ocean. *Nature, London*, **325**, 803–804.

Parker, W.K. and Jones, T.R. (1856). On some Foraminifera of the North Atlantic and Arctic Oceans, including Davis Straits and Baffin Bay. *Philosophical Transactions of the Royal Society of London* **155**, 325–441.

Patterson, D.J., Larsen, J. and Corliss, J.O. (1982). The ecology of heterotrophic flagellates and ciliates living in marine sediments. *Progress in Protistology* **3**.

Pfannkuche, O. (1985). The deep-sea meiofauna of the Porcupine Seabight and abyssal plain (NE Atlantic): population structure, distribution, standing stocks. *Ocèanologica Acta* **8(3)**, 343–353.

Pfannkuche, O. (1992). Organic carbon flux through the benthic community in the temperate abyssal northeast Atlantic. *In* "Deep-Sea Food Chains and the Global Carbon Cycle" (G.T. Rowe and V. Pariente, eds), pp. 183–198, Kluwer, Dordrecht.

Pfannkuche, O. (1993a). Benthic standing stock and metabolic activity in the bathyal Red Sea from 17°N to 27°N. *P.S.Z.N.I., Marine Ecology* **14**, 67–79.

Pfannkuche, O. (1993b). Benthic response to the sedimentation of particulate organic matter at the BIOTRANS station, 47°N, 20°W. *Deep-Sea Research* **II**, **40**, 135–149.

Pfannkuche, O. and Thiel, H. (1987). Meiobenthic stocks and benthic activity on the NE-Svalbard Shelf and in the Nansen Basin. *Polar Biology* **7**, 253–266.

Pfannkuche, O. and Thiel, H. (1988). Chapter 9. Sample processing. *In* "Introduction to the Study of Meiofauna" (R.G. Higgins and H. Thiel, eds), pp. 134–145. Smithsonian Institute Press, Washington DC.

Pfannkuche, O., Theeg, R. and Thiel, H. (1983). Benthos activity, abundance and biomass under an area of low upwelling off Morocco, Northwest Africa. *"Meteor" Forschungsergebnisse* **36**, 85–96.

Pfannkuche, O., Beckman, W., Christiansen, B., Lochte, K., Rheinheimer, G., Thiel, H. and Weikert, H. (1990). BIOTRANS, Biologisches vertikaltransport und energiehaushault in der bodennahen Wasserschicht der Tiefsee. *Berichte aus dem Zentrum für Meeres- und klimaforschung der Universität Hamburg* **10**, 159 pp.

Rachor, E. (1975). Quantitative Untersuchungen über das Meiobenthos der nordostatlantischen Tiefsee. *"Meteor" Forschungsergebnisse D* **21**, 1–10.

Renaud-Mornant, J. (1989). Espèces nouvelles de Florarctinae de l'Atlantique Nord-est et du Pacifique Sud (Tardigrada, Arthrotardigrada). *Bulletin du Muséum national d'Histoire naturelle, Paris*, 4th series **11**, A, 571–592.

Renaud-Mornant, J. and Gourbault, N. (1990). Evaluation of abyssal meiobenthos in the eastern central Pacific (Clarion–Clipperton fracture zone). *Progress in Oceanography* **24**, 317–329.

Rice, A.L. and Lambshead, P.J.D. (1993). Patch dynamics in the deep-sea benthos: the role of a heterogeneous supply of organic matter. Proceedings of British Ecological Society/American Society of Limnology and Oceanography Symposium on "Aquatic ecology: scale pattern and process", University College, Cork, April 1992.

Rice, A.L., Billett, D.S.M., Fry, J. John, A.W.G., Lampitt, R.S., Mantoura, R.F.C. and Morris, R.J. (1986). Seasonal deposition of phytodetritus to the deep-sea floor. *Proceedings of the Royal Society of Edinburgh* **88B**, 265–279.

Rice, A.L., Billett, D.S.M., Thurston, M.H. and Lampitt, R.S. (1991). The Institute of Oceanographic Sciences Biology Programme in the Porcupine Seabight: background and general introduction. *Journal of the Marine Biological Association of the UK* **71**, 281–310.

Richardson, M.J., Wimbush, M. and Mayer, L. (1981). Exceptionally strong near-bottom flows on the continental rise off Nova Scotia. *Science* **213**, 887–888.

Riemann, F. (1974). On hemisessile nematodes with flagelliform tails living in marine soft bottoms and on micro-tubules found in deep-sea sediments. *Microfauna des Meeresbodens* **40**, 1–15.

Robinson, M.K., Bauer, R.A. and Schroeder, E. (1979). Atlas of North Atlantic–Indian Ocean monthly mean temperatures and mean salinities of the surface layer. *U.S. Naval Oceanographic Office, Reference Publication* **18**, 1–17.

Romano, J.-C. and Dinet, A. (1981). Relation entre l'abondance du méiobenthos et la biomasse des sédiments superficiels estimée par la mesure des adénosines 5'phosphate (ATP, ADP, AMP). In "Géochimie organique des sédiments marins profonds, ORGON IV, golfe d'Aden, mer d'Oman", pp. 159–180. CNRS, Paris.

Rona, P.A. (1980). The Central North Atlantic Ocean basin and continental margin: geology, geophysics, geochemistry and resources, including the Trans-Atlantic Geotraverse (TAG). *NOAA Atlas 3*. NOAA Environmental Research Laboratories.

Rutgers van der Loeff, M.M. (1991). Oxygen in pore waters of deep-sea sediments. *Philosophical Transactions of the Royal Society of London* **A331**, 69–84.

Rutgers van der Loeff, M.M. and Lavaleye, M.S.S. (1986). Sediments, fauna and the dispersal of radionuclides at the N.E. Atlantic dumpsite for low-level radioactive waste. Report of the Dutch DORA program. Netherlands Institute for Sea Research, pp. 1–134.

Schnitker, D. (1980). Quaternary deep sea foraminifers and bottom water masses. *Annual Review of Earth and Planetary Sciences* **8**, 343–370.

Schriever, G., Bussau, C. and Thiel, H. (1991). DISCOL—Precautionary environmental impact studies for future manganese nodule mining and first results on meiofauna abundance. *Proceedings and Advances of the Marine Technology Conference* **4**, 47–57.

Shirayama, Y. (1983). Size-structure of deep-sea meiobenthos in the western Pacific. *International Revue des gesamten Hydrobiologie* **68**, 799–810.

Shirayama, Y. (1984a). The abundance of deep-sea meiobenthos in the Western Pacific in relation to environmental factors. *Oceanologica Acta* **7**, 113–121.

Shirayama, Y. (1984b). Vertical distribution of meiobenthos in the sediment profile in bathyal, abyssal and hadal deep-sea systems in the Western Pacific. *Oceanologica Acta* **7**, 123–129.

Sibuet, M. (1984). Quantitative distribution of echinoderms (Holothurioidea, Asteroidea, Ophiuroidea, Echinoidea) in relation to organic matter in the sediment, in deep sea basins of the Atlantic Ocean. "Proceedings Fifth International Echinoderm Conference, Galway, 24–29 September 1984" (B.F. Keegan and B.D.S. O'Connor, eds), pp. 99–108. A.A. Balkema, Rotterdam.

Sibuet, M., Monniot, C., Desbruyeres, D., Dinet, A., Khripounoff, A., Rowe, G. and Segonzac, M. (1984). Peuplements benthiques et characteristiques trophiques du milieu dans la plaine abyssale de Demerara dans l'ocean Atlantique. *Ocèanologica Acta* **7**, 345–358.

Sibuet, M., Lambert, C.E., Chesselet, R. and Laubier, L. (1989). Density of the major size groups of benthic fauna and trophic input in deep basins of the Atlantic Ocean. *Journal of Marine Research* **47**, 851–867.

Smart, C.W., King, S.C., Gooday, A.J., Murray, J.W. and Thomas, E. (1994). A benthic foraminiferal proxy of pulsed organic matter paleofluxes. *Paleoceanography*, in press.

Snider, L.J., Burnett, B.R. and Hessler, R.R. (1984). The composition and distribution of meiofauna and nanobiota in a central North Pacific deep-sea area. *Deep-Sea Research* **31**, 1225–1249.

Soetaert, K., Heip, C. and Vincx, M. (1991a). The meiobenthos along a Mediterranean deep-sea transect off Calvi (Corsica) and in an adjacent canyon. *P.S.Z.N.I., Marine Ecology* **12**(3), 227–242.

Soetaert, K., Heip, C. and Vincx, M. (1991b). Diversity of nematode assemblages along a Mediterranean deep-sea transect. *Marine Ecology Progress Series* **75**, 275–282.

Soltwedel, T. (1993). Meiobenthos und biogene Sedimentkomponenten im tropischen Ost-Atlantik. *Berichte aus dem Zentrum für Meeres- und Klimaforschung der Universität Hamburg* **13**, 178 pp.

Sorensen, J. and Wilson, T.R.S. (1984). A headspace technique for oxygen measurements in deep-sea sediment cores. *Limnology and Oceanography* **29**, 650–652.

Stein, R. (1991). "Accumulation of Organic Carbon in Marine Sediments". *Lecture Notes in Earth Sciences* **34**, 217 pp. Springer, Berlin.

Streeter, S. (1973). Bottom water and benthonic foraminifera in the North Atlantic-Glacial-Interfacial contrasts. *Quaternary Research* **3**, 131–141.

Thiel, H. (1966). Quantitative Untersuchungen über die Meiofauna des Tiefseebodens. *Veröffenlichungen des Institut für Meeresforschungen Bremerhaven* (Sonderband) **2**, 131–148.

Thiel, H. (1972a). Meiofauna und Struktur der benthischen Lebensgemeinschaft des Iberischen Tiefseebeckens. *"Meteor" Forschungsergebnisse D* **12**, 36–51.

Thiel, H. (1972b). Die Bedeutung der Meiofauna in küsternfernen benthischen Lebensgemeinschaften verschiedener geographischer Regionen. *Verhandlungsbericht der Deutschen zoologische Gesellschaft* **65**, 42–57.

Thiel, H. (1975). The size structure of the deep-sea benthos. *International Revue des gesamten Hydrobiologie* **60**, 575–606.

Thiel, H. (1978). Benthos in upwelling regions. In "Upwelling Ecosystems" (R. Boje and M. Tomczak, eds), pp. 124–138. Springer, Berlin.

Thiel, H. (1979a). First quantitative data on the deep Red Sea benthos. *Marine Ecology Progress Series* **1**, 347–350.

Thiel, H. (1979b). Structural aspects of deep-sea benthos. *Ambio Special Report* **6**, 25–31.

Thiel, H. (1982). Zoobenthos of the CINECA area and other upwelling regions. *Rapport et Procès-Verbaux des Réunions du Conseil International pour l'Exploration de la Mer* **180**, 323–334.

Thiel, H. (1983). Meiobenthos and nanobenthos of the deep-sea. In "Deep-sea Biology" (G.T. Rowe, ed), pp. 167–230. Wiley, New York.

Thiel, H. (1993). Benthos size classification. *Deep-Sea Newsletter* **20**, 8–11.

Thiel, H., Pfannkuche, O., Theeg, R. and Schriever, G. (1987). Benthic metabolism and standing stock in the central and northern Red Sea. *P.S.Z.N.I., Marine Ecology* **8**, 1–20.

Thiel, H., Pfannkuche, O., Schriever, G., Lochte, K., Gooday, A.J., Hemleben, C., Mantoura, R.F.C., Turley, C.M., Patching, J.W. and Rieman, F. (1989–

1990). Phytodetritus on the deep-sea floor in a central oceanic region of the northeast Atlantic. *Biological Oceanography* **6**, 203–239.

Thistle, D. (1983). The role of biologically produced habitat heterogeneity in deep-sea diversity maintenance. *Deep-Sea Research* **30**, 1235–1245.

Thistle, D. and Eckman, J.E. (1988). Response of harpacticoid copepods to habitat structure at a deep-sea site. *Hydrobiologia* **167/168**, 143–149.

Thistle, D. and Sherman, K. (1985). The nematode fauna of a deep-sea site exposed to strong near-bottom currents. *Deep-Sea Research* **32**, 1077–1088.

Thistle, D., Yingst, J.Y. and Fauchald, K. (1985). A deep-sea benthic community exposed to strong near-bottom currents on the Scotian Rise (Western Atlantic). *Marine Geology*, **66**, 91–112.

Tietjen, J.H. (1971). Ecology and distribution of deep-sea meiobenthos off North Carolina. *Deep-Sea Research* **18**, 941–957.

Tietjen, J.H. (1976). Distribution and species diversity of deep-sea nematodes off North Carolina. *Deep-Sea Research* **23**, 755–768.

Tietjen, J.H. (1984). Distribution and species diversity of deep-sea nematodes in the Venezuela Basin. *Deep-Sea Research* **31(2)**, 119–132.

Tietjen, J.H. (1989). Ecology of deep-sea nematodes from the Puerto Rico Trench area and Hatteras Abyssal Plain. *Deep-Sea Research* **36**, 1567–1577.

Tietjen, J.H. (1992). Abundance and biomass of metazoan meiobenthos in the deep-sea. *In* "Deep-Sea Food Chains and the Global Carbon Cycle" (G.T. Rowe and V. Pariente, eds), pp. 45–62. Kluwer, Dordrecht.

Tietjen, J.H., Deming, J.W., Rowe, G.T., Macko, S. and Wilke, R.J. (1989). Meiobenthos of the Hatteras Abyssal Plain and Puerto Rico Trench: abundance, biomass and associations with bacteria and particulate fluxes. *Deep-Sea Research* **36**, 1567–1577.

Tyler, P.A. and Zibrowius, H. (1992). Submersible observations of the invertebrate fauna on the continental slope southwest of Ireland (NE Atlantic ocean). *Oceanologica Acta* **15**, 211–226.

Udintsev, G.B. (1989–1990). "International Geological–Geophysical Atlas of the Atlantic Ocean". IOC (of UNESCO), Min. Geol. USSR, Ac. Sci. USSR, GUGK USSR Moscow.

Vangreisheim, A. (1985). Hydrologie et circulation profonde. *In* "Peuplements profonds du Golfe de Gascogne" (L. Laubier and C. Monniot, eds), pp. 43–70. IFREMER, Paris.

van Harten, D. (1990). Modern abyssal ostracod faunas of the eastern Mid-Atlantic Ridge area in the North Atlantic and a comparison with the Mediterranean. *In* "Ostracoda and Global Events" (R. Whatley and C. Maybury, eds), pp. 321–328. Chapman & Hall, London.

Vanreusel, A., Vincx, M., Van Gansbeke D. and Gijselinck, W. (1992). Structural analysis of the meiobenthos communities of the shelf break area in two stations of the Gulf of Biscay (N.E. Atlantic). *Belgian Journal of Zoology* **122**, 185–202.

Vivier, M.H. (1978). Influence d'un déversement industriel profond sur la nématofaune (canyon de Cassidaigne, Méditerranée). *Téthys* **8(4)**, 307–321.

Wallace, H.E., Thomson, J., Wilson, T.R.S., Weaver, P.P.E., Higgs, N.C. and Hydes, D.J. (1988). Active diagenetic formation of metal-rich layers in the northeast Atlantic sediments. *Geochemica et Cosmochimica Acta* **52**, 1557–1569.

Weaver, P.P.E. and Kuijpers, A. (1983). Climate control of turbidite deposition on the Madeira Abyssal Plain. *Nature, London* **306**, 360–363.

Weaver, P.P.E., Searle, R.C. and Kuijpers, A. (1986). Turbidite deposition and the origin of the Madeira Abyssal Plain. In "North Atlantic Palaeoceanography" (C.P. Summerhayes and N.J. Shackleton, eds), pp. 131–143. Geological Society Special Publication 21.

Weston, J.F. (1985). Comparison between recent benthic foraminiferal faunas of the Porcupine Seabight and western approaches continental slope. Journal of Micropaleontology 4, 165–183.

Weston, J.F. and Murray, J.W. (1984). Benthic foraminifera as deep-sea water mass indicators. In "Benthos '83, 2nd International Symposium on Benthic Foraminifera, Paris, April 1983" (H.J. Oerbli, ed.), pp. 605–610.

Wieser, W. (1953). Beziehungen zwischen Mundhölengestalt, Ernährungsweise und Vorkommen bei freilebenden marinen Nematoden. Archive der Zoologie (2) 4, 439–484.

Wigley, R.L. and McIntyre, A.D. (1964). Some quantitative comparisons of offshore meiobenthos and macrobenthos south of Martha's Vineyard. Limnology and Oceanography 9, 485–493.

Wilson, T.R.S. and Wallace, H.E. (1990). The rate of dissolution of calcium carbonate from the surface of deep-ocean turbidite sediments. Philosophical Transactions of the Royal Society of London A 331, 41–49.

Wilson, R.R., Smith, J.R., Rosenblatt, K.L. (1985). Megafauna associated with bathyal seamounts in the central North Pacific Ocean. Deep-Sea Research 32, 1243–1254.

Wilson, T.R.S., Thomson, J., Hydes, D.J., Colley, S., Culkin, F. and Sorensen, J. (1986). Oxidation fronts in pelagic sediments: diagenetic formation of metal-rich layers. Science 232, 972–975.

Woods, D.R. and Tietjen, J.H. (1985). Horizontal and vertical distribution of meiofauna in the Venezuela Basin. Marine Geology 68, 233–241.

Worthington, L.V. (1976). "The North Atlantic Circulation". Johns, Hopkins University Press, Baltimore.

The Biology of Oniscid Isopoda of the Genus *Tylos*

A. C. Brown and F. J. Odendaal

Department of Zoology, University of Cape Town, South Africa 7700

1. INTRODUCTION

There appears to be no doubt that terrestrial Crustacea in general evolved from marine ancestors (Bliss, 1968; Edney, 1968) and that in most cases their invasion routes were via the marine intertidal zone. This

ADVANCES IN MARINE BIOLOGY VOL. 30
ISBN 0–12–026130–8

transition from sea to land usually involved extensive adaptations (Hurley, 1968), especially with regard to locomotion, respiration, water balance, ionic regulation and heat relationships. Both isopods and amphipods have sometimes been considered ill-adapted to life on land (Williamson, 1951; Edney, 1954; Kuenen, 1959), as they show little modification from aquatic forms, yet in fact both groups are abundant in a very wide range of terrestrial habitats, including even deserts. This seems to have been achieved more by behavioural than by morphological or physiological adaptations. It is thought that isopods invaded the land rather earlier than did amphipods and are in consequence better adapted to terrestrial habitats (Hurley, 1968).

Living terrestrial isopods all belong to the suborder Oniscidea (formerly Oniscoidea, a name now reserved for a superfamily within the suborder — see Figure 1). This suborder was divided into two series by Vandel (1943): the Ligienne, including the genera *Oniscus* and *Armadillidium*, and comprising the truly terrestrial isopods (although at least one species has returned to a semi-aquatic existence (Dalens, 1989)) and the Tylienne. The latter were considered to be intermediate, semi-terrestrial forms in that they have not achieved complete independence from the sea. The Tylienne comprised two families, according to Vandel (1943) — the Tylidae and the Stenoniscidae. The only genus within the Tylidae is *Tylos*.

After an intensive study of the anatomy and morphology of the oniscids, Vandel (1943) came to the conclusion that they had a polyphyletic origin. He derived the semi-terrestrial Tylienne series from the marine Valvifera, with which they have structurally much in common, while the origin of the Ligienne he considered less certain but tended towards favouring the Cirolanidae as ancestral stock.

Holdich *et al.* (1984) separated the Tylidae from other oniscids to an even greater extent than Vandel (1943) had done, by placing them in their own infra-order, the Tylomorpha, all other families being assigned to the infra-order Ligiamorpha. This is unlikely to reflect true phylogenetic relationships, however, and Schmalfuss (1989) not only abandons infra-orders but places the Tylidae firmly within a new superfamily, the Scyphacoidea, together with the families Scyphacidae and Actaeciidae (see Figure 1). His classification of the Oniscidea is currently accepted by most isopod experts (B. Kensley, personal communication).

The genus *Tylos* Latreille 1828 is widespread on sandy beaches around the world but is most characteristic of the mid-latitudes. It typically occurs not far above high water mark, although some populations may invade the dunes behind the beach and sometimes even colonize dune slacks on a semi-permanent basis (McLachlan *et al.*, 1987). The genus currently comprises nearly 30 species. All are psammophilic and all

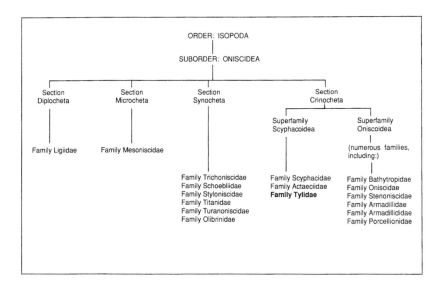

Figure 1   Classification of the Oniscidea according to Schmalfuss (1989). (The family Tylidae, to which *Tylos* belongs, is shown in bold type.)

burrow into the substratum. Their most obvious link with the sea, apart from being dependent on it for a food supply, is a marked tidal rhythm; the animals characteristically feed at night in the intertidal zone (Figure 2), following the tide down the shore and up again, to reburrow above the tidal limit. Although they can survive some hours of submersion (Kensley, 1974), the adults never willingly enter the water and there is no free-living aquatic larval stage; the term "semi-terrestrial" is therefore apt.

Some members of the genus *Tylos* have on occasion been referred to as "giant isopods". They are, indeed, larger than the majority of fully aquatic isopods, although they do not compare in size with some aquatic species from very high latitudes (e.g. *Saduria*, *Glyptonotus*) or some deep-water forms (e.g. *Bathynomus*). Maximum measurements for *Tylos granulatus* Krauss, on the South African west coast, are approximately 55 mm in length and 26 mm in maximum width; the wet weight of such an individual approaches 10 g. Most species are, however, smaller than this.

The importance of *Tylos* in the ecology of sandy beaches varies with species and with site. On some sandy beaches in southern California and northern Mexico, *T. punctatus* is virtually the only macrofaunal scavenger present and, in terms of biomass, outweighs all other resident inverte-brate macrofauna by a factor of 20 (Hayes, 1969). The beaches at Carlsbad, Torrey Pines and Punta Banda support populations of at least 20 000 isopods (210 g dry weight) per metre of beach frontage, with peak

Figure 2 *Tylos granulatus* Krauss feeding on kelp on a South African west
coast beach. (Photo: C.L. Griffiths.)

populations of three times this figure (Hayes, 1974). On some sandy
beaches in Mozambique, *Tylos* is again by far the most dominant
scavenger, numbers in some areas approaching those given above
(Brown, unpublished). On the other hand, on some beaches which even
to the experienced eye would appear to be ideal habitats for *Tylos*, the
animal is rare or absent. On others, such as the beaches around Sydney,
Australia, it is replaced by an oniscid isopod of a different genus (Dexter,
1983), although *Tylos australis* occurs elsewhere in New South Wales
(Lewis and Bishop, 1990).

Early publications on *Tylos* were naturally most concerned with
taxonomy and with distribution, and this work continues (see Brandt,
1883; Budde-Lund, 1885; Holmes and Gay, 1909; Barnard, 1924, 1940;
Boone, 1934; Vandel, 1943; Verhoef, 1949; Schultz, 1970, 1974, 1983;
Schultz and Johnson, 1984; Lewis and Bishop, 1990). There have from
time to time been interesting developments in these studies, including a
debate as to whether the generic name *Tylos* is valid as applied to this
group of crustaceans (Holthius, 1951; Gordon, 1952). There also remain
some taxonomic problems; for example, it has on several occasions been
suggested that individuals of *Tylos punctatus* Holmes & Gay from the
Pacific coast of North America are taxonomically distinct from those
encountered in the Gulf of California (Hamner *et al.*, 1969; Hayes, 1977).
The matter does not appear to have been successfully resolved. *Tylos*

granulatus Krauss, common on the west coast of southern Africa and occurring as far east as Port Elizabeth (Kensley, 1978), has been confused with *Tylos granulatus* Miers from Japan (e.g. see Imafuku, 1976). These species are, however, clearly taxonomically distinct, even though they may well display certain morphological similarities. The Japanese *T. granulatus* Miers is smaller than the South African species and also appears to differ in aspects of its behaviour.

A special problem in any historical account of *Tylos* concerns *T. latreillei* Aud. and Sav. Two subspecies, *T. latreillei europaeus* and *T. latreillei sardous*, were erected by Arcangeli (1939). Both occur on the French coast and elsewhere in the Mediterranean (Mead, 1968), while the former also occurs on the European Atlantic coast (Dexter, 1990). More recently, these subspecies have been elevated to specific rank (see Giordani-Soika, 1954). Both typically burrow just above high water mark (Peres and Picard, 1955) but, according to Giordani-Soika (1954) and Mead (1968), the two species favour different habitats, *T. europaeus* inhabiting significantly finer sand than *T. sardous*. Much of the literature, even as late as the 1970s, refers only to *T. latreillei*, without mention of subspecies or their promotion to species. To complicate matters, Giordani-Soika (1954) erected three subspecies of *T. sardous*. In this review we ignore such subspecies and refer to *T. latreillei* wherever we are in doubt as to whether *T. europaeus* or *T. sardous* is indicated. The reader should bear in mind that only two species are involved, not three.

It is not our intention to pursue taxonomic considerations in this review and our interest centres on the biology and behaviour of the animals, especially in so far as they represent a condition intermediate between the truly marine and the fully terrestrial. This interest is heightened by the fact that within the genus a wide gradient of behaviour is revealed, from populations extremely dependent on the sea to populations that have adopted a virtually terrestrial way of life and display an attenuation of the tidal rhythm typical of the genus. Furthermore, this range is due not only to differences between species but more particularly to behavioural plasticity within the species, so that a complex web of interrelated factors and behavioural adaptations is revealed.

2. RHYTHMIC ACTIVITY AND BEHAVIOUR

Tylos is essentially a nocturnal genus (Ondo, 1953; Pardi, 1955; Hamner *et al.*, 1968, 1969; Kensley, 1972, 1974; Iga and Kitamura, 1975; Imafuku, 1976; Hayes, 1969; Sieben, 1982; Odendaal *et al.*, 1994). Some diurnal activity may, however, be in evidence in some species. This is true, for example, of *T. europaeus*, particularly in those areas where it occurs in damp, shady, rock crevices rather than in sand (Pardi, 1955). Both *T. europaeus* and *T. sardous* emerge from the sand during the day if the sea inundates their burrows (Mead and Mead, 1974), the impression being given that this is not uncommon. Ondo (1954) and Imafuku (1976) recorded some daytime activity in the Japanese *T. granulatus*, this behaviour being mostly restricted to juveniles, while Yuasa (1973) reported the migration of large numbers of the same species onto a breakwater during the day, to escape typhoon waves. Tongiorgi (1969) described a similar phenomenon in *T. europaeus*. Authors have associated the predominantly nocturnal habit of *Tylos* with the avoidance of diurnal predators such as birds (Hamner *et al.*, 1969; Kensley, 1974), or with the maintenance of favourable ambient temperatures (Holanov and Hendrickson, 1980; Matthewson, 1991) and humidity levels (Imafuku, 1976; Holanov and Hendrickson, 1980).

Most *Tylos* populations spend their periods of inactivity in temporary burrows near the high tide mark (Hamner *et al.*, 1968, 1969; Mead, 1968; Kensley, 1974; Hayes, 1977; Chelazzi and Ferara, 1978). They typically emerge at night during ebb or at low tide, for about 2–3 h, following the tide down the slope to forage below the upper tidal level (Ondo, 1952; Hamner *et al.*, 1969; Kensley, 1974, Odendaal *et al.*, 1994). Exceptions may occur; according to Geppetti and Tongiorgi (1967), a Mediterranean population of *T. latreillei* moved *up* the beach after emergence, foraged, and then moved back down the slope to burrow. These authors claimed that this reversal of the more typical response was related to the food which on these beaches tended to lie well above the high water mark, in contrast to the kelp which forms the staple diet of *T. punctatus* and *T. granulatus* Krauss (Hamner *et al.*, 1969; Kensley, 1974; Odendaal *et al.*, 1994), kelp being heavy and bulky so that it is deposited below the driftline. In any case, Mediterranean tides are of very low amplitude compared with changes of sea level induced by the weather, so that upslope and downslope responses may be relatively unimportant here (Hamner *et al.*, 1968). Some Japanese populations of *T. granulatus* Miers spread out over the surface during their activity period, some going up, some down the slope (Imafuku, 1976).

After feeding, the animals typically return to the approximate position of the next high water mark, where they again burrow into the sand, so that their burrows are found in a narrow band extending along the beach. This ability to adjust their position may prevent them from being washed away by the next high tide and also keeps them in relatively moist sand and within range of their food (Kensley, 1974). According to Imafuku (1976), *T. granulatus* at Cape Banshozaki in Japan hides between pebbles and under driftwood and wrack during the day. Hamner *et al.* (1969) also note that some individuals of *T. punctatus* hide under flotsam or wrack during the hours of daylight. While this behaviour has not been observed for the South African *T. granulatus* Krauss in the field, de Villiers and Brown (1994) report that, given dry sand in the laboratory, many animals choose to shelter under debris rather than attempt burial.

Distribution in a narrow band along the high water mark is not invariable and, at least in some species, may be modified. Thus populations of *T. capensis* on part of the shore of Algoa Bay, South Africa, have assumed a terrestrial existence in slacks of the Alexandria dune field (McLachlan, 1986; Matthewson, 1991), almost certainly induced by a lack of food in the intertidal zone. No tidal rhythm of emergence is present in these populations, in contrast to conspecifics living elsewhere along the South African south coast, although they remain nocturnal, thus avoiding diurnal predation and high daytime temperatures.

Both *T. granulatus* and *T. capensis* typically display a circatidal rhythm of nocturnal emergence from their burrows; this is essentially a 24.8 h lunar-day rhythm which ensures that the animals emerge close to the time of low water (Kensley, 1974). Thus, in phase with the tides, the population emerges later each night. This means that after about 2 weeks of repeated nocturnal emergence, dawn would catch the animals still out of their burrows if they chose to emerge. This is not allowed to happen, however, and the animals give that morning tide a miss, lying buried until the following, early evening, tide (Figure 3). This switch back may be seen as resulting from a 15-d, or semi-lunar, rhythm imposed on the tidal, lunar-day rhythm (Kensley, 1974). Similar rhythms are generally in evidence in most other *Tylos* species as well.

In addition to these rhythms of emergence, *T. capensis* and *T. granulatus* show an activity rhythm, with a peak of activity at neap tides and much less activity at springs (Kensley, 1974; Branch and Branch, 1981), although in some populations this may be obscured on any particular occasion by light intensity. *T. granulatus* at Groenrivier show considerable spring-tide activity at new moon, when it is dark, but little or no activity at full moon (Odendaal, unpublished). Although terrestrial populations of *T. capensis* in the Alexandria dune slacks have suppressed the tidal rhythm of emergence, a semi-lunar rhythm of activity continues to be in evidence (Matthewson, 1991), ensuring that the animals are not swept away by exceptionally high spring tides.

That these rhythms have a strong endogenous component is beyond question. Tidal rhythms of emergence are evident in the laboratory in the absence of environmental cues (Kensley, 1974; Brown and Trueman,

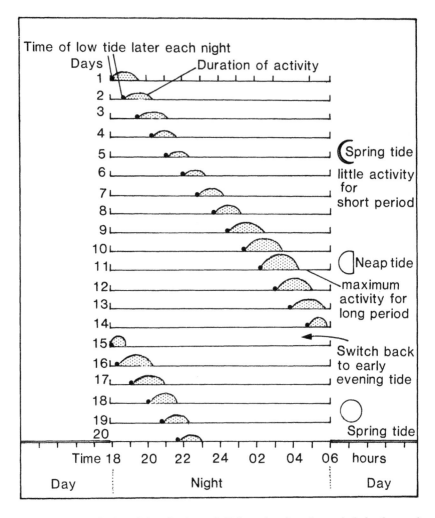

Figure 3   Typical activity rhythm of *Tylos*, showing the switch back on the fifteenth night and the activity peak at neap tides. (After Branch and Branch, 1981; based on Kensley, 1972.)

1994b) and Marsh and Branch (1979) demonstrated marked circadian and circatidal rhythms of oxygen consumption under constant dim light conditions in the laboratory (see Section 5). However, it is recognized that endogenous clocks have to be "set" — synchronized and reinforced — by changing environmental stimuli (Brown, 1961; Hardy, 1970) and that endogenous rhythms and exogenous cues commonly complement one another in determining behaviour. For example, the ghost crabs, *Ocy-*

pode ceratophthalmus and *O. kuhlii*, living, like *Tylos*, above the driftline on sandy shores, display similar endogenous semi-circalunar and diurnal rhythms (Barrass, 1963; Hughes, 1966); nevertheless, Jones (1972) has shown that the emergence of these crabs is largely dictated by the level of water in the burrow. This cannot be true, however, of *Tylos*, in which the burrows are normally too shallow to be reached by the rising water. Moreover, de Villiers and Brown (1994) have shown that, while a rising water table may cause *T. granulatus* to migrate upwards through the sand column so as to avoid being submerged, it does not trigger emergence from the sand.

Imafuku (1976) maintained the Japanese *T. granulatus* Miers in the laboratory under various regimes of light, temperature and humidity. He found that, given a 12-h cycle of bright illumination alternating with 12 h of darkness or dim light, the isopods were active only during the dark period even when this was out of phase with the natural cycle. Under constant dim light, at temperatures of 25 or 30°C, the activity rhythm could be controlled by fluctuations of humidity, while at constant humidity it adapted to fluctuations in temperature between 20 and 30°C, the animals being active at the lower humidity or temperature. It should be stressed, however, that his animals were from a pebble beach and were always at the surface of the substratum, so that cycles of burrowing and emergence did not occur. Imafuku makes no mention of tidally correlated activity patterns.

Tylos granulatus Krauss at Groenrivier in South Africa are not active when the moon is out (Odendaal *et al.*, 1994) and burrow at the first sign of the moon rising, even if the tide is favourable. Kensley (1974) makes no mention of this phenomenon and it should be noted that his detailed study of *T. granulatus* was undertaken at different sites. *T. punctatus* is certainly active in moonlight (Holanov and Hendrickson, 1980) and there is no suggestion that moonlight depresses its activity. *T. europaeus* not only emerges in moonlight but also during the day and orientates both to the sun and the moon (Pardi, 1954). The different responses by different populations to the presence of the moon or daylight may well reflect different predatory regimes. While Kensley (1974) discovered no predators in his study of populations of *T. granulatus*, Odendaal *et al.* (1994) reported predation on this species at Groenrivier by the yellow mongoose, *Cynictis pencillata*, which is often nocturnally active. It is thus possible that the Groenrivier population has evolved an aversion to moonlight as a predator avoidance mechanism. *T. capensis* does not appear to avoid moonlight and according to Matthewson (1991), its activity peak at neap tide implies that more moonlit nights are available for mate recognition and foraging than at springs, when it is light only during full moon and dark at new moon. However, this argument is only valid if sight plays an important part in these activites, and there is no evidence to suggest this.

An interesting aspect of emergence and reburrowing in *T. granulatus* is the extremely marked synchrony of these events within the population. At Groenrivier, Odendaal *et al.* (1994) observed the sudden emergence of many animals. Once the animals reach the surface, they pause for several minutes with only their heads showing (see also Kensley, 1974), so that immediately before the start of their excursions, numerous individuals may be seen waiting in their exit holes. Within a minute or two they all emerge and move around on the beach. After about 2 h all the animals reburrow in a similarly synchronized fashion. This is the case even if the tide is out and several hours of apparently suitable time are left for foraging before dawn. We have been unable to ascertain whether such synchrony is found in other species as well. Hamner *et al.* (1968) note only that *T. punctatus* emerges at night but returns to the high tide mark before dawn, being active for more than 3.5 h. Marked synchrony in *T. europaeus* or *T. sardous* on Mediterranean sandy beaches appears unlikely from the descriptions of such observers as Arcangeli (1952) and Tongiorgi (1969), the latter noting activity periods of many hours; the small Mediterranean tides presumably fail to impose any necessity for a short activity time window. Imafuku (1976) does not mention any marked synchrony of activity in his non-burrowing population of *T. granulatus* Miers at Cape Banshozaki.

3. LOCOMOTION: CRAWLING AND BURROWING

Tylos is typically active only during low tide at night, when it crawls over the sand in search of food, sometimes covering considerable distances. A second form of locomotion, burrowing, punctuates these excursions. A third type of locomotion, surfing, is practised by the juveniles of at least some species, the animal rolling itself up into a ball and allowing the swash to transport it up the beach (Kensley, 1974). Adults have not been observed to show this behaviour but, if caught by a wave, flatten themselves against the beach and allow the water to pass over them before migrating upshore. It might be thought that the juvenile response is simply one of protection when unexpectedly caught by a wave; however, the differing response from that of the adults suggests otherwise. Moreover, the difference in response tends to separate the juveniles from the adults and leads to the transport of the former to areas where carrion is most likely to be encountered (Kensley, 1974) (see Section 5).

3.1. Crawling

Crawling in *Tylos granulatus* Krauss has been briefly examined by Brown and Trueman (1994b). They observed a rate of locomotion of some $2.5 \, \mathrm{cm \, s^{-1}}$ for large, adult animals in the laboratory, over flat, damp sand but point out that this is a great deal slower than rates attained in the field as, for instance, when the animal is actively searching for food. Odendaal *et al.* (1994) have recorded average rates of no less than $13 \, \mathrm{cm \, s^{-1}}$ for adults of the same species on the beach at Groenrivier at night.

Movement trajectories for *T. granulatus* at Groenrivier have been studied by Odendaal *et al.* (1994), using night vision equipment. The study of such trajectories, and particularly the analysis of individual move lengths and directions, has become a useful tool in elucidating mechanisms influencing the spatial patterns of animals (for reviews see Turchin, 1989; Turchin *et al.*, 1991). The observation of *Tylos* under natural conditions is, however, not easy, as the animals are disturbed by any movement or light. *T. granulatus* was seen to crawl over the sand during its activity period, stopping frequently when encountering depressions, such as footprints, or potential food items, and sometimes without any visible stimulus. Their movement trajectories can therefore be approximated by a series of stopping points, joined by straight lines (see Root and Kareiva, 1984; Turchin *et al.*, 1991). Twenty *Tylos* were followed during their nocturnal activity period, one at a time, by two investigators. One carried a TRS-80 computer on which the time and type of all visible behavioural events were recorded; the other investigator marked the position of stopping points in the trajectory with numbered wire flags. Despite all precautions and the fact that there was no moon, several animals appeared to sense the presence of observers and thus 12 trajectories were discarded as being possibly influenced by sensitivity to human presence. The remaining eight were used to calculate the speed of an average *T. granulatus* and the distance moved during an activity period.

The eight usable trajectories (two of which are shown in Figure 4) yielded a total of 59 moves, with an average of 7.34 ± 2.97 moves per trajectory. The average move length was $1.35 \pm 1.73 \, \mathrm{m}$. The average speed at which the animals moved on the beach, including the frequent stops, was $14 \pm 11 \, \mathrm{mm \, s^{-1}}$. Over a 2-h period of activity an individual can therefore cover a distance of some $80 \, \mathrm{m}$. The average cruising speed between stops was $130 \pm 61 \, \mathrm{mm \, s^{-1}}$; if a *Tylos* travelled at this speed for 2 h without stopping, it would cover $936 \, \mathrm{m}$.

All the pereiopods are used in a similar manner during crawling, except that the terminal pair (pair 7) are somewhat less involved than the others (Brown and Trueman, 1994b). The limbs on either side of the body

A. C. BROWN AND F. J. ODENDAAL

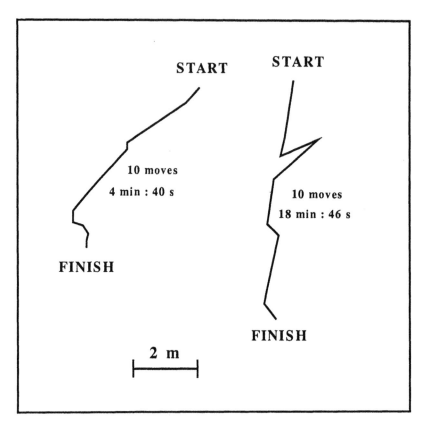

START   START

10 moves
4 min : 40 s
                        10 moves
                        18 min : 46 s

FINISH

                FINISH

2 m

Figure 4   Movement trajectories of two individuals of *Tylos granulatus* Krauss on the beach at Groenrivier (after Odendaal *et al.*, 1994). The path taken by each animal is assumed to be straight between stopping points.

display a metachronal rhythm, the wave passing along one side in about 0.2 s and being 180° out of phase with that of the other side of the body. On the surface of the sand, *T. granulatus* employs Manton's (1952) midgear ratio of 1:1, with approximately equal time for fore-stroke and back-stroke. This is comparable with observations on the aquatic isopod *Chiridotea* (Griffith and Telford, 1985).

3.2. Burrowing

According to Kensley (1974) and Newell (1979), the anterior three pairs of pereiopods are the effective digging appendages during burrowing, these passing the sand on to the posterior four pairs, which in turn push it backwards to form a little mound. The animal then rotates, according to these authors, in an anticlockwise direction, through an angle of some 45° before again starting to burrow. The isopod thus descends into the sand in a spiral fashion, while the displaced sand forms a cone above the entrance to the burrow. As it burrows deeper, the sand continues to be simply pushed behind it, blocking the burrow. The animal eventually rolls up into a ball, in which position it rests until its next emergence.

Burrowing in *T. granulatus* has recently been studied in greater detail by Brown and Trueman (1994b), using video and electronic recording techniques. The animal commences burrowing by pushing its head into the sand (Figure 5a). Pereiopods 1–3 then commence digging movements, while the body is supported by the head and pereiopods 4–7. The head is meanwhile moved up and down, four or five times in the first 3 s of burial, consolidating the burrow walls. The most anterior pair of pereiopods commence digging first, at a rate of 2–4 strokes per second, followed by those on the second and third thoracic segments. The pereiopods extend for 7 mm beyond the lateral tergal folds of the pereion during excavation of sand (in an animal 35 mm long). Sand is passed backwards, in the midline, from the first three pairs of pereiopods, forming a loose ball which is compacted and pushed backwards by limbs 4–6 (Figure 5b) at a rate of 1–2 strokes per second. The bolus of sand is finally pushed powerfully behind the burrow by the synchronous action of pereiopod pair 7 (Figure 5c). The whole body passes into the sand in about 12 s. During this digging period, sand accumulates in small heaps at the surface but as digging proceeds, the excavated sand remains in the burrow behind the animal, partially blocking it. In burrowing, pereiopods 1–3 display approximately equal times for fore- and back-stroke, as they do in crawling, but the posterior four pairs move into a lower gear of about 1:2.

The differences in function between the anterior three and posterior four pairs of pereiopods are reflected only slightly in their morphology (Figure 6). All are of approximately equal length and are robust, this feature increasing somewhat through the series from pereiopod 1 to pereiopod 7. The proportionate lengths of the pereiopod joints are not markedly different from those of typical aquatic Isopoda. The dactylopodites of the first pair of pereiopods are slender and elongate, those further back less so. A terminal claw is typically present on each of the pereiopods but is sometimes lacking on the last two pairs in older individuals, giving these appendages a more peg-like appearance. Figure 6 shows the structure of pereiopod 1 and the distal portion of pereiopod 7 of *T. granulatus* Krauss. Kensley (1974) figured pereiopods 2 and 5 of the same species.

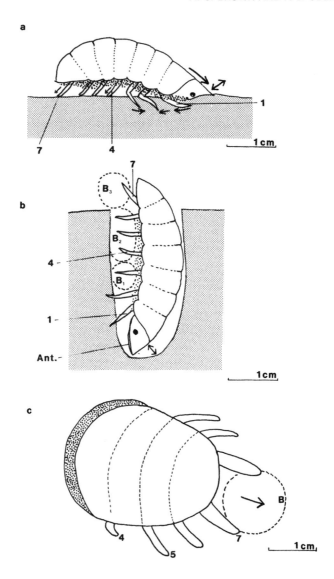

Figure 5   Diagrams of *Tylos granulatus* during burial, drawn from single frames of a video tape record. (a) At initial entry into the sand (hatched); arrows indicate action of the pereiopods, 1–3 excavating a hole, while 4–7 support the body and push the head forward into the sand (large arrow). The raising and lowering of the head to consolidate the burrow is indicated ($\leftrightarrow$). (b) *Tylos* burrowing beneath the sand against the glass side of an aquarium. Successive positions of a bolus of sand (B1–3) in the mid-line between the pereiopods are indicated, as the sand is brought from the head region to the surface. While the

Cals (1974) investigated the cuticular structures on the pereiopods of *T. latreilli* (*sic*) by means of scanning electron microscopy and compared them briefly with those found in *Periplaneta* (Insecta). The structures he describes and figures for *T. latreillei* are found also on the pereiopods of *T. granulatus* Krauss (Stenton-Dozey and Webb, unpublished), although he does not remark on the complex, apparently movable spines which are a feature of the pereiopods of the latter species.

Rotation of *T. granulatus* in the burrow was observed on numerous occasions by Brown and Trueman (1994b). The amount of rotation varied between 10 and 90°, although generally it was about 45°, as previously noted by Kensley (1974). However, rotation was found to be either clockwise or anticlockwise, one direction or the other usually being maintained throughout burial. Burrowing is a discontinuous process (Figure 7a,b) and takes place in a step-wise manner which brings to mind the burrowing of soft-bodied invertebrates (Brown and Trueman, 1994b).

The effectiveness of burrowing in *T. granulatus* may be compared with that of other invertebrates by determination of the Burrowing Rate Index (BRI) (Stanley, 1970, modified by Vermeij and Zipser, 1986 — see Brown and Trueman, 1994a):

$$\text{BRI} = \frac{\sqrt[3]{\text{wet weight (g)}}}{\text{time for burial (min)}}$$

For *T. granulatus*, BRI = 3 (Brown and Trueman, 1994b). This is much better than most burrowing molluscs (Trueman and Brown, 1992) and is comparable with that of the aquatic whelk *Bullia digitalis* (BRI = 4.6) (Trueman and Brown, 1989), which inhabits the same beaches. As animals with BRI > 1 are considered to be rapid burrowers (Vermeij and Zipser, 1986; Trueman and Brown, 1992), *Tylos granulatus* must be considered a very powerful burrowing animal. Unlike *Bullia* and other fully aquatic animals, it cannot utilize the thixotropic properties displayed by sand saturated with water (see Brown and Trueman, 1991) but burrows into the mechanically far more difficult medium of relatively dry or damp sand. The weight-specific energy for burial (Brown *et al.*, 1989) is similar in *Tylos* and *Bullia*.

---

burrow is continually being excavated by legs 1–3, the head is raised and lowered to consolidate the burrow walls (↔). (c) Dorsal aspect of *Tylos* when about halfway into the sand, just preceding rotation of the body, with pereiopods 4–6 pushing backwards and a bolus (B) of sand being ejected by pereiopods 7 functioning synchronously. 1–7, pereiopods; Ant, second antenna. (After Brown and Trueman, 1994b.)

(a)

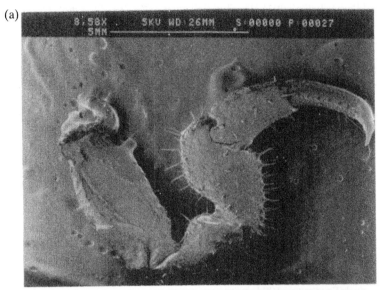

(b)

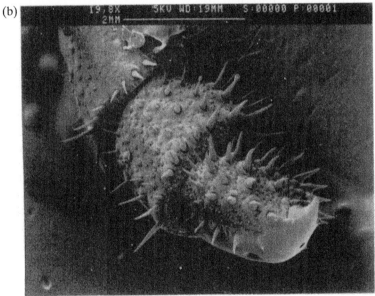

Figure 6   Scanning electron micrographs of (a) pereiopod 1 and (b) the distal region of pereiopod 7 of *Tylos granulatus* Krauss. The larger spines are set in ball-and-socket joints but whether they can be moved by the animal or are sensory has not been ascertained.

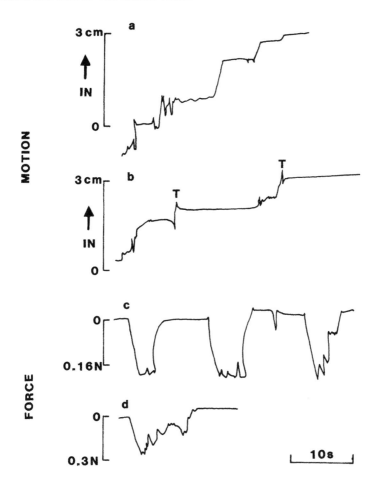

Figure 7 Recordings of the movements of *Tylos granulatus* Krauss while excavating a burrow, made by means of an isotonic transducer (a,b), and of the force with which it is pushed into the burrow, using an isometric transducer (c,d). (a) Record showing motion into the sand (in, arrow) in five steps, until the animal was just completely buried. (b) Extract of a trace showing turning (T), with a characteristic overshoot movement as the animal moves into the sand. (c) Series of pushes into the sand, saturating the trace. Tension was not sustained and deflections occur as the limbs slip in the sand. (After Brown and Trueman, 1994b.)

Brown and Trueman (1994b) have calculated the mechanical energy requirement for burial of a *Tylos granulatus* of wet weight 5.7 g (dry tissue weight 1.9 g) as 0.28 J for a depth of 1 m. This is probably a slight underestimate and must, in any case, be multiplied by at least five to give an estimate of the total metabolic cost to the animal. A realistic assessment of the metabolic cost of burrowing to a depth of 50 cm, which may be about average for this species, is thus between 0.75 and 1 J (Odendaal *et al.*, 1994). This would appear to be less than 2% of the animal's total energy expenditure over a 24-h period.

Different depths of burrowing have been reported for different species and for the same species at different localities or in different seasons. *T. punctatus* has been shown to burrow to a depth of between 10 and 50 cm in summer, but deeper in winter, during hibernation (Hayes, 1969, 1974). It is considered that the depth of burrowing is largely determined by the moisture content of the substratum, or perhaps more correctly the saturation deficit of the interstitial air (Hayes, 1977; Holanov and Hendrickson, 1980; de Villiers and Brown, 1994) (see Section 6). This is probably true of other oniscids as well, and Brusca (1966) reported that *Alloniscus perconvexus* burrows to a depth dependent on sand moisture content. There may also be some correlation in talitrid amphipods (Williams, 1983a).

Moisture content is not, however, the only factor determining depth of burrowing. Hayes (1977) conducted simple experiments in which groups of *T. punctatus* were placed in three "burrowing tubes" over sand of three different grain sizes but of equal moisture content. He demonstrated a greater average depth of burrowing in sand of 0.7–1.4 mm grain diameter than in finer grades. This does not, however, imply that the coarser the sand the deeper the burrowing but rather that there is an optimum grain size for deep burrowing. De Villiers and Brown (1994) have found that a layer of coarse, shelly sand well below the surface on Yserfontein Beach formed an effective barrier to the burrowing of *T. granulatus*.

Brown (1959) made some casual observations of the fauna at the mouth of the Orange River, on South Africa's west coast, and noted that what appeared to be a single *T. granulatus* burrow often proved to house several animals. Each animal tended to have its own chamber but all the chambers were at approximately the same depth, communicating directly with one another (Brown, unpublished). Multichambered burrows have not been found in other localities, although Kensley (1974) notes that a number of *T. granulatus* may successively begin to burrow at the same point. It is possible that such communal burrows may, by taking advantage of the water inevitably lost by individuals, reduce desiccation by increasing humidity within the burrow. Another advantage to the animals could be a saving in the energy cost of burrowing. Brown (1959) observed that *Tylos* had a habit of commencing burrowing in depressions or where the sand had already been disturbed. Indeed, entrances to new burrows were commonly found in the footprints left by investigators the day before. These observations have been confirmed by subsequent work

(Brown and Trueman, 1994b; Odendaal *et al.*, 1994), both in the field and in the laboratory.

Another common occurrence, at least in the case of the South African *T. granulatus*, is burrowing into exit holes resulting from the emergence of individuals at the beginnning of their activity cycle (Odendaal *et al.*, 1994). This behaviour has a marked effect on the longshore distribution of the animals (see Section 8), although it does not seem to result in a significant saving of energy (Odendaal *et al.*, 1994). Loose sand remains in the burrow after an animal has emerged. Burrowing under these circumstances appears to be somewhat different from that described by Brown and Trueman (1994b) and the resulting surface deposit can be distinguished visually from that formed when an animal digs a new burrow (Odendaal, unpublished). Less sand is apparent on the surface and it forms a low, star-shaped deposit rather than the mound which is otherwise apparent. This phenomenon requires further investigation.

Most writers have assumed that, when emerging, *Tylos* simply retraces its steps up the burrow formed during burial. This is not necessarily always the case, however, and Kensley (1974) noted that buried *T. granulatus* dig upwards "either following the path made by their entry or may follow a fresh path", while at Lambert's Bay animals entering the sand one after the other, at the same point, were found to separate once in the sand and eventually to emerge from different exit holes.

Although the animals invariably burrow above high water mark, they cannot make burrows into totally dry sand. De Villiers and Brown (1994) have found that, given completely dry sand in the laboratory, *T. granulatus* may attempt to burrow but cease this activity as the burrow collapses behind them, thus coming to rest just below the surface. Many animals do not even attempt to burrow into dry sand but hide under any debris that may be available to them.

4. SENSORY PHYSIOLOGY AND ORIENTATION

Sensory responses to changing environmental conditions, and particularly to light, to the position of the sun or moon, to moisture and to beach slope, have been studied in *Tylos europaeus* (as *T. latreillii* (*sic*)) (Pardi, 1954, 1955), in *T. sardous* (Mead and Mead, 1974) and in *T. punctatus* (Hamner *et al.*, 1968). Some relevant information, both published and unpublished, also exists for *T. granulatus* and *T. capensis* on South African beaches and for *T. granulatus* Miers in Japan.

One of the major problems facing sandy-beach invertebrates which undertake tidal migrations up and down the slope is overall maintenance of position on the shore (Newell, 1979; Brown, 1983; Brown and McLachlan, 1990). Not only must they migrate in the right direction at the right time but their migratory excursions must not carry them too far up or down the slope. Endogenous rhythms, as displayed by so many sandy-shore animals, including *Tylos*, may ensure emergence and reburial at appropriate times but cannot provide orientational cues or tell the animal where it is located within the physical system. The animals must therefore be receptive to directional stimuli such as the position of the sun or moon, polarized light patterns or the slope of the beach, as well as responding to factors which indicate their position on the shore, such as whether the sand is wet or relatively dry.

In a series of experiments on almost flat Mediterranean beaches, Pardi (1954, 1955) showed that *T. europaeus* orientates to both the sun and the moon in a manner very similar to that displayed by talitrid Amphipoda such as *Talitrus* (Pardi and Papi, 1953; Papi and Pardi, 1953; Pardi and Ercolini, 1986). On the beaches studied by Pardi (1955), *Tylos europaeus*, although predominantly nocturnal, commonly emerges during the day, largely in response to tidal inundation of its burrows, so that orientation to both celestial bodies is appropriate. When the animal finds itself under water or on wet sand, it employs orientation to the sun or moon to migrate rapidly up the beach, virtually at right-angles to the shore-line. On dry sand, however, the response is reversed, the animal moving towards the sea. The animals possess a time clock that allows them to compensate for the position of the sun throughout the day. In laboratory experiments under artificial light and with no view of the sky, they assume an orientation to the light source which corresponds to that which they would have adopted towards the sun on their home beach at the same time. Furthermore, *T. europaeus* transported to a beach facing a different direction display an orientational response to the sun appropriate to their home beach and thus move in the "wrong" direction. In addition to these observations, there is some indication that the animals can detect polarized light patterns (Pardi, 1955; Herrnkind, 1972).

Mead and Mead (1974) attempted to repeat the above experiments on solar orientation on the closely related *T. sardous*, using an orientation wheel with wet or dry sand. They obtained quite different results, however, and concluded that the visual orientation of their animals was not based on solar orientation at all but rather on "perception of height or of relative distance of the different sectors of the horizon". Slope was not taken into account in their experiments.

Hamner *et al.* (1968), using a flat orientation wheel whose rim prevented the isopods from seeing any topographical features, obtained results which indicated that the responses of *T. punctatus* to the sun represented a simple negative phototaxis and was unrelated to the

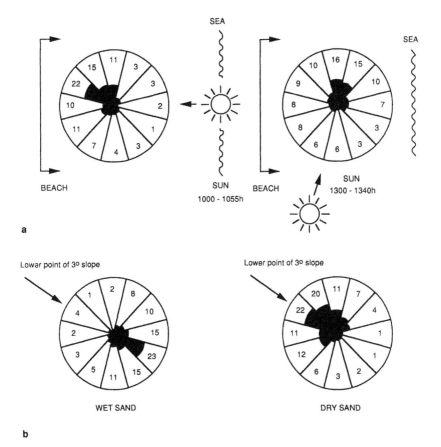

Figure 8   Diagrams showing the orientation of *Tylos punctatus*. (a) Orientation to the sun at two different times of day in a flat orientation wheel which excluded other directional stimuli. The response of the animals is essentially a negative phototaxis, unrelated to the orientation of the beach. (b) Orientation on a plate covered with wet and dry sand respectively and inclined at an angle of 3°. The animals were negatively geotactic on wet sand, positively geotactic on dry sand. These responses would allow the animals to migrate up and down the beach without reference to other directional cues. In both (a) and (b), the numbers indicate percentage response. (After Newell, 1979; data from Hamner *et al.*, 1968.)

orientation of the beach (Figure 8). On the other hand, the responses to slope displayed by *T. punctatus* are of clear significance in the tidal migrations of the isopod. An orientation wheel tilted at 3°, or even as little as 1°15′, resulted in a response from the animals (Hamner *et al.*, 1968); moreover, the response was reversed, as in the case of *T.*

europaeus, according to whether the sand was wet or dry (Figure 8). On dry sand, the isopods moved consistently down the slope, regardless of the direction of the sun, but moved up the slope on wet sand. This behaviour will clearly allow *Tylos* to move up and down the beach quite independently of other environmental cues and could also account for the fact that the centre of gravity of the population moves down into the intertidal sands at neap tides and follows it upshore again as the tides approach springs.

This does, however, raise the question as to how the animals behave during or soon after rain, when the surface of the sand may be wet from the water's edge to the top of the beach. No-one has seriously addressed this problem, although *T. latreillei* on Mediterranean beaches is said to move shorewards up to 50 m not only in response to stormy seas but also during extended, heavy rain (Tongiorgi, 1962, 1969; Herrnkind, 1972). Casual observation on South African beaches over many years suggests that *T. granulatus* tends to remain within its burrows during heavy rain (Brown, unpublished), while Kensley (1974) mentions that neither light rain nor wind appears to have any effect on the behaviour of *T. capensis* or *T. granulatus*. According to Imafuku (1976) rainfall results in a temporary increase in activity in a non-burrowing population of *T. granulatus* Miers at Cape Banshozaki, while strong wind suppresses activity.

The responses to slope displayed by *Tylos* are thought not to depend on individual learning but to be inherent, for they require no compass reference (as does orientation to the sun or moon) and are therefore adaptive to the offspring on any beach (Herrnkind, 1972). Certainly their orientational abilities are considerable and consistent. Hamner *et al.* (1968) observed that, despite the lengthy excursions undertaken by *T. punctatus*, they had never found an individual buried in an inappropriate area of the beach. We have found this to be true also of *T. granulatus* Krauss on South African beaches.

Vision is clearly of great importance to the behaviour of *Tylos*, quite apart from whether or not it orientates to the sun or moon, and the eyes are accordingly prominent and well developed. *T. granulatus* easily detects observers on the beach at night, a fact which has hampered investigation of its behaviour at Groenrivier in South Africa (Odendaal *et al.*, 1994). The animals also apparently use vision in deciding whether to emerge from their burrows at the start of an activity period. As they reach the surface they pause, often for several minutes, with only the head and eyes showing (Kensley, 1974; Odendaal *et al.*, 1994). This has been interpreted as a testing of light intensity but of course the animals may use this pause to detect other environmental variables as well. The degree of photonegativity of the responses varies considerably not only from species to species but apparently between populations of the same species. Thus populations of *T. europaeus* and *T. sardous* on Mediterranean beaches often emerge during the day in response to changing conditions (Pardi, 1955; Mead, 1968), while *T. granulatus* at Groenrivier will not emerge even in bright moonlight (Odendaal *et al.*, 1994). Populations of *T. granulatus* at Bloubergstrand and of *T. capensis* in Algoa Bay appear to be intermediate, in that, while emerging only at night, they are not deterred by full moonlight (Kensley, 1974; Sieben, 1982; Matthewson, 1991). The Japanese *T. granulatus* Miers is essentially nocturnal and displays photonegativity to bright light; nevertheless it has been found to be photopositive in dim light (Imafuku, 1976). The possibility that *Tylos* detects physical features of the beach or skyline by means of vision has not been investigated except for the observations of Mead and Mead (1974) mentioned above.

It may also be noted that responses to light may change according to circumstances. Most terrestrial isopods, like *Tylos*, are photonegative in their responses, but this may be temporarily reversed under certain conditions such as starvation, heat stress or desiccation (Carthy, 1958). Such factors will cause even the highly photonegative *T. granulatus* to emerge from its burrows even in bright sunshine (Brown, unpublished). Presumably such a reversal of response also occurs when Mediterranean populations of *Tylos* emerge from the sand during the day, when their burrows are inundated (see Section 2).

Iga (1972) and Iga and Kitamura (1975) have described a photopositive response in an essentially nocturnal population of Japanese *T. granulatus*. They found that their animals, studied in the dark at night, moved about "irregularly" but in relatively straight lines in dim light. Young animals of between 5 and 9 mm body length, when subjected to a single dim light source overhead, moved towards this light source and adopted a circular path around it. The radius of this "circus movement" increased with the height of the light source but not with variation in light intensity. It is not clear what significance, if any, these reactions have in the lives of the animals in the field.

Responses to moisture and humidity also require further investigation. While responses to wet and dry sand are now well documented and it is clear that depth of burrowing is largely determined by the moisture content of the substratum (see Section 6), responses to humidity as such have not yet been reported in the literature, although they have been studied in several fully terrestrial oniscids (Gunn, 1937; Edney, 1954; Cloudsley-Thompson, 1956, 1958). *T. granulatus* has been found to move up a humidity gradient in dim light at 15°C, when given no sand in which to burrow (Brown, unpublished). This preliminary series of experiments also revealed that the animal moved down a gradual temperature gradient when this lay between 15 and 30°C but moved at random in a gradient of 5–20°C. However, as humidity was not controlled during these temperature experiments, the results need to be treated with caution. Also relevant is the finding that terrestrial isopods in general move faster in dry air than in moist air (Fraenkel and Gunn, 1940), thus reducing periods of stress due to desiccation, although of course the animals may lose more water per unit time by moving faster. Stenton-Dozey and Webb (unpublished) have identified structures on the pereiopods of *T. granulatus* which may well be humidity sensors.

Chemosensory phenomena in *Tylos* have been totally ignored, although olfaction may well play a part in foraging for food (Kensley, 1974) and in sexual attraction. The second antennae may be implicated in olfaction, as they are in *Ligia* (Hewitt, 1907), these structures being very well developed. Fischbach (1954) and Kuenen and Nooteboom (1963) have demonstrated that terrestrial oniscids have a sensitive olfactory sense, responses including the recognition of other members of the species and possibly of other oniscids.

Pardi *et al.* (1988) have shown that the sandy-beach talitrid amphipod *Talorchestia martensii* utilizes a magnetic compass which allows orientation to the normal axis of their home beach and is used in conjunction with their sun compass response. Ugolini and Pardi (1992) have further shown that, in *T. martensii* from equatorial regions, the sun compass is secondary to the magnetic compass and seems to be based on an imperfect endogenous clock. The sun compass is, in fact, only of value to the amphipod in the early morning and late afternoon. Whether *Tylos* also employs a magnetic compass and places less reliance on a possible sun compass in the tropics than in temperate regions are matters which, like so many others, remain to be investigated.

5. NUTRITION, RESPIRATION AND ENERGETICS

5.1. Nutrition

Tylos has generally been regarded as essentially herbivorous. It should, however, more correctly be classified as an omnivorous scavenger and indeed some species clearly prefer animal matter to plant material (Kensley, 1974; Newell, 1979). Nevertheless, on the majority of beaches where *Tylos* populations occur, algal detritus is much more readily available than is carrion, so that the adult diet consists predominantly of stranded kelp and wrack. It has been suggested that the surfing behaviour of the juveniles, of some species at least, tends to carry them to the drift line, where stranded animal material is more likely to occur (Kensley, 1974). Observations at Groenrivier, on the contrary, indicate that it is rare for the juveniles of *T. granulatus* to surf and that this behaviour is most unlikely to play a significant role in food-finding. This could be yet another instance of different populations of a species behaving differently at different sites.

Hayes (1974) found that *T. punctatus* preferred kelp to fresh zooplankton. This species selected fronds of the kelp *Macrocystis pyrifera* in preference to other algal material. *M. pyrifera* stipes were not eaten, however. On some Mexican beaches on the Sea of Cortez, the primary food source of *Tylos punctatus* is stranded eelgrass, *Zostera marina* (G. Simmons, pers. commun.).

Kensley (1974) performed food choice experiments on adult *Tylos granulatus* from Bloubergstrand, on the South African west coast, presenting the animals with five potential food items. The preference was for dead fish, followed by the green alga *Ulva*, followed by *Laminaria*, *Gigartina* and *Macrocystis*. He also tested juveniles of the same species, substituting the small red alga *Placomium* for *Laminaria*, as this is far more common at the drift line. As with the adults, fish was eaten in preference to the other foods but *Ulva* and *Placomium* were taken in roughly equal amounts; cabbage was seldom eaten and *Macrocystis* not at all. Kensley (1974) considers it likely that *Tylos* is attracted to its food largely by olfactory stimulation.

Tylos latreillei, on Mediterranean beaches, has been said to feed on "decomposed matter" (Arcangeli, 1953) but has also been observed to feed on living amphipods (Matsakis, 1956) and dead fish (Vandel, 1960). While other species have also been seen to feed on dead fish, cnidarians and even dead seals and cats (Hamner *et al.*, 1969), *T. latreillei* is the only one reported as feeding on living Crustacea. *Tylos* may sometimes be found feeding on rather unlikely objects and Hamner *et al.* (1969) observed *T. punctatus* in the Gulf of California eating paper Coca-Cola cups.

Adults of *T. granulatus* on the west coast of South Africa congregate on kelp fronds (*Ecklonia* and *Laminaria*), reaching densities of up to $300/0.2 \, m^2$ there (Kensley, 1974). A large adult consumes some 110 mg of kelp material (20–25% of its body weight) in a meal lasting 2–3 h. Their activities are particularly apparent around the top of the intertidal slope, as they prefer thalli which are relatively dry and from which the mucilage has disappeared; feeding near the top of the intertidal also gives them a longer period in which to feed before the tide reaches the food. There may be not only daily but also seasonal variations in the feeding cycle, the extreme being reached by *Tylos punctatus* on southern Californian beaches, where the animals do not feed at all during the winter months but lie dormant in deep burrows (Hayes, 1974).

The structure of the alimentary canal of *T. granulatus* has been investigated by Barnard (1925) and in greater detail by Kensley (1974). Not surprisingly, there are marked similarities with the alimentary system of *Ligia* as described by Nicholls (1930), although some differences are apparent. The mouth parts are adapted for the rapid cutting of pieces of food, while the gut, and especially the mid-gut, is very capacious, allowing the animal to ingest a maximum amount in a short period of time. The powerful gastric mill present in the "stomach" is presumably also an important factor in dealing with large amounts of food.

Simple tests were performed by Kensley (1974) on the secretions of the hepatopancreas. Although these were relatively crude, he was able to demonstrate the presence of carbohydrases, proteases and lipases, as is consistent with an omnivorous diet. He also showed that the pH of the gut was not uniform; the hepatopancreas gave a reading of 6.3, that of the gastric mill and anterior part of the mid-gut 6.9, while the posterior mid-gut gave a reading of 7.4 and the hind-gut 6.8. The posterior mid-gut is the probable site of maximum protease activity. These results are similar to those obtained on *Ligia* (Nicholls, 1930).

Assimilation efficiencies of *T. punctatus* fed on *Macrocystis* fronds ranged from 53 to 72%, with a mean of 64% (Hayes, 1974). This value is quite high for a sandy-beach detritivore, the amphipod *Talorchestia capensis* on the same beaches having an efficiency of little more than 50% (Muir, 1977). On the other hand, it is much lower than that of a sandy-beach scavenging carnivore such as *Bullia*, with an efficiency of 88% (Stenton-Dozey and Brown, 1988). However, the value compares well with those of other terrestrial isopods. Hubble *et al.* (1965) recorded assimilation efficiencies of between 53 and 75% in *Armadillidium vulgare* and Reichle (1967), using different techniques, estimated an efficiency of 64% for the same species.

Some terrestrial isopods are coprophagous, which allows them to concentrate trace metals such as copper (Wieser, 1966; Hayes, 1970b). It is not clear, however, how widespread this habit is among species of *Tylos*.

5.2. Respiration

Rates of oxygen consumption have been studied in *T. punctatus* (Hayes, 1969, reported briefly in Hayes, 1974) and in the South African *T. granulatus* (Stoch, 1975; Marsh and Branch, 1979). *T. granulatus* presents a low rate of oxygen consumption throughout the day but displays a

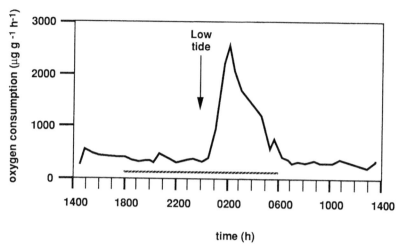

Figure 9   Typical respiratory rhythm of an individual of *Tylos granulatus* Krauss, showing a marked peak following low tide at night, an equally marked decline as dawn approaches, and a low inactive rate of oxygen consumption during the day, while the animal is buried. The horizontal bar denotes hours of darkness. (After Marsh and Branch, 1979; Brown and McLachlan, 1990.)

dramatic increase of between 300 and 700% during low tide at night, the peaks having a 24.8 h periodicity (Marsh and Branch, 1979) (Figure 9). Its periods of high oxygen uptake in the laboratory, under constant low light conditions, thus correspond to its periods of activity in the field. In addition to this endogenous circatidal rhythm, a semi-lunar rhythm is in evidence, the height of the peak and the percentage of individuals showing a peak increasing from spring tide to neaps. The "switch back" of the activity rhythm displayed every 15 d, the animals missing a tide as dawn approaches (see Section 2), is also shown with regard to the peaks of oxygen consumption in the laboratory. Tidal rhythms of oxygen uptake have been noted in a number of sandy-beach animals, both aquatic and semi-terrestrial, displaying endogenous activity rhythms (Brown, 1983) but none as marked as that found in *T. granulatus* Krauss. It follows that the endogenous rhythm is particularly strong in this species.

Marsh and Branch (1979) give the "basal" (i.e. inactive) rate of oxygen consumption for a standard animal of 0.1 g ash-free dry weight as 320 μl O_2 g^{-1} h^{-1}, rising to some 960 μl during the nocturnal activity period at spring tide and to 2240 μl at neaps. However, while the inactive rate was very uniform among individuals, the height of the active rate peaks was variable. Logarithmic regressions of body weight against oxygen consumption gave slopes that were relatively steep ($\bar{x}$ 0.86) but comparable with those reported by Hayes (1969) for *T. punctatus*.

The results presented by Hayes (1969) for *T. punctatus* differ from those of Marsh and Branch (1979) in that he did not discover peaks of oxygen uptake coinciding with peaks of activity in the field. However, his experiments were performed differently, under artificial regimes of light and dark, and his first two light periods occurred during the night, while the intervening artificial dark period fell during the day. Marsh and Branch (1979) suggest that possibly the somewhat higher rate of oxygen uptake recorded by Hayes during these two light periods may have represented nocturnal peaks dampened by the presence of artificial light. Stoch (1975) showed that exposure to light damps the respiratory peaks in *T. granulatus*. This finding is also strengthened by the discovery that, in some areas, *T. granulatus* refuses to emerge from its burrows in moonlight (Odendaal *et al.*, 1994). *T. punctatus* may also show a suppression of the activity rhythm in response to light. However this may be, we believe it likely that *T. punctatus* will, in view of its similar behaviour to *T. granulatus*, be found to display a marked rhythm of oxygen uptake. This may not be the case in *T. europaeus* or *T. sardous* from the Mediterranean, however, in view of their apparently much weaker activity rhythms.

An interesting observation regarding *T. punctatus* is that experimentally starved animals reduce their rate of oxygen consumption, at 18°C, from about 117 to only 19 $\mu l\,g^{-1}\,h^{-1}$ (Hayes, 1969). This is a very low rate indeed and is presumably similar to that of hibernating animals; if so, it represents a very considerable saving in energy during the winter months.

5.3. Energetics

Hayes (1974) estimated the rate of energy assimilation by *T. punctatus* to be 29.6 cal (125 J)/kcal isopod tissue/day. Of this, an estimated 15.8 cal (66 J) were expended in metabolism, while 10.2 cal (42.6 J) were expended in growth. We suspect, however, that his figure for metabolic expenditure may be too low in view of his failure to identify endogenous peaks of oxygen uptake. In any case it is apparent that a high proportion of the energy budget of *T. punctatus* is employed in respiration.

Using similar calculations for *T. granulatus* Krauss, Marsh and Branch (1979) concluded that the energy expended in respiration by this species is only about 9.74 cal/kcal isopod tissue/day, based on 21 h of inactive rate and 3 h at the maximum (or active) rate. Recalculating the energy expenditure of *T. granulatus* from the data presented by Marsh and Branch (1979) leads to the conclusion that the average energy expended by a medium-sized adult (5.7 g wet weight), including metabolic heat losses, is about 58 J over a 24-h period (Odendaal *et al.*, 1994). It is not possible at the present time to proceed to the drawing up of an activity budget, as has been prepared, for example, for the sandy-beach whelk *Bullia* (Brown, 1982b), as only the cost of burrowing has been measured (see Section 3) and this turns out to be less than 2% of the animal's energy expenditure even if it emerges on every night tide. Further work, and in particular an estimate of the cost of crawling, remains to be undertaken.

6. WATER AND HEAT RELATIONSHIPS

Among the chief problems facing an aquatic organism invading the land are those concerned with water balance and temperature relationships. Terrestrial and semi-terrestrial isopods obtain most of their water by feeding on moist food. Many can also imbibe "free" water through both the mouth and anus (Spencer and Edney, 1954), although they will drown if submerged for an extended period (Kensley, 1974; Matthewson, 1991; Brown, unpublished). The degree of protection against water loss afforded by the isopod cuticle does not approach that found in insects (Bursell, 1955; Edney, 1957; Hurley, 1968) and the cuticle does not appear to play any major role in restricting water loss in *Tylos capensis* (Matthewson, 1991). *Tylos* does have a heavier exoskeleton than other oniscids that have been studied but this is presumably a protective adaptation against sand abrasion (Hayes, 1974). There could, however, be some involvement of endocuticular lipids, which have apparently not been investigated in *Tylos* but are known to occur and restrict water loss in some other Oniscidea (Edney, 1968; Moore and Francis, 1986). Matthewson (1991) compared rates of water loss in living and dead *T. capensis*, under controlled conditions of temperature and humidity, and found that the latter did not transpire faster than the former, indicating that there is almost certainly no significant active control of water loss in the living animal.

Imafuku (1976) studied *T. granulatus* Miers on a Japanese beach, where the animals hid between pebbles and under debris during the day. Exposed to daytime conditions in the absence of a particulate substratum or debris, the animals rapidly lost weight and died. Imafuku also reported on the high-temperature tolerances of his animals, which, in the field, must frequently contend with ambient temperatures well in excess of 30°C. Humidity was found to play a vital role in survival, even at temperatures as low as 24°C (Figure 10).

Indeed, all the ecophysiological work which has been conducted on the water relationships of oniscid isopods points to the conclusion that the

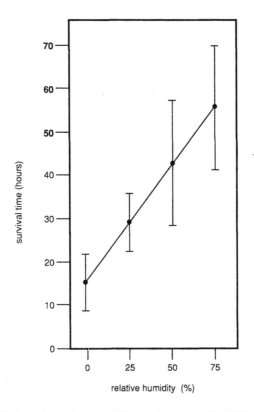

Figure 10   Effect of relative humidity on the survival of *Tylos granulatus* Miers at 24°C (after Imafuku, 1976). The animals were between 10 and 20 mm in length and humidity was controlled by means of sulphuric acid. Means and standard deviations are indicated for four different humidities.

animals are critically dependent on behavioural mechanisms to reduce water loss, seeking out microhabitats of high humidity for at least part of every day (Gunn, 1937; Waloff, 1941; Cloudsley-Thompson, 1956; Warburg, 1964; Edney, 1968; Holanov and Hendrickson, 1980). Brusca (1966) reported that the oniscid *Alloniscus perconvexus* burrows until it reaches sand with a moisture content of between 2.9 and 5.5%. Kensley (1974) suggested that the South African *Tylos granulatus* burrows far enough into the sand to remain in moist conditions while the surface layers above it dry out during the heat of the day and Hayes (1977) came to a similar conclusion in the case of *T. punctatus*, suggesting that the animals may display specific moisture preferences.

Holanov and Hendrickson (1980) tested this hypothesis on *T. punctatus* inhabiting a sandy beach in the northern Gulf of California. The day-time distribution of the animals within the substratum was determined, as well as the moisture content of the sand at 5-cm intervals. Results indicated that *T. punctatus* burrows downwards until it reaches conditions of 1.5–2.0% moisture, regardless of temperature, and that depth of burrowing is thus controlled by the need to reach saturation humidities. These authors further showed that animals held in sand with a 1.1% moisture content (i.e. 5% saturation of the sand), at 33°C for 12 h all survived, whereas 14 of 25 animals held at the same temperature but with a moisture content of only 0.1% died. A relationship between substratum moisture content and depth of burrowing in *T. punctatus* has been confirmed by Garthwaite *et al.* (1985).

De Villiers and Brown (1994) have recently shown, using sand columns in the laboratory, that *Tylos granulatus* Krauss also burrows to depths dependent on the moisture content of the sand (Figure 11), although the preferred moisture content is apparently somewhat higher, and the range wider, than was found by Holanov and Hendrickson (1980) for *T. punctatus*.

Transpiration through the integument is the major source of water loss in terrestrial and semi-terrestrial isopods (Lindqvist, 1972; Hadley and Warburg, 1986). Water is also lost through the pleopods, with the faeces and due to excretion, including the release of ammonia (Wieser *et al.*, 1969), which may account for up to 30% of the metabolic wastes in some oniscids (Dresel and Moyle, 1950). Matthewson (1991) subjected *Tylos capensis* of various size classes to a series of temperature/relative humidity combinations and measured rates of water loss. Her results do not present any surprises. Water loss increased gradually with increasing temperature in dry air, the highest rate of loss occurring at combinations of high temperature, low humidity and small isopod size. As Matthewson points out, semi-terrestrial isopods tend to be significantly larger than their aquatic counterparts and this could be considered an adaptation to life in air, as a relatively smaller surface area is thus presented to the environment, with a consequent reduction in transpiration rates. This is not to deny that there may be other advantages to semi-terrestrial isopods in being relatively large. In general *T. capensis* loses water more rapidly than do fully terrestrial isopods of similar size subjected to similar conditions, but less rapidly than the talitrid amphipod *Talorchestia capensis* occupying the same areas as *Tylos capensis* (Matthewson, 1991).

An interesting suggestion concerning water balance in oniscids was made by den Boer (1961), who worked mainly with *Porcellio scaber*. He showed that the nightly excursions of this species are regulated in frequency, duration and direction by the water content of the animals and the humidity of the air, and postulated that such excursions are essential for the evaporation of surplus water accumulated during day-time

A. C. BROWN AND F. J. ODENDAAL

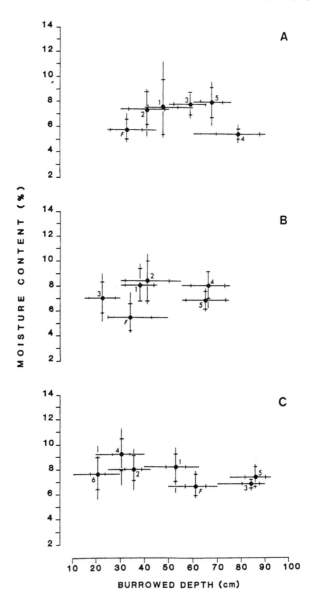

Figure 11 Sand moisture/burrowing depth relationships in *T. granulatus* Krauss, from three separate series of experiments (A–C) (*n* = 7, 6 and 7 respectively). *F* indicates the depth from which the animals were collected in the field and the moisture content of the substratum at that depth. Numbers 1–6 are consecutive reburrowing results for each of the three groups of animals, transferred to columns with different moisture/depth profiles. All results are given as the mean + sᴅ and range. (After de Villiers and Brown, 1994.)

inactivity in microhabitats of high humidity. The postulate is supported by the fact that at high humidities *Porcellio* gains weight, presumably by the uptake of water through the cuticle, a circumstance which den Boer attributes to an inability of the osmoregulatory system to excrete water at a rate sufficient to balance its uptake in near-saturated air. The argument is intriguing in that it stands the usual water-balance story on its head. The situation in *Tylos granulatus* is currently being investigated (Brown and Odendaal, unpublished).

Water loss clearly has implications for osmotic and ionic balance; indeed it has been suggested that physiological adaptations to retain ions may have been of greater evolutionary significance than resistance to desiccation in terrestrial Amphipoda (Lazo-Wasem, 1984; Moore and Francis, 1985, 1986; Friend and Richardson, 1986; Morritt, 1988) and this probably applies to the Isopoda as well. Kuemmel (1981) has presented evidence suggesting that the respiratory surfaces of the pleopods have an osmoregulatory function in oniscid isopods. Most studies on water balance in non-aquatic isopods and amphipods have, however, considered only desiccation and rates of water loss (Warburg, 1965, 1968; Mayes and Holdich, 1975; Dubinsky *et al.*, 1979; Coenen-Stass, 1981), although there is some more recent work on osmotic and ionic responses in terrestrial isopods (Price and Holdich, 1980; Coenen-Stass, 1985; Warburg, 1987; Matthewson, 1991).

Matthewson (1991) measured the osmolality and concentrations of sodium, potassium and chloride in the haemolymph of four size classes of *Tylos capensis*, both immediately after capture and after they had been subjected to a series of temperature/humidity combinations (Figure 12). She reported a summer average osmolality of 1070 mOsm in untreated animals, rising to 1270 mOsm in spring, values which compare well with those recorded for the supralittoral isopod *Ligia oceanica* by Parry (1953) and Price and Holdich (1980). Osmolality in *Tylos* haemolymph increased in roughly linear fashion with increasing temperature and decreasing relative humidity (Figure 13), the slopes for small individuals being steeper than those for larger animals. The slopes for damaged or dead animals did not differ significantly from those for live, undamaged individuals. Chloride concentrations were markedly temperature-dependent but did not correlate significantly with either relative humidity or body weight. Sodium concentrations also increased with increasing temperature, while potassium values, although increasing to a certain level, appeared, according to Matthewson, to be subject to some control by the living animal.

In general the results gained by Matthewson (1991) confirm that the regulation of water balance and ionic concentrations in the body fluids of *Tylos* depends on behavioural rather than physiological mechanisms.

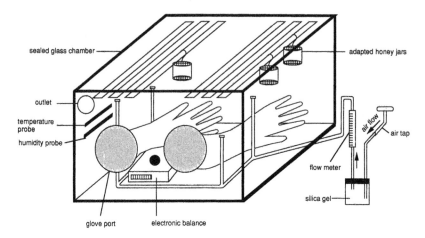

Figure 12   The experimental chamber used by Matthewson (1991) to investigate the effects of a range of controlled temperatures and humidities on *Tylos capensis*. The chamber was 700 mm in length, 445 mm wide and 420 mm high and was situated in a temperature-controlled room. The animals were housed within the chamber in plastic honey jars with the bottoms sawn off and the two open ends covered in plastic mesh to ensure passive air flow through each jar and around the animals. Relative humidity was maintained in the chamber by blowing in air from an air tap through silica gel and via a clamp and flow meter; the air entered the chamber through four air stones, equally spaced beneath the suspended jars so that the animals were subjected to minimal air convection. A top pan balance, situated inside the chamber, was used to determine the weight of the jars and hence the weight loss of the animals. (After Matthewson, 1991.)

However, this dependence is not necessarily as limiting as one might think, as is shown, for example, by the occurrence of permanent colonies of *T. capensis*, a species normally found between the drift line and the foredunes, in dune slacks, some distance from the sea, on beaches near Port Elizabeth in the Eastern Cape Province of South Africa (McLachlan, 1986; McLachlan *et al.*, 1987; Matthewson, 1991).

7. REPRODUCTION, GROWTH AND SURVIVORSHIP

7.1. Reproduction

Observations relevant to reproduction in *Tylos* have been made by Mead (1965, 1967, 1976), Hamner *et al.* (1969), Kensley (1974), Hayes (1977), Odendaal (unpublished) and other workers.

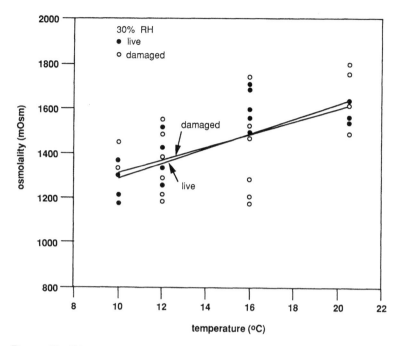

Figure 13 Linear regressions of haemolymph osmolality of *Tylos capensis* plotted against temperature at 30% relative humidity for live and damaged isopods. (After Matthewson, 1991.)

Ovigerous females of *Tylos granulatus* were first described by Barnard (1940); subsequently, Kensley (1974) noted that egg formation in both *T. granulatus* and *T. capensis* commences in spring and that ovigerous females are present in the population for only 3 months of the year. Size frequency analysis of these two species supports the view that oviposition and release of the young take place in midsummer over a 3-month period.

Hayes (1977) analysed samples of *T. punctatus* from five beaches in California and Mexico. He noted that all the eggs carried by a particular female appeared to be at an identical stage of development, suggesting that the entire brood had been fertilized at one time. Size frequency analyses supported the hypothesis that the animals breed only once a year. Two- and 3-year old females were found to produce roughly equal numbers of eggs and there appeared to be no breeding females older than 3 years. Hamner *et al.* (1969) agreed that both 2- and 3-year-old females could breed but suggested that the contribution of the 2-year olds was significantly less than that of 3-year olds. They observed the first gravid females at Estero de Punta Banda in May, with the first juveniles that had not yet moulted released from the brood pouch in August. This implies an incubation period of about 3 months and does not leave time for a second brood before the onset of hibernation, usually in October. *T. punctatus* incubates its eggs in the sand above high water mark, where it is presumably warmer than lower down, for about 9 weeks and produces an average of 13.6 young. Larger females tend to produce larger broods. Hamner *et al.* (1969) thought that few females breed before they are 3 years old and that most or all die soon thereafter.

Hamner *et al.* (1969) point out that the reproductive rate of *T. punctatus* is very low for an oniscid isopod; from four to 20 young are produced in each brood, only one brood is produced a year, and most females appear to breed only once during their lifespan, at 3 years of age. Mead (1965) found that ovigerous females of *T. sardous*, also presumably breeding once a year, carried only up to 14 embryos, while *T. capensis* averages 11 embryos (McLachlan and Sieben, 1991). In contrast, the terrestrial isopod *Armadillidium vulgare* produces two broods a year, of 20–75 eggs each and at least some of the females breed at 1 year of age (Paris and Pitelka, 1962). *T. europaeus*, on the Mediterranean coast of France, resembles other *Tylos* species in producing few young per brood (Hatchett, 1974); Vandel (1960) reports a mean of 20 eggs/brood, with a range of 10–38. Hayes (1977) found egg mortality in *T. punctatus* to be 0.67% at Punta Banda and 1.87% at Carlsbad, while Hamner *et al.* (1969) reported an approximate egg mortality of 1% at Punta Banda. Verhoef (1949) noted that juvenile *T. latreillei* are unusually large and suggested that this might be correlated with small broods. Juvenile *T. punctatus* are between 3 and 5 mm long when they emerge from the brood pouch (Hamner *et al.*, 1969), while emerging juveniles of *Armadillidium vulgare* are only between 2 and 3 mm.

It would appear likely that ovigerous females tend to stay within their burrows. McLachlan and Sieben (1991) failed to collect any ovigerous *T. capensis* in their pit traps and the present writers have not found ovigerous *T. granulatus* crawling on the sand surface.

Kensley (1974) described the reproductive structures of *T. granulatus* and these appear to be similar in other members of the genus. The male has a pair of copulatory stylets, which are modifications of the endopodites of the second pair of pleopods. Each is concave on its inner surface and the two stylets can be brought together so that their inner surfaces form an effective tube for the transfer of sperm to the female. The male reproductive opening lies on pereon segment seven, immediately opposite a right-angle bend in the stylets. The female reproductive opening is between the bases of the fifth pereiopods. The oostegites which form the marsupium are not present in all adult females but appear after a moult, when the female has eggs ripening in the ovaries. Barnard (1940) noted that the oostegites remain flat in females carrying embryos, the brood forcing the sternum upwards to form an internal sac. Barnard considered that this leads to a degeneration of the female's internal organs, an opinion which was not, however, confirmed by Kensley (1974). Mead (1965) also reports on an internal sac in *T. sardous*.

The only published accounts of *Tylos* mating behaviour are those of Mead (1967, 1976), who observed mating in both *T. europaeus* and *T. sardous* and found no important differences between these two species. Mating appears to be linked closely to the annual parturial moult of the females, copulation usually taking place either between the posterior and anterior ecdyses or within 48 h of moulting, although mating after a longer period was also noted on occasion.

Courtship in both *T. europaeus* and *T. sardous* is brief (Mead, 1967). The male approaches the female and stimulates her by touching her with his antennae, rubs himself against her and may bump her, becoming increasingly more agitated as he does so. These activities last no more than a few minutes before copulation occurs. During copulation, the male inserts the endopodites of his second pleopods, individually and alternately, into the single reproductive opening of the female. Each insertion lasts about 3 min. The male mounts the female *after* copulation has terminated, grasping her posterior tergites with his anterior pereiopods. While mounted, the male slaps the tergites of the female, at irregular intervals, with his antennae. After this activity has continued for an hour or more, it may become interrupted for short periods; however, the postcopulatory behaviour may continue intermittently for up to 24 h. This postcopulatory behaviour may clearly prevent other males from mating with the female, although alternative interpretations are also possible.

Odendaal *et al.* (unpublished) observed *T. granulatus* on the beach at Groenrivier, South Africa, almost every night during the course of a 1-month study and several nights a month over a period of 5 years, yet virtually no interactions or sexual activity between individuals was observed. However, on 27 September, 1992, a visit by two investigators to Island Point, some 8 km from Groenrivier, led to the observation of numerous pairs of *T. granulatus* mounted in piggy-back fashion (Odendaal, unpublished) (Figure 14). When a pair was deliberately separated and the female moved off, the male would follow her closely, crawling exactly on her trail, then again attempting to mount. It seemed possible that the male was being attracted by a pheromone released by the female; pheromonal stimulation of sexual activity has been suggested for a number of isopods and amphipods, and Hammoud *et al.* (1975) produced evidence to suggest that copulation in *Gammarus pulex* is induced by the pheromonal activity of female ecdysterone. Certainly fully terrestrial oniscids have an acute sense of smell (Fischbach, 1954) and can detect the presence of other individuals by means of olfaction (Kuenen and

Figure 14   *Tylos granulatus* Krauss displaying mating behaviour. This is believed to be the first photograph to be published showing sexual activity in any species of the genus. The male is mounted back-to-front on the female and is stroking her with his antennae. This is probably postcopulatory grooming. A third individual has been attracted to the pair and is seen burrowing in the background. (Photo: C. Velasquez.)

Nooteboom, 1963). Whether the mounted pairs of *Tylos granulatus* were *in copula* or indulging in postcopulatory behaviour was not ascertained. However, in photographs taken at the time, the majority of pairs are seen to be mounted back to front (see Figure 14), a position in which copulation clearly cannot occur, although the males were stroking the females with their antennae. We thus incline to the view that postcopulatory mounting and grooming is a feature of this species, as in Mediterranean species of the genus.

Occasionally a larger group of up to seven individuals was seen interacting. Many animals were observed waiting in the mouths of their exit holes; every now and then, an individual would make a grab at, and sometimes hold on to, an animal that passed by its hole. The investigators subsequently proceeded to the beach at Groenrivier, where they encountered similar activities on the part of the isopods. Here they recorded over 50 mounted pairs of *T. granulatus*. Several individuals were trying to get into a single burrow and up to eight animals, attempting to mount others, were found entangled in a group. This behaviour is not dissimilar to the "explosive breeding systems" described for insects and frogs having short breeding seasons (Wells, 1977; Thornhill and Alcock, 1983;

Odendaal *et al.*, 1985). A number of animals that were just protruding from their exit holes were found to have a second individual attached to them. Despite the apparent free-for-all, the top individual of a mounted pair was invariably found to be male, the lower partner female.

7.2. Growth and Survivorship

Although for most isopods, and indeed for most Crustacea, size is measured in terms of length, the fact that *Tylos* tends to roll up into a ball makes it more convenient to measure the maximum width of the animal. This is justified by the fact that length bears a linear relationship to width in all those species of *Tylos* which have been examined in this regard (Hamner *et al.*, 1969; McLachlan and Sieben, 1991; Brown, unpublished). This linear relationship is not unique to *Tylos* but is a feature of other oniscid isopods as well (Paris and Pitelka, 1962; Schultz, 1965).

Estimates of growth and life-span exist for *Tylos capensis* Krauss in the Eastern Cape of South Africa (McLachlan and Sieben, 1991) and for *T. punctatus* in southern California (Hayes, 1974). *T. capensis* from dune slacks in Algoa Bay revealed up to five cohorts, based on width frequency histograms. Growth curves indicated an age of 4–5 years at the maximum recorded width (14 mm). Captive animals moulted approximately every month, with a mean increment of 0.46 mm/moult and it may therefore take 20–30 moults to reach maximum size (McLachlan and Sieben, 1991). Kensley (1974) thus appears to have been mistaken in suggesting only four or five moults during the life cycle of this species.

Growth rates of *T. punctatus* (Figure 15) were estimated from collections made at Carlsbad and Estero de Punta Banda (Hayes, 1974). Four cohorts were generally distinguishable. As in the case of *T. capensis*, width measurements were used, the width/volume ratio being constant as growth is isometric throughout the animal's life-span. These are much smaller animals than *T. capensis*, the maximum body width being something under 5.5 mm. Nevertheless, it is estimated that it takes about 4 years for the animals to reach this size. Hamner *et al.* (1969) also found that some males of *T. punctatus* lived for at least 4 years. The animal grows only during a 5-month period, in summer, hibernating in winter. No significant differences were found between mean sizes or growth rates of males and females of this species and the growth rate declined with age.

The only calculations of mortality rate or survivorship appear to be those of Hayes (1977) for *T. punctatus* in southern California, and of McLachlan and Sieben (1991) for *T. capensis* in dune slacks in the Eastern Cape Province of South Africa. The latter authors give a figure

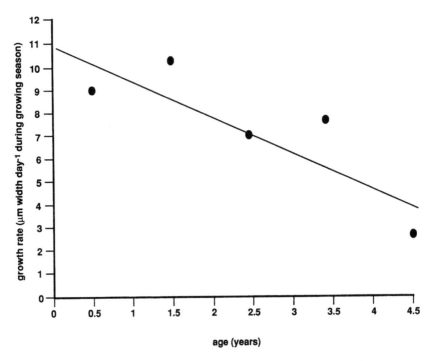

Figure 15   Age-specific growth rates for the five cohorts of *Tylos punctatus* identified during the 1966 growing season. The points represent the slopes of least-square regression lines fitted to the five growth curves obtained. Ages of the cohorts are represented by the midpoints of the year classes (i.e. the 1966 cohort is taken as 0.5 years old, the 1965 cohort as 1.5 years old, and so on). (After Hayes, 1974.)

for the age-specific mortality rate of $Z = 0.77/y$ (Figure 16). However, they point out that this value needs to be treated with some caution, as pit trapping favours the larger individuals (Hayes, 1974). Hayes (1977) throws light on the difficulties of assessing mortality rates in a mobile population and arrives at a best estimate of survivorship for his population of *T. punctatus* at Punta Banda of 39% year^{-1} ($M = 0.61$). Again, caution is indicated but we may conclude that it is usual for over half the population to die each year. This is by no means uncommon for populations of any short-lived species of animal. Mortality is due chiefly to physical causes rather than to predation.

McLachlan and Sieben (1991) give an overall P/B ratio for the population of *T. capensis* in the Alexandria dunefield of 1.38 year^{-1}. Highest energy values were reported for adults during the breeding season (February–March) and the lowest values for juveniles. The overall mean energy value was $7.8 \, \text{kJ} \, \text{g}^{-1}$.

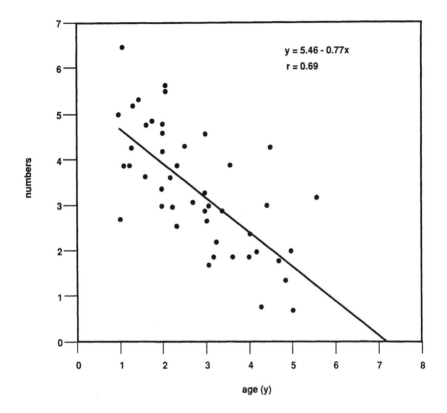

$$y = 5.46 - 0.77x$$
$$r = 0.69$$

Figure 16   Mortality curve for *Tylos capensis* in dune slacks in Algoa Bay, South Africa. (After McLachlan and Sieben, 1991.)

8. POPULATION DENSITIES, DISTRIBUTION AND DISPERSAL

8.1. Densities

A number of estimates have been made of the size and densities of *Tylos* populations in several different parts of the world. Although one can accept estimates such as those of Hayes (1969, 1974), indicating 20 000 isopods per metre of beach frontage at Carlsbad, Torrey Pines and Punta Banda, as being a realistic order of magnitude, accurate assessments are extremely difficult to achieve. This is particularly true where attempts have been made to determine sex ratios or the proportion of juveniles in the population. There are several reasons for this difficulty. First, the number of mounds or entrance holes is not a reflection of the number of animals actually buried in the sand (Brown, 1959; Kensley, 1974; Odendaal et al., 1994). Secondly, if quadrats of sand are excavated, juveniles are easily overlooked unless all the sand is passed through a sieve (S. Eekhout, pers. commun.), a mammoth task if a number of quadrats are to be excavated to a depth of 1 m or more. Thirdly, the population density is seldom uniform along the beach (see below), so that there can be no such thing as a representative quadrat. Fourthly, juveniles commonly burrow lower down the beach than do the adults (Mead, 1968; Kensley, 1974) and in some cases the females likewise tend to be separated from the males (Mead, 1968; Hamner et al., 1969). Fifthly, pit trapping of animals during their activity periods has proved to be inadequate (Hayes, 1970a) and lastly gravid females may show less tendency to emerge from the sand than other members of the population (Hamner et al., 1969; McLachlan and Sieben, 1991; Brown, unpublished). We are thus faced at best with population estimates which may incorporate a 50–100% error.

8.2. Distribution on the Beach

Tylos spends most of its time either within the sand column or under cover, not far above the high tide mark, typically emerging for a few hours at night. The centre of gravity of the population moves towards and away from the sea during each fortnightly tidal cycle. This behaviour has led to three aspects of the distribution of buried animals on the beach which have been studied by several workers: vertical distribution in the sand, shore-normal distribution (i.e. at right-angles to the sea) and longshore distribution.

Different species of Tylos burrow to different depths, as do populations of the same species on different beaches and the same populations at different times of year. Within a given species, depth of burrowing is largely determined by the moisture content of the sand (see Section 6) but undoubtedly other factors, such as particle-size stratification of the substratum, also play a role. Kensley (1974) found that T. granulatus burrowed to depths of up to 500 mm and T. capensis to 300 mm at the sites he investigated. However, at other sites T. granulatus burrows to 1 m or more (Odendaal et al., 1994; Brown, unpublished), while at Yserfontein many individuals of this species were found only 10–20 cm below the surface (Brown and Trueman, 1994b). Adult T. punctatus also burrow as deep as 1 m at certain times (Hamner et al., 1969). Both Hamner et al. (1969) and Hayes (1977) noted that juvenile T. punctatus burrow to shallower depths than do the adults but that the depth of burrowing changed from night to night. During the winter months, T.

punctatus burrow particularly deeply, high on the shore, to hibernate (Hamner *et al.*, 1969).

Although grain size influences depth of burrowing (see Section 3), Hayes (1977) was unable to correlate the presence or absence of *T. punctatus* with particle size composition of the substratum and concluded that grain size itself is not important in determining the distribution of the isopods. Kensley (1974) also noted that *T. capensis* and *T. granulatus* burrow into a wide variety of sand types and are not generally limited by particle size. This was also the conclusion reached by Imafuku (1976) for the Japanese *T. granulatus* Miers; he confirmed this in the laboratory, showing that the isopods did not select a particular particle-size range, although they always selected wet rather than dry sediment of any particle size. These findings contrast with those of Giordani-Soika (1954) and Mead (1968), who found that the distribution of *T. europaeus* and *T. sardous* on Mediterranean beaches was largely controlled by particle-size preferences.

The shore-normal distribution of buried animals changes with the position of high water mark through the tidal cycle, these changes thus being most marked on beaches with a considerable tidal range and least on beaches where the tidal range is small, such as are found in the Mediterranean. On beaches having a gentle slope at and above high water mark, the band in which the animals burrow is often as much as 100 m in width, while on other beaches the population may be compressed into an area only 10 m or less wide (Kensley, 1974; Odendaal *et al.*, 1994). These observations are complicated by the fact that beach slopes may change radically from time to time, particularly on beaches exposed to heavy wave action (Brown and McLachlan, 1990), so that the width of the zone inhabited by *Tylos* grows wider or narrower according to the changing beach profile. The changing distribution of *Tylos* populations perpendicular to the shore has been explained in terms of changing high water mark, which in turn affects the area in which food is deposited, and factors such as slope and humidity (Hamner *et al.*, 1969; Hayes, 1970a; Kensley, 1972; Imafuku, 1976).

As is the case with so many animals inhabiting sandy shores (Brown and McLachlan, 1990), the longshore distribution of *Tylos* tends to be extremely discontinuous in most populations studied, areas of high density alternating with stretches of beach in which no individuals can be found (Kensley, 1974; Hayes, 1977; Odendaal *et al.*, 1994; Brown, unpublished). Various explanations have been put forward to account for this phenomenon in a number of different sandy-beach animals, including longshore differences in the penetrability of the substratum (Griffiths and Griffiths, 1983; Brown *et al.*, 1989; Brown, unpublished), the effect of waves and currents on surfing forms (Brown, 1982a; Brown and McLachlan, 1990), aggregation correlated with food distribution (Ansell and Trevallion, 1969; Brown, 1982a), aggregation resulting from mating behaviour (Perry, 1980) and simple gregarious tendencies (Boaden and Erwin, 1971). These various explanations are not mutually exclusive. In any case it now seems likely that no single explanation can account for the discontinuous distribution of all or even most psammophiles and that different factors are involved in different species.

None of the above possible explanations of longshore discontinuity appears to apply to buried *Tylos granulatus*, however, although its patchy distribution is extremely marked in all those populations which have been studied. A special study of longshore distribution in this species has been made by Odendaal *et al.* (1994) at Groenrivier, on the South African west coast. Within the band of 5–10 m width in which the animals occur, large, dense clusters are separated from one another by very low-density areas. High- and low-density areas are clearly visible from surface observation of mounds and exit holes (Figure 17) and this highly non-random distribution has been confirmed by actual sampling of the animals. The position, shape and intensity of the aggregations differ from night to night.

Odendaal *et al.* (1994) were unable to find any convincing relationship between spatial patterns of buried *T. granulatus* and their food and developed an alternative hypothesis to explain them, based on the knowledge that the non-random movement of conspecifics will, under certain circumstances, result in spatial aggregation (Odendaal *et al.*, 1988; Turchin, 1989). Night-time observation of the population revealed that, when *T. granulatus* returns to the vicinity of high water mark after foraging, they have a tendency to re-use existing holes vacated at the start of the activity period and also tend to burrow at surface irregularities such as the cone-shaped mounds caused by the digging of other individuals. The hole-and-mound hypothesis developed by Odendaal *et al.* (1994) states that an area with a large number of exit holes and mounds will tend to "capture" more *Tylos* than an area with few of these disturbances. Experiments carried out by these authors show that significantly more holes are, in fact, re-used in high-density than in low-density areas. Clearly, the higher the density of exit holes the more likely it is that a returning animal will encounter a hole.

In contrast to the distribution of buried populations, the distribution of emerged, active isopods can clearly be related to the distribution of suitable food. This is immediately apparent from casual observation (Kensley, 1974; Brown, unpublished) and has been confirmed on a more quantitative basis by G. Simmons (pers. commun.) and his coworkers. They studied the distribution and abundance of active *T. punctatus* in relation to the distribution of its primary food resource, stranded eelgrass

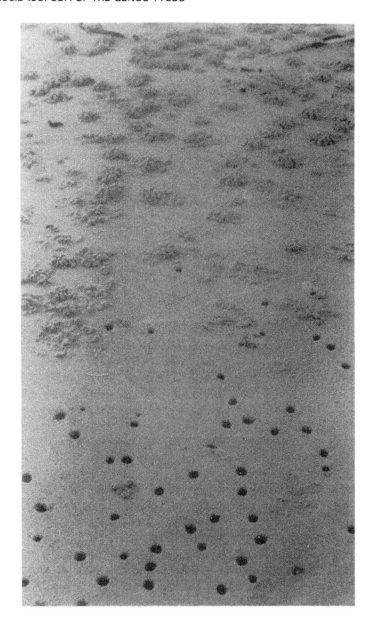

Figure 17 Mounds and exit holes (in the foreground) of *Tylos granulatus* Krauss on the beach at Groenrivier, on the west coast of South Africa. The mounds nearest the camera have been made by individuals using existing entrance burrows and are low and star-shaped. (Photo: F.J. Odendaal.)

Zostera marina, on a beach in Mexico. They constructed patches of this wrack, which they varied in size, shape, age and situation relative to the drift line, and then assessed the density of isopods, nightly, on these patches. They also attempted to determine whether patch shape influenced emigration from the patches. In addition, they surveyed transects perpendicular and parallel to the shoreline to determine any relationship between isopod distribution and prevalence of eelgrass.

They were able to show that at night active *T. punctatus* is most abundant where there is most wrack but that there was no significant correlation between isopod density and the size, shape or age of the patches, at least at the scale they considered. The degree of moisture was, however, important in the distribution and they were able to attract isopods to wrack patches well above the drift line by keeping these wet with seawater.

8.3. Dispersal

A problem which has not been seriously addressed as far as *Tylos* is concerned is the dispersal of the species. Aquatic organisms and animals that fly can spread easily from beach to beach, while animals such as the isopod *Ligia* will crawl at night from one rocky outcrop to another over sand, particularly if kelp or wrack is present on the beach (Brown, unpublished). However, rock devoid of sand appears to present a barrier to at least some species of *Tylos*, there is no aquatic larval stage and the behaviour of the adults leads them to avoid direct contact with the water as far as possible. Nevertheless, they may be found on isolated pocket beaches and on stretches of sand remote from other beaches. The only explanation which presents itself is that an occasional individual may be caught by a wave or washed out of its burrow during a storm and then survive long enough at sea to reach another beach. The animals drown if submerged for any length of time, as the pleopod surface is not extensive enough to support aquatic respiration (Arcangeli, 1957; Kensley, 1974; Ebbe, 1981) but it is possible that some may be washed out to sea while clinging to a piece of kelp or wrack or may accidentally find such a potential raft subsequently, and so survive the journey to another beach. Juveniles that surf, rolled into a ball (Kensley, 1974), may perhaps be more likely to undertake such forced journeys than the strictly non-surfing adults. A partial confirmation of this theory of dispersal is that living individuals of both *T. punctatus* and *T. granulatus* Krauss have been taken at sea, clinging to debris, sometimes at a considerable distance from shore (Menzies, 1952; Odendaal, unpublished). Kensley (1974) has shown that individuals of *T. granulatus* denied a raft will survive submersion in seawater at 17°C for a number of hours and that juveniles survive longer, up to 12 h, due at least partly to a more advantageous pleopod area/body volume ratio. Also the juveniles, unlike the adults, trap an air bubble between the pereiopods when they roll up into a ball, which may give them access to more oxygen in addition to making them buoyant. Brown (unpublished) placed a young adult *T. granulatus* on a small raft of wrack in a tank of seawater at 15°C and found, after an hour, that it was gripping the wrack firmly with pereiopods 1–3, while pereiopods 6 and 7 were fully extended, pushing the abdomen upwards, so that the pleopods were largely out of the water. The animal was removed after 48 h and placed on sand, into which it burrowed after a few minutes.

Hayes (1977) was of the opinion that *T. punctatus* in California is frequently washed out to sea, sometimes in large numbers during storms. He found that not only the juveniles, but also the adults, of this species tend to roll up into a ball when submerged, frequently trapping an air bubble which causes them to float. Such animals lived for more than a week in an aerated aquarium. Imafuku (1976) found that submerged *T. granulatus* Miers also survived longer, at 20–21°C, than does *T. granulatus* Krauss (at 15°C). He found the salinity of the seawater to be of considerable importance, the animals surviving much longer in 100% seawater than at lower salinities.

Colonization of a beach by intermittent sea transport must be expected to take a long time and this is confirmed by our experience that where a population has been eliminated, recolonization is extremely slow, if it occurs at all. Sea dispersal might also account for the essentially discontinuous distribution of *Tylos*, many beaches which appear suitable being in fact devoid of them (Hayes, 1977; Brown, unpublished). A beach would depend for colonization on an appropriate current regime (Menzies, 1952).

Hayes (1977) was inclined to attribute the distribution of *T. punctatus* on beaches in southern California to sea dispersal in combination with the random occurrence of catastrophic beach erosion such as reported by King (1959), who drew attention to instances of beach erosion of two vertical metres or more in a single day. Hayes (1977) points out that most erosion occurs in winter, when *T. punctatus* is hibernating, so that the animals can do little to escape it.

Land dispersion also occurs under appropriate circumstances. When the mouth of the Orange River was investigated by Brown (1959), it was open and *T. granulatus* Krauss was found on its southern bank. In 1966 the mouth became blocked by a sand bar, some 3 km in length. In 1968 the site was visited by Kensley (1974), who found the entire bar colonized by this species.

9. THE IMPACT OF MAN ON *TYLOS*

Tylos is very vulnerable to human activities, although the animals have not been exploited for food as far as we are aware. Three different types of activity have been identified as severely reducing *Tylos* and other semi-terrestrial populations (Brown and McLachlan, 1990): pollution, destruction of habitats and human invasion of the back beach and dunes for recreational activities.

Being semi-terrestrial, *Tylos* is caught between sea and land; it is thus susceptible to pollution from either direction. Marine pollution may contaminate the isopod's food, foul its burrows, and affect the animal by direct contact. Crude oil pollution may achieve all three effects and must be considered a most serious threat. Oil/dispersant mixtures are even more lethal than crude oil on its own and in the past have caused great destruction when dispersants were sprayed too near to, or in some cases onto, the shore (Brown, 1985).

The flexible behaviour of the animals does afford them some chance of survival, however, even under conditions of extreme pollution. Crude oil from the ballast tanks of a tanker was washed ashore at Bloubergstrand (on South Africa's west coast) during the period that Kensley (1974) was studying the population of *T. granulatus* there. A spring tide occurred the following day, resulting in a beach covered with oil up to the HWS mark, the upper slope being covered in a thick layer. Kensley reports that, of the few animals which emerged that night, almost all were killed, probably due to a clogging of the pleopod respiratory surfaces. The following night a large number emerged and many of these moved upshore, well above the drift line, to bury themselves amongst the vegetation of the low foredunes. Those that did bury themselves lower down, through the now much thinner layer of oil, appeared not to be seriously affected and no oil could be found adhering to the bodies of those subsequently dug up, except for traces on the tips of the pereiopods. Presumably burrowing through the sand had served to remove any oil sticking to the body. Contamination of the stranded kelp, on which the animals would normally have fed, remained a threat, however, and at least some of the animals that had migrated into the dunes resorted to eating dune vegetation. After about 3 weeks the animals returned to their normal position on the beach and resumed their normal behaviour, having survived a pollution event to which any less adaptable animal would have succumbed.

Pollution from the land includes airborne pollution from factories and agricultural land, as well as pollution of the ground water by seepage and run-off. In addition to such pollution, the drawing-off of ground water for agricultural or domestic purposes may lower the water table (Brown and

McLachlan, 1990) and thus decrease humidity within the sand and in the burrows, although these effects have not been studied specifically with respect to *Tylos* populations.

Recreational activities confined to the intertidal zone of a sandy beach have relatively little impact on the invertebrate fauna (Brown and McLachlan, 1990). However, walking, and particularly the movement of vehicles, above or through the drift line, has serious detrimental effects on the system (Van der Merwe, 1988). While it is unlikely that walking over the surface has significant impact on *Tylos* populations, off-road vehicles may run the animals down while they are active at night or crush them within their burrows during the day, as shearing and compacting effects extend to a depth of some 20 cm or more (Godfrey *et al.*, 1978). Odendaal (unpublished) has observed some of the effects on *T. granulatus* of marine diamond mining activities on the South African west coast. Many animals are run over by heavy vehicles pulling boats, etc. What makes the situation worse is the observation at Groenrivier that *Tylos* has a marked tendency to burrow in the tracks left by such vehicles and as the same tracks are used over and over again by these vehicles the effect on the population is magnified.

Van der Merwe and Van der Merwe (1991) conducted experiments in which *Tylos capensis* were allowed to burrow in sand within 200-mm deep frames with mesh bottoms, sunk flush with the sand surface, and then driven over by a 4×4 Toyota LDV (weighing 1420 kg). They demonstrated a linear relationship between the number of vehicle passes and the percentage of *Tylos* crushed (Figure 18). Clearly the proportion killed must depend largely on the depth at which the animals are buried and this in turn depends on the moisture content of the sand. Thus animals in damp sand, being more shallow, will be at greater risk than more deeply buried animals in drier sand. While regulations are now in force in many parts of the world limiting the use of off-road vehicles, the total impact on *Tylos* (and other semi-terrestrial psammophiles) increases as recreational activities involving the shore escalate in developed countries (what Dower (1965) termed "the fourth wave"), and with increasing shore-mining activities in countries such as South Africa.

Destruction of habitats, or the physical obstruction of free movement up and down or along the beach, also has serious implications for *Tylos* populations and may in many cases eliminate them. Bathing beaches commonly have extensive developments immediately to shorewards, including roads, parking lots, pavilions, shops, stalls, paved walkways and other facilities. These are generally constructed on the back beach and/or levelled foredunes, so destroying the habitats of semi-terrestrial animals. Sea walls and similar barriers, which effectively interrupt the slope and prevent detritus washing up from the sea, also make it

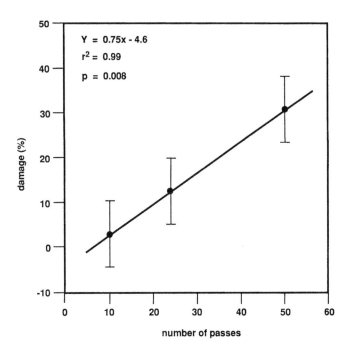

Figure 18   Regression between the number of vehicles passing over an area of beach and the percentage of *Tylos capensis* damaged. 95% confidence limits are indicated. (After Van der Merwe and Van der Merwe, 1991.)

impossible for *Tylos* populations to survive, while the building of tarred roads through or in front of the foredunes spells disaster for the animals and indeed for the ecosystem as a whole (Brown and McLachlan, 1990). Brown (unpublished) noted a thriving population of *T. granulatus* on the back beach at Hout Bay, South Africa, in the 1950s; however, within a few months of the construction of a road and a parking lot in front of the foredunes, this population had totally disappeared and has not returned.

The removal of kelp and carrion from bathing beaches, to render them more attractive, also leads to a dramatic decrease in the populations of semi-terrestrial psammophiles, as has been observed by Brown (unpublished) on the sandy coastline of False Bay, near Cape Town. The effects of mobile sand-cleaning machines, which suck up the surface sand, pass it through filters and return it to the beach, are more problematical. While these machines effectively eliminate talitrid amphipod populations, it is unlikely that *Tylos*, lying buried in relatively deep burrows during the day, will be sucked up by the mechanism; however, as the vehicles are heavy, they are likely to crush the animals within their burrows.

Development of the coastal zone for recreation or housing has implications in addition to those associated with the physical structures themselves. The presence of lights, even some distance away, may inhibit the emergence of *Tylos*. It is possible that noise has a similar effect, although this has not been investigated. On the other hand, the animals are so adaptable that it may be argued that noise and light pollution will not seriously inhibit them for long.

Use of the animal as a bait organism and the effect of this on *Tylos* populations does not appear to have been investigated, although in regions such as Mozambique the practice appears to be widespread. In contrast, very few fishermen along the South African coast claim to have used *Tylos* as bait and those that have state that it is not very effective (Brown, unpublished).

Along the South African west coast, north of Lambert's Bay, beach mining for diamonds and sublittoral diamond pumping have been undertaken for some years. Regrettably there are no faunistic records from these beaches dating from before the commencement of such operations, so that to link them with the absence of *Tylos* must at best be circumstantial. However, the mining operations typically lead to dramatic changes in beach structure, while heavy vehicles move back and forth along the back beach. Consulting work undertaken in these areas (Brown, unpublished) indicates that the meiofauna and many components of the macrofauna recolonize a beach rapidly after mining operations have ceased, while the fauna at and above the drift line, and *Tylos* in particular, does not.

Our observations on South African sandy shores over the years certainly indicate a serious decline in *Tylos* populations, most marked where surfaces have been hardened and recreational facilities constructed, or where beach mining has been in progress. A decline is also apparent on other beaches, not seriously altered morphologically but quite heavily utilized by man, such as Long Beach, Kommetjie. Only on remote, relatively inaccessible shores does *Tylos* continue to flourish in the vast populations that were at one time common throughout the region.

10. DISCUSSION

Sandy beaches, and particularly those exposed to heavy wave action, present extremely harsh conditions to the biota. Apart from the aquatic fauna having to face the force of the waves, both aquatic and semi-terrestrial animals must adapt to a highly unstable substratum, whose profile may be drastically altered during storms, tonnes of sand being removed and washed out to sea. This instability also precludes the presence of attached plants throughout the intertidal zone and to well above the highest level reached by the water, thereby depriving the fauna of both a primary food supply and shelter. The macrofauna of a high energy sandy beach thus tends to consist of relatively few species, mostly filter feeders and scavengers, highly adapted to their harsh environment both morphologically and behaviourally (Brown, 1983; Brown and McLachlan, 1990).

A few groups of Crustacea have managed to overcome the rigours of direct exposure to wave action by leaving the water and moving up the shore to colonize the area above high water mark. Chief among these are the ocypodid crabs, talitrid amphipods and oniscid isopods of the genus *Tylos*. While a large number of other oniscids have continued moving landwards until all links with the sea have been lost, *Tylos* typically maintains its dependence on the marine environment, feeding intertidally on material washed up by the sea and organizing its time by means of tidally linked endogenous activity rhythms that, in addition to saving energy, provide maximum protection against high temperatures and desiccation, against predation, against submergence and against being washed out to sea. It is of interest that, despite its continued dependence on the sea, the animal has lost the ability to breathe under water and will drown if submerged for any length of time. The term "semi-terrestrial" is thus highly appropriate as applied to this genus.

Endogenous diurnal, circatidal and semicircalunar rhythms are well known among aquatic intertidal isopods (Jones and Naylor, 1970; Enright, 1972; Fincham, 1973; Dexter, 1977; Hastings, 1981) and the rhythms displayed by *Tylos* may well be derived from similar rhythms that were present in its aquatic, intertidal ancestors. These rhythms in *Tylos* would, however, appear to be particularly intense. This view is consistent with previous findings that burrowing shore Crustacea in general possess more consistent, precise and persistent endogenous rhythms than do mobile non-burrowing forms (Atkinson and Naylor, 1973; Enright, 1975; Williams, 1980, 1983b). One advantage of this is that it permits a proportion of the population to remain buried during one or more activity periods, yet emerge approximately in synchrony with other members of the population at the beginning of a subsequent period of activity (Naylor, 1976, 1988). This may be of considerable importance to *Tylos*, as it is known that not all individuals emerge every night and that in some populations there may be almost no surface activity on some nights. The 15-d "switch back", when the entire population misses a tide, also requires a particularly strong internal clock.

Recent work confirms previous views that the undoubted success of *Tylos* as a semi-terrestrial crustacean resides less in morphological and physiological adaptations than in behavioural flexibility. Indeed, structurally the animals do not differ markedly from the pattern of typical, aquatic isopods and they remain relatively unspecialized. Neither has the cuticle become highly resistant to water loss or the excretory mechanism adapted to convert ammonia into some less toxic form. The pereiopods are highly efficient at burrowing, and yet their overall structure, though robust, conforms to that of other Isopoda and is not remarkable in any way.

The position is very different, however, when we consider behaviour. Endogenous rhythms of activity, set by predictable environmental variables, play a crucial role in determining appropriate behaviour in *Tylos*, as in other semi-terrestrial Crustacea, but are of little value on their own in subtending the survival and continuing essential functions of the animal. Of equal importance is the ability to orientate to environmental cues, such as beach slope, moisture and light. If their endogenous rhythms originated in similar ancestral rhythms, directional orientation to environmental cues must have evolved since they ceased being fully aquatic, and may thus be regarded as having been superimposed on the endogeny. In these orientational abilities they resemble the semi-terrestrial talitrid amphipods rather than aquatic isopods. This is obviously the result of convergent evolution. A rigorous comparison between mechanisms of orientation in *Tylos* and talitrids would make a rewarding research project but for the moment we have evidence of similarities in orientational behaviour in some species and also data which suggest important differences in others. Orientation in talitrids has been studied intensively but that of *Tylos* has not.

Relatively few species of *Tylos* have been studied in any detail — less than one quarter of those presently enjoying specific rank. Nevertheless, despite the relative paucity of information, one is struck by the range of behaviour found within the genus, a range due not only to differences between species but more particularly to a behavioural plasticity within the species. *T. capensis* on the South African shore, for example, normally takes full advantage of conditions on the beach and its behaviour is adapted to tidal ebb and flow. However, it is not fully committed to this mode of life and some populations have adapted successfully to an existence higher up the shore, in dune slacks (McLachlan and Sieben, 1991), presumably initiated by a shortage of suitable food intertidally. Another remarkable instance of behavioural flexibility is provided by the observations of Kensley (1974) on the behaviour of *T. granulatus* following an oil spill: the animals moved up the shore into the dunes, ate dune vegetation, and only returned to their former habitat and behaviour after 3 weeks, when the threat of contamination had been substantially reduced.

Hazlett (1988) has pointed out that "the outstanding feature of behaviour is its flexibility in interfacing organisms with an environment which is variable in time and space". Moreover such flexibility becomes increasingly critical the greater and more unpredictable the environmental changes with which the animal has to contend. Endogenous rhythms, however complex and sophisticated, can clearly only interface the animal with environmental elements which are predictable in time and space. The animal is entirely dependent on behavioural plasticity to survive irregular events, especially if they are also of unpredictable magnitude, such as storms (or oil spills). Exposed sandy beaches present particularly dramatic changes. Hazlett (1988) comes to the conclusion that the essential adaptation which has been selected for in such unpredictably harsh environments is the preservation of behavioural flexibility, the loss of which would effectively preclude the animal from survival. While not at present wishing to argue this point for the sandy-beach fauna as a whole, we believe that Hazlett's concept is totally true with regard to *Tylos*.

Several additional points of interest arise from a consideration of this behavioural plasticity. It has brought home to the present writers how easily and rapidly behavioural responses may be modified as compared with morphological or physiological adaptations. While no claim of originality is made for this concept, it is certainly worth stating again in the context of oniscid biology. Many previous writers have remarked on the success of the oniscid isopods *despite* their lack of morphological and physiological adaptations to a terrestrial or semi-terrestrial mode of life. We are of the opinion, however, that their success largely *depends* on this lack of specialization, coupled with an ability to modify behaviour according to circumstances. The success of *Tylos* is not in question, bearing in mind that those species that have been studied breed but once a year and produce remarkably few young. They appear to be threatened only by human activity.

Differences in behaviour between species and populations also render it difficult to make generalizations about the biology of the genus and some of the generalizations that have been made in this review must be treated with caution. This lack of behavioural uniformity is likely to become even more apparent when further species and populations have been studied. At the moment, all we know about some species of *Tylos* is a statement as to where the animal was originally found, together with the morphological description given when the species was erected. Of the 16 species of *Tylos* recorded from what may be termed Indo-West Pacific regions (Roman, 1970; Lewis and Bishop, 1990), not one has been accorded an ecological, a physiological or a behavioural study and several have not been found again since they were originally described (Grave and Simon, 1992). In other tantalizing cases, species are mentioned in passing as part of an ecological overview of a sandy beach but no behavioural information presented. For example, Dexter (1990) has recorded *T. europaeus* (as *T. europeus*) from beaches in Portugal, the tidal amplitude there being extreme compared with that in the Mediterranean, where the animal has been studied in some detail. One of the features of this species in the Mediterranean is a behaviour which is different from that of species of *Tylos* from exposed shores, and the suggestion has been made above that this may be due to the weak tidal regime. Indeed, in a number of Crustacea differences in rhythmic patterns of behaviour between Mediterranean populations and those on European Atlantic shores have been linked to differences in tidal amplitude (Naylor, 1988). This could easily be confirmed or refuted by studying populations of *T. europaeus* on the Atlantic coast.

Despite the relative paucity of the information available, it is clear that in *Tylos* we encounter a genus unremarkable in its anatomical, physiological or biochemical adaptations to a harsh environment, but linking this lack of specialization to a behavioural flexibility which is truly astonishing. This flexibility extends to modification or suppression of the normal endogenous rhythms and to the reversal of orientational responses under appropriate circumstances. Moreover, the animals display what may be termed a synchronous flexibility; a reversal or modification of response commonly involves most of the population, not only when a reasonably long period is provided for adaptation (as for the *T. capensis* populations in the Alexandria Dunefield) but even when a potentially catastrophic, unpredictable event requires virtually immediate response (as in the case of the oil spill recorded by Kensley (1974)). We have no doubt that further work on *Tylos* will bring to light numerous additional examples of this remarkable behavioural plasticity.

ACKNOWLEDGEMENTS

We thank the South African Foundation for Research Development for providing us with funds for travel and research which made this review possible. Dr J.M.E. Stenton-Dozey, Research Officer to the first author, prepared diagrams for publication. We are grateful to Dr Brian Kensley, Chairperson of the Department of Invertebrate Zoology at the Smithsonian Institution, Washington, for commenting on the currently accepted classification of the Oniscidea and for bringing certain references to our attention. We also thank Dr G. Simmons for allowing us access to his unpublished report on the night-time distribution of active *Tylos punctatus* on a beach in Mexico, and Dr Deborah Dexter for replying promptly to our queries. Professor Anton McLachlan, of the University of Port Elizabeth, kindly criticized a draft of the review. We thank Mrs Rosalind Brown for her help in translating passages from Italian and Dr Naota Ohsaki, of Kyoto University, for translating from the Japanese.

REFERENCES

Ansell, A.D. and Trevallion, A. (1969). Behavioural adaptations of intertidal molluscs from a tropical sandy beach. *Journal of Experimental Marine Biology and Ecology* **4**, 9–35.
Arcangeli, A. (1939). *Tylos latraillei* Aud. et Sav., suoi biotopi, sua area di diffusione. *Bollettino del Museo di Zoologia dell' Universita di Torino* **46**, 151–157.
Arcangeli, A. (1952). Appunti sopra il genre *Tylos* Latr. *Bollettino del Istituto di Zoologia dell' Universita di Torino* **3**, 133–141.
Arcangeli, A. (1953). Precisazioni ecologische sui generi *Ligia* Fabr. e *Tylos* Latr. (Crustacei Isopodi terrestri). *Atti della Accademia delle Scienze di Torino* **87**, 200–212.
Arcangeli, A. (1957). Rapporti del sistema respiratorio degli exopoditi dei pleipodi del genre *Tylos* Latr. con la ecologia della stresso (Crostacei Isopodi terrestri). *Att della Accademia delle Scienze di Torino* **91**, 94–113.
Atkinson, R.J.A. and Naylor, E. (1973). Activity rhythms in some burrowing decapods. *Helgoländer wissenschaftlichen Meeresuntersuchungen* **24**, 192–201.
Barnard, K.H. (1924). Contributions to a knowledge of the fauna of South West Africa. 3. Crustacea, Isopoda Terrestria. *Annals of the South African Museum* **20**, 27–36.
Barnard, K.H. (1925). The digestive canal of isopod crustaceans. *Transactions of the Royal Society of South Africa* **12**, 27–36.
Barnard, K.H. (1940). Contributions to the crustacean fauna of South Africa. No. 11. Further additions to the Tanaidacea, Isopoda and Amphipoda, together with keys for the identification of the hitherto recorded marine and fresh-water species. *Annals of the South African Museum* **32**, 381–543.
Barrass, R. (1963). The burrows of *Ocypode ceratophthalmus* (Pallas) (Crustacea, Ocypodidae) on a tidal wave beach at Inhaca Island, Mozambique. *Journal of Animal Ecology* **32**, 73–85.
Bliss, E.D. (1968). Transition from water to land in decapod crustaceans. *American Zoologist* **8**, 355–392.
Boaden, P.J.S. and Erwin, D.G. (1971). *Turbanella hyalina* versus *Protodrilus symbioticus*: a study in interstitial ecology. *Vie et Milieu* Suppl. **22**, 479–492.
Boone, L. (1934). New and rare Cuban and Haitian terrestrial Isopoda. *Bulletin of the American Museum of Natural History* **66**, 567–598.
Branch, G.M. and Branch, M. (1981). "The Living Shores of Southern Africa." C. Struik, Cape Town.

Brandt, J.F. (1883). Conspectus monographiae Crustaceorum Oniscodorum Latreilli. *Bulletin of the Society of Naturalists, Moscow* **6**, 171–193.

Brown, A.C. (1959). The ecology of South African estuaries. Part IX: notes on the estuary of the Orange River. *Transactions of the Royal Society of South Africa* **35**, 463–473.

Brown, A.C. (1982a). The biology of sandy-beach whelks of the genus *Bullia. Oceanography and Marine Biology, Annual Review* **20**, 309–361.

Brown, A.C. (1982b). Towards an activity budget for the sandy-beach whelk *Bullia digitalis* (Dillwyn). *Malacologia* **22**, 681–683.

Brown, A.C. (1983). The ecophysiology of sandy beach animals — a partial review. *In* "Sandy Beaches as Ecosystems" (A. McLachlan and T. Erasmus, eds), pp. 297–301. W. Junk, The Hague.

Brown, A.C. (1985). The effects of crude oil pollution on marine organisms — a literature review in the South African context: conclusions and recommendations. *South African National Scientific Programmes Report No. 99*, 33 pp.

Brown, A.C. and McLachlan, A. (1990). "Ecology of Sandy Shores." Elsevier, Amsterdam.

Brown, A.C. and Trueman, E.R. (1991). Burrowing of sandy-beach molluscs in relation to penetrability of the substratum. *Journal of Molluscan Studies* **57**, 134–136.

Brown, A.C. and Trueman, E.R. (1994a). The burrowing rate index. *Journal of Molluscan Studies* **60**, 354–355.

Brown, A.C. and Trueman, E.R. (1994b). Burrowing behaviour and cost in the sandy-beach oniscid isopod *Tylos granulatus* Krauss. *Crustaceana*, in press.

Brown, A.C., Stenton-Dozey, J.M.E. and Trueman, E.R. (1989). Sandy beach bivalves and gastropods: a comparison between *Donax serra* and *Bullia digitalis. Advances in Marine Biology* **25**, 179–247.

Brown, F.A. Jr (1961). Physiological rhythms. *In* "The Physiology of Crustacea", Vol. 2. Academic Press, New York.

Brusca, G.J. (1966). Studies on the salinity and humidity tolerances of five species of isopods in a transition from marine to terrestrial life. *Bulletin of the Southern California Academy of Science* **65**, 146–154.

Budde-Lund, G. (1885). "Crustacea Isopoda Terrestria, per Familias et Genera et Species". Nielsen and Lydiche, Helcinae.

Bursell, E. (1955). The transpiration of terrestrial isopods. *Journal of Experimental Biology* **32**, 238–255.

Cals, P. (1974). Mise en evidence par le microscope electronique a balayage, de champs morphogenetiques polarises, exprime par les cellules epidermiques normales dans l'appendices locomoteur des arthropodes *Tylos latreillei* (Audouin) (Crustace isopode) et *Periplaneta americana* (Insecta dictophere). *Comptes Rendus hebdomadaires des Sèances de l'Academie des Sciences de Paris* (Series D) **279B**, 663–666.

Carthy, J.D. (1958). "An Introduction to the Behaviour of Invertebrates." Allen & Unwin, London.

Chelazzi, G. and Ferara, F. (1978). Researches on the coast of Somalia. The shore and dune system at Sar Uanle. 19. Zonation and activity of terrestrial isopods (Oniscoidea). *Monitore Zoologica Italiano* N.S. Suppl. **8**, 189–219.

Cloudsley-Thompson, J.L. (1956). Studies in diurnal rhythms. VII. Humidity responses and nocturnal activity in woodlice (Isopoda). *Journal of Experimental Biology* **33**, 576–582.

Cloudsley-Thompson, J.L. (1958). Water relations and diurnal rhythms in woodlice. *Annals of Applied Biology* **46**, 117–119.

Coenen-Stass, D. (1981). Some aspects of the water balance of two desert woodlice: *Hemilepistus aphganicus* and *Hemilepistus reaumuri* (Crustacea, Isopoda, Oniscoidea). *Comparative Biochemistry and Physiology* **70A**, 405–419.

Coenen-Stass, D. (1985). Effects of desiccation and hydration on the osmolality and ionic concentration in the blood of the desert woodlouse *Hemilepistus reaumuri* (Crustacea, Isopoda, Oniscoidea). *Comparative Biochemistry and Physiology* **81B**, 717–721.

Dalens, H. (1989). Sur un nouveau genre d'oniscoide "aquatique" provenant de sud-est Asiatique: *Thailandoniscus annae* (Isopoda, Oniscidea, Styloniscidae). *Spixiana* **12**, 1–6.

De Villiers, C. and Brown, A.C. (1994). Sand moisture as a factor determining depth of burrowing in the oniscid isopod *Tylos granulatus* Krauss. *South African Journal of Zoology*, in press.

Den Boer, P.J. (1961). The ecological significance of the activity patterns in the woodlouse *Porcellio scaber* Latr. (Isopoda). *Archivs Nèerlandaise de Zoologie* **14**, 283–409.

Dexter, D.M. (1977). Natural history of the Pan-American sand beach isopod *Excirolana braziliensis* (Crustacea: Malacostraca). *Journal of Zoology, London* **183**, 103–109.

Dexter, D.M. (1983). Community structure of intertidal sandy beaches in New South Wales, Australia. *In* "Sandy Beaches as Ecosystems" (A. McLachlan and T. Erasmus, eds), pp. 461–472. W. Junk, The Hague.

Dexter, D.M. (1990). The effect of exposure and seasonality on sandy beach community structure in Portugal. *Ciencia Biologica Ecology and Systematics* **10**, 31–50.

Dower, M. (1965). "Fourth Wave: the Challenge to Leisure." Civic Trust, London.

Dresel, E.I.B. and Moyle, V. (1950). Nitrogenous excretion of amphipods and isopods. *Journal of Experimental Biology* **27**, 210–225.

Dubinsky, Z., Steinberger, Y. and Shachak, M. (1979). The survival of the desert isopod *Hemilepistus reaumuri* (Audouin) in relation to temperature (Isopoda, Oniscoidea). *Crustaceana* **36**, 147–154.

Ebbe, B. (1981). Beitrag zur Morphologie Ultrastruktuur und Funktion der Respirationsapparates von *Tylos granulatus* Krauss (Isopoda, Oniscoidea). *Zoologische Jahrbücher (Anatomie)* **105**, 551–578.

Edney, E.B. (1954). Woodlice and the land habitat. *Biological Reviews* **29**, 185–219.

Edney, E.B. (1957). "The Water Relations of Terrestrial Arthropods." Cambridge University Press, Cambridge.

Edney, E.B. (1968). Transition from water to land in isopod crustaceans. *American Zoologist* **8**, 309–326.

Enright, J.T. (1972). A virtuoso isopod: circalunar rhythms and their tidal fine structure. *Journal of Comparative Physiology* **77**, 41–162.

Enright, J.T. (1975). Orientation in time: endogenous clocks. *In* "Marine Ecology, Vol. 2(2): Physiological Mechanisms" (O. Kinne, ed.), pp. 917–944. Wiley-Interscience, New York.

Fincham, A.A. (1973). Rhythmic swimming behaviour in the New Zealand sand beach isopod *Psuedaga puntata* Thomson. *Journal of Experimental Marine Biology and Ecology* **11**, 229–237.

Fischbach, E. (1954). Licht, Schwere und Geruchssinn bei Isopoden. *Zoologisches Jahrbuch Abt Zool Physiol* **65**, 141–170.

Fraenkel, G.S. and Gunn, D.L. (1940). "The Orientation of Animals." Clarendon Press, Oxford.

Friend, J.A. and Richardson, A.M.M. (1986). Biology of terrestrial amphipods. *Annual Review of Entomology* **31**, 25–48.

Garthwaite, R.L., Hochberg, F.G. and Sassaman, C. (1985). The relationship of sand moisture to burrowing depth of the sand-beach isopod *Tylos punctatus* Holmes and Gay. *Bulletin of the Southern California Academy of Sciences* **84**, 23–37.

Geppetti, L. and Tongiorgi, P. (1967). Nocturnal migrations of *Talitrus saltator* (Montagu) (Crustacea: Amphipoda). *Monitore Zoologica Italiano* (N.S.) **1**, 37–40.

Giordani-Soika, A. (1954). Ecologia, sistematica, biogeografia ed evoluzione del *Tylos latreillei* Auct. (Isop. Tylidae). *Bollettino del Museo Civico di Storia Naturale di Venezia* **7**, 63–83.

Godfrey, P.J., Leatherman, S.P. and Buckley, P.A. (1978). Impact of off-road vehicles on coastal ecosystems. In "Coastal Zone '78", pp. 581–600. American Society of Coastal Engineers, New York.

Gordon, I. (1952). Support for Professor A. Vandel's proposal for the use of the plenary powers to preserve the generic name *Tylos* (Latreille M.S.) Audouin (1826) (Class Crustacea, Order Isopoda). *Bulletin of Zoological Nomenclature* **6**, 173.

Grave, S. de and Simon, A. (1992). Rediscovery and range extension of *Tylos opercularis* Budde-Lund, 1885 (Isopoda, Oniscoidea). *Crustaceana* **62**, 222–223.

Griffith, H. and Telford, M. (1985). Morphological adaptations to burrowing in *Chiridotea coeca* (Crustacea: Isopoda). *Biological Bulletin* **168**, 296–311.

Griffiths, C.L. and Griffiths, R.J. (1983). Biology and distribution of the rove beetle *Psammathobledius puntatissima* (Le Conte) (Coleoptera: Staphylinidae). *Hydrobiologia* **101**, 203–214.

Gunn, D.L. (1937). The humidity reactions of the woodlouse, *Porcellio scaber* (Latreille). *Journal of Experimental Biology* **14**, 178–186.

Hadley, N.F. and Warburg, M.R. (1986). Water loss in three species of xeric adapted isopods: correlations with cuticular lipids. *Comparative Biochemistry and Physiology* **85A**, 669–672.

Hammoud, W., Comte, J. and Ducruet, J. (1975). Recherche d'une substance sexuellement attractive chez les gammares du group *Pulex*. *Crustaceana* **28**, 152–157.

Hamner, W.M., Smythe, M. and Mulford, E.D. (1968). Orientation of the sand-beach isopod *Tylos punctatus*. *Animal Behaviour* **16**, 405–409.

Hamner, W.M., Smythe, M. and Mulford, E.D. (1969). The behaviour and life history of a sand-beach isopod, *Tylos punctatus*. *Ecology* **50**, 442–453.

Hardy, G.S. (1970). Circadian rhythms. *Tuatara* **18**, 124–131.

Hastings, M.H. (1981). Semi-lunar variations of endogenous circatidal rhythms of activity and respiration in the isopod *Eurydice pulchra*. *Marine Ecology Progress Series* **4**, 85–90.

Hatchett, B.P. (1974). Biology of Isopoda of Michigan, *Ecological Monographs* **17**, 47–49.

Hayes, W.B. (1969). Ecological studies on the high-beach isopod *Tylos punctatus* Holmes and Gay. Ph.D. thesis, University of California, San Diego.

Hayes, W.B. (1970a). The accuracy of pitfall trapping for the sand-beach isopod *Tylos punctatus*. *Ecology* **51**, 514–516.

Hayes, W.B. (1970b). Copper concentrations in the high-beach isopod *Tylos punctatus*. *Ecology* **51**, 721–723.

Hayes, W.B. (1974). Sand-beach energetics: importance of the isopod *Tylos punctatus*. *Ecology* **55**, 838–847.

Hayes, W.B. (1977). Factors affecting the distribution of *Tylos punctatus* (Isopoda, Oniscoidea) on beaches in southern California and northern Mexico. *Pacific Science* **31**, 165–186.

Hazlett, B.A. (1988). Behavioural plasticity as an adaptation to a variable environment. In "Behavioural Adaptation to Intertidal Life" (G. Chelazzi and M. Vanini, eds), pp. 317–332. Plenum Press, New York.

Herrnkind, W.F. (1972). Orientation in shore-living arthropods, especially the sand fiddler crab. In "Behavior of Marine Animals" (H.E. Winns and B.L. Olla, eds), pp. 1–59. Plenum Press, New York.

Hewitt, C.G. (1907). *Ligia. L.M.B.C. Memoirs of Typical British Marine Plants and Animals* **16**, 37 pp.

Holanov, S.H. and Hendrickson, J.R. (1980). The relationship of sand moisture to burrowing depth of the sand-beach isopod *Tylos punctatus* Holmes and Gay. *Journal of Experimental Marine Biology and Ecology* **46**, 81–88.

Holdich, D.M., Lincoln, R.J. and Ellis, J.P. (1984). The biology of terrestrial isopods: terminology and classification. *Symposia of the Zoological Society of London* **53**, 1–6.

Holmes, S.J. and Gay, M.E. (1909). Four new species of isopods from the coast of California. *Proceedings of the U.S. National Museum* **36**, 375–379.

Holthius, L.B. (1951). On the objection, from the carcinological point of view, of accepting the name *Tylos* Meigen, 1800 (Class Insecta, Order Diptera) and the consequent rejection of the name *Tylos* (Latreille M.S.) Audoin, 1826 (Class Crustacea, Order Isopoda). *Bulletin of Zoological Nomenclature* **6**, 128.

Hubble, S.P., Sikora, A. and Paris, O.H. (1965). Radiotracer, gravimetric, and calorimetric studies on ingestion and assimilation rates of an isopod. *Health Physics* **11**, 1485–1501.

Hughes, D.A. (1966). Behavioural and ecological investigations of the crab *Ocypode ceratophthalmus* (Crustacea, Ocypodidae). *Journal of Zoology, London* **150**, 129–143.

Hurley, D.E. (1968). Transition from water to land in amphipod crustaceans. *American Zoologist* **8**, 327–353.

Iga, T. (1972). Circular movement of *Tylos granulatus*. *Zoological Magazine* **81**, 408 (in Japanese).

Iga, T. and Kitamura, S. (1975). The circus movement of a sand-beach isopod, *Tylos granulatus* Miers. *Memoirs of the Faculty of Literature and Science of Simane University of Natural Science* **9**, 89–102 (in Japanese).

Imafuku, M. (1976). On the nocturnal behaviour of *Tylos granulatus* Miers (Crustacea: Isopoda). *Publications of the Seto Marine Laboratory* **23**, 299–340.

Jones, D.A. (1972). Aspects of the ecology and behaviour of *Ocypode ceratophthalmus* (Pallas) and *O. kuhlii* de Haan (Crustacea: Ocypodidae). *Journal of Experimental Marine Biology and Ecology* **8**, 31–43.

Jones, D.A. and Naylor, E. (1970). The swimming rhythm of the sand beach isopod *Eurydice pulchra*. *Journal of Experimental Marine Biology and Ecology* **4**, 188–199.

Kensley, B. (1972). Behavioural adaptations of the isopod *Tylos granulatus* Krauss. *Zoologica Africana* **7**, 1–4.

Kensley, B. (1974). Aspects of the biology and ecology of the genus *Tylos* Latreille. *Annals of the South African Museum* **65**, 401–471.

Kensley, B. (1978). "Guide to the Marine Isopods of Southern Africa." South African Museum, Cape Town.

King, C.A.M. (1959). "Beaches and Coasts." Edward Arnold, London.

Kuemmel, G. (1981). Fine structural indications of an osmoregulatory function of the "gills" in terrestrial isopods (Crustacea, Oniscoidea). *Cellular and Tissue Research* **214**, 663–666.

Kuenen, D.J. (1959). Excretion and water balance in some land isopods. *Entomologia Experimentalis et Applicata* **2**, 287–294.

Kuenen, D.J. and Nooteboom, H. (1963). Olfactory orientation in some land isopods (Oniscoidea, Crustacea). *Entomologia Experimentalis et Applicata* **6**, 133–142.

Lazo-Wasem, E.A. (1984). Physiological and behavioural ecology of the terrestrial amphipod, *Arcitalitrus sylvaticus* (Haswell, 1880). *Journal of Crustacean Biology* **4**, 343–355.

Lewis, F. and Bishop, L. (1990). *Tylos australis*: a new species of Tylidae (Isopoda: Oniscidea), a family previously not recorded in Australia. *Invertebrate Taxonomy* **3**, 747–757.

Lindqvist, O.V. (1972). Components of water loss in terrestrial isopods. *Physiological Zoology* **45**, 316–324.

Manton, S.M. (1952). The evolution of arthropodan locomotory mechanisms. Pt. 2. General introduction to the locomotory mechanisms of the Arthropoda. *Journal of the Linnean Society (Zoology)* **42**, 93–117.

Marsh, B.A. and Branch, G.M. (1979). Circadian and circatidal rhythms of oxygen consumption in the sandy-beach isopod *Tylos granulatus* Krauss. *Journal of Experimental Marine Biology and Ecology* **37**, 77–89.

Matsakis, J. (1956). Observations ethologiques sur les *Tylos* (Isopodes Oniscoides) du Rousillon. *Vie et Milieu* **7**, 107–109.

Matthewson, A.C. (1991). Ecophysiology of two crustaceans in a coastal dune field. Unpublished thesis, University of Port Elizabeth, South Africa.

Mayes, K.R. and Holdich, D.M. (1975). Water exchange between woodlice and moist environments, with particular reference to *Oniscus asellus*. *Comparative Biochemistry and Physiology* **51A**, 295–300.

McLachlan, A. (1986). "Sandy Beach Research at UPE 1975–1986." *Institute for Coastal Research, University of Port Elizabeth, Report No. 14.*

McLachlan, A. and Sieben, P.R. (1991). Growth and production of *Tylos capensis*. *Crustaceana* **61**, 43–48.

McLachlan, A., Ascaray, C. and du Toit, P. (1987). Sand movement, vegetation succession and biomass spectrum in a coastal dune slack in Algoa Bay, South Africa. *Journal of Arid Environments* **12**, 9–25.

Mead, F. (1965). Sur l'existence d'un sac incubateur interne chez l'isopode terrestre *Tylos latreillei* Audouin. *Comptes Rendus hebdomadaires des Seances de l'Academie des Sciences de Paris* **160**, 2336–2337.

Mead, F. (1967). Observations sur l'accouplement et la chevauchee nuptiale chez l'isopode *Tylos latreillei* Audouin. *Comptes Rendus hebdomadaires des Seances de l'Academie des Sciences de Paris* **264D**, 2154–2157.

Mead, F. (1968). Observations sur l'ecologie de *Tylos latreillei* Audouin (Isopode Tylidae) et sur comportement en milieu naturel. *Vie et Milieu* **19C**, 345–362.

Mead, F. (1976). La place de l'accouplement dans le cycle de reproduction des isopodes terrestre (Oniscoidea). *Crustaceana* **31**, 27–41.

Mead, M. and Mead, F. (1974). Etude de l'orientation chez l'isopode terrestre *Tylos latreillei s.sp sardous*. *Vie et Milieu* **23C**, 81–93.

Menzies, R.J. (1952). The occurrence of a terrestrial isopod in plankton. *Ecology* **33**, 303.

Moore, P.G. and Francis, C.H. (1985). On the water relations and osmoregulation of the beach hopper *Orchestia gammarellus* (Pallas) (Crustacea, Amphipoda). *Journal of Experimental Marine Biology and Ecology* **94**, 131–150.

Moore, P.G. and Francis, C.H. (1986). Environmental tolerances of the beach hopper *Orchestia gammarellus* (Pallas) (Crustacea, Amphipoda). *Marine Environmental Research* **19**, 115–129.

Morritt, D. (1988). Osmoregulation in littoral and terrestrial talitroidean amphipods (Crustacea) from Britain. *Journal of Experimental Marine Biology and Ecology* **123**, 77–94.

Muir, D.G. (1977). The biology of *Talorchestia capensis* (Amphipoda: Talitridae), including a population energy budget. Unpublished thesis, University of Cape Town, Cape Town.

Naylor, E. (1976). Rhythmic behaviour and reproduction in marine animals. *In* "Adaptation to Environment: Essays on the Physiology of Marine Animals" (R.C. Newell, ed.), pp. 393–429. Butterworths, London.

Naylor, E. (1988). Clock-controlled behaviour in intertidal animals. *In* "Behavioural Adaptation to Intertidal Life" (G. Chelazzi and M. Vanini, eds), pp. 1–14. Plenum Press, New York.

Newell, R.C. (1979). "Biology of Intertidal Animals", 3rd ed. Ecological Surveys, Faversham, Kent.

Nicholls, A.G. (1930). Studies on *Ligia oceanica*. II. The process of feeding, digestion and absorption, with a description of the foregut. *Journal of the Marine Biological Association of the United Kingdom* **17**, 675–705.

Odendaal, F.J., Iwasa, Y. and Ehrlich, P.R. (1985). Duration of female availability and its effect on butterfly mating systems. *American Naturalist* **125**, 673–678.

Odendaal, F.J., Turchin, P. and Stermitz, F.E. (1988). An incidental-effect hypothesis explaining aggregation of males in a population of *Euphydryas anicia* (Nymphalidae). *American Naturalist* **132**, 735–749.

Odendaal, F.J., Eekhout, S., Brown, A.C. and Branch, G.M. (1994). An incidental-effect hypothesis explaining aggregations in a semi-terrestrial isopod inhabiting on sandy beaches. *Journal of Zoology, London*, in press.

Ondo, Y. (1952). A behavioural and ecological study of *Tylos granulatus* Miers. *Zoological Magazine* **61**, 55. (In Japanese.)

Ondo, Y. (1953). Daily rhythmic activity of *Tylos granulatus* Miers II. On some environmental elements to induce the nocturnal activity of the animals. *Journal of the Faculty of Education of Tottori University for Natural Sciences* **4**, 20–23. (In Japanese, with English summary.)

Ondo, Y. (1954). Daily rhythmic activity of *Tylos granulatus* Miers. III. Modification of rhythmic activity in accord with its growing stage. *Japanese Journal of Ecology* **4**, 1–3. (In Japanese, with English summary.)

Papi, F. and Pardi, L. (1953). On the lunar orientation of sandhoppers (Amphipoda, Talitridae). *Biological Bulletin* **124**, 97–105.

Pardi, L. (1954). Über die Orientierung von *Tylos latreillei* Aud. and Sav. (Isopoda terrestria). *Zeitschrift fur Tierpsychologie* **11**, 175–181.

Pardi, L. (1955). L'orientamento diurno di *Tylos latreilli* (Aud. & Sav.). *Bollettino del Museo di Universite di Torino* **4**, 167–196.

Pardi, L. and Ercolini, A. (1986). Zonal recovery mechanisms in talitrid crustaceans. *Bollettino di Zoologia* **53**, 139–160.

Pardi, L. and Papi, F. (1953). Ricerche su l'orientamento di *Talitrus saltator* (Montagu) (Crustacea-Amphipoda). I. L'orientamento durante il giorno in una populazione del litorale tirrenico. *Zeitschrift für Vergleichende Physiologie* **35**, 459–489.

Pardi, L., Ugolini, A., Faqi, A.S., Scapini, F. and Ercolini, A. (1988). Zonal recovering in equatorial sandhoppers: interaction between magnetic and solar orientation. *In* "Behavioural Adaptations to Intertidal Life" (G. Chelazzi and M. Vanini, eds), pp. 79–92. Plenum Press, New York.

Paris, O.H. and Pitelka, F.A. (1962). Population characteristics of the terrestrial isopod *Armidillidium vulgare* in California grassland. *Ecology* **39**, 229–248.

Parry, G. (1953). Osmotic and ionic regulation in the isopod crustacean *Ligia oceanica* L. *Journal of Experimental Biology* **30**, 567–574.

Peres, J.M. and Picard, J. (1955). Biotopes et biocoenoses de la Mediterranée occidentale compares a ceux de la Manche et de l'Atlantique nord-oriental. *Archives de Zoologie expérimentale et générale* **92**, 1–72.

Perry, D.M. (1980). Factors influencing aggregation patterns in the sand crab *Emerita analoga* (Crustacea: Hippidae). *Oecologia* **45**, 379–384.

Price, J.B. and Holdich, D.M. (1980). Changes in osmotic pressure and sodium concentration of the haemolymph of woodlice with progressive desiccation. *Comparative Biochemistry and Physiology* **66A**, 155–161.

Reichle, D.E. (1967). Radioactive turnover and energy flow in terrestrial isopod populations. *Ecology* **48**, 351–366.

Roman, M.-L. (1970). Les Oniscoides halophiles de Madagascar (Isopods, Oniscoidea). *Beaufortia* **26**, 107–152.

Root, R.B. and Kareiva, P.M. (1984). The search for resources by cabbage butterflies (*Pieris rapae*): ecological consequences and adaptive significance of Markovian movements in a patchy environment. *Ecology* **65**, 147–165.

Schmalfuss, H. (1989). Phylogenetics in Oniscidea. *Monitore Zoologico Italiana, Monographie* **4**, 3–27.

Schultz, G.A. (1965). The distribution and general biology of *Hyloniscus riparius* (Koch) (Isopoda, Oniscoidea) in North America. *Crustaceana* **8**, 131–140.

Schultz, G.A. (1970). A review of the species of the genus *Tylos* Latreille from the New World (Isopoda Oniscoidea). *Crustaceana* **19**, 297–305.

Schultz, G.A. (1974). Terrestrial isopod crustaceans (Oniscoidea) mainly from the West Indies and adjacent regions. 1. *Tylos* and *Ligia*. *Uitgaven natuurwetenschappelijke Studiekring voor Suriname* **45**, 162–173.

Schultz, G.A. (1983). Two species of *Tylos* Audouin from Chile, with notes on species of *Tylos* with three flagellar articles (Isopoda: Oniscoidea: Tylidae). *Proceedings of the Biological Society of Washington* **96**, 675–683.

Schultz, G.A. and Johnson, C. (1984). Terrestrial isopod crustaceans from Florida (USA) (Oniscoidea: Tyliodae, Ligidae, Halophilosciidae, Philosciidae and Rhyscotidae. *Journal of Crustacean Biology* **4**, 154–171.

Sieben, P. (1982). Biology of *Tylos capensis* (Krauss). Unpublished student project, Zoology Department, University of Port Elizabeth, South Africa.

Spencer, J.D. and Edney, E.B. (1954). The absorption of water by woodlice. *Journal of Experimental Biology* **31**, 491–496.

Stanley, S.M. (1970). Relation of shell form to life habits of the Bivalvia (Mollusca). *Geological Society of America Memoirs* **125**, 1–296.

Stenton-Dozey, J.M.E. and Brown, A.C. (1988). Feeding, assimilation and scope for growth in the scavenging sandy beach neogastropod *Bullia digitalis*. *Journal of Experimental Marine Biology and Ecology* **119**, 253–268.

Stoch, C. (1975). Daily rhythms of O_2-consumption in the isopod *Tylos granulatus* (Latreille), and the effects that light has on the rhythms. Unpublished student project in the Niven Library, University of Cape Town, Cape Town.

Thornhill, R. and Alcock, J. (1983). The Evolution of Insect Mating Systems." Harvard University Press, Boston, USA.

Tongiorgi, P. (1962). Sulle relazioni tra habitat ed orientamento astronomico in alcune specie del gen. *Arctosa* (Araneae-Lycosidae). *Bollettino Zoologia* **28**, 683–689.

Tongiorgi, P. (1969). Ricerche ecologiche sugli artropodi di una spiaggia sabbiosa del litorale tirrenico: III. Migrazioni e ritmo di attivita locomotoria nell'isopode *Tylos latreilli* (Aud. and Sav.) e nei tenebrionidi *Phaleria provincialis* Fauv. e *Halammobia pellucida* Herbst. *Redia* **51**, 1–19.

Trueman, E.R. and Brown, A.C. (1989). The effect of shell shape on the burrowing performance of species of *Bullia* (Gastropoda: Nassariidae). *Journal of Molluscan Studies* **55**, 129–131.

Trueman, E.R. and Brown, A.C. (1992). The burrowing habit of marine gastropods. *Advances in Marine Biology* **28**, 389–431.

Turchin, P. (1989). Population consequences of aggregative movement. *Journal of Animal Ecology* **58**, 75–100.

Turchin, P., Odendaal, F.J. and Rausher, M.D. (1991). Quantifying insect movement in the field. *Environmental Entomology* **20**, 955–963.

Ugolini, A. and Pardi, L. (1992). Equatorial sandhoppers do not have a good clock. *Naturwissenschaften* **78**, 279–281.

Vandel, A. (1943). Essai sur l'origine, l'evolution et la classification des Oniscoidea (isopodes terrestres). *Bulletin Biologique de la France et de la Belguique Suppl.* **30**, 1–136.

Vandel, A. (1960). "Isopodes Terrestres (premiere partie)", *Faune de France* **64**. Lechevallier, Paris.

Van der Merwe, D. (1988). "The Effects of Off-road Vehicles (ORVs) on Coastal Ecosystems — a Review." Institute for Coastal Research, University of Port Elizabeth, South Africa.

Van der Merwe, D. and Van der Merwe, D. (1991). Effects of off-road vehicles on the macrofauna of a sandy beach. *South African Journal of Science* **87**, 210–213.

Verhoef, K.W. (1949). *Tylos*, eine terrestrich-maritime Ruckwanderer-Gattung der Isopoden. *Archiv für Hydrobiologie, Stuttgart* **423**, 329–340.

Vermeij, J. and Zipser, E. (1986). Burrowing performance of some tropical Pacific gastropods. *Veliger* **29**, 200–206.

Waloff, N. (1941). The mechanisms of humidity reactions of terrestrial isopods. *Journal of Experimental Biology*, **18**, 115–135.

Warburg, M.R. (1964). The responses of isopods towards temperature, humidity and light. *Animal Behaviour* **12**, 175–186.

Warburg, M.R. (1965). The evaporative water loss of three isopods from semi-arid habitats in South Australia. *Crustaceana* **9**, 302–308.

Warburg, M.R. (1968). Simultaneous measurement of body temperature and water loss in isopods. *Crustaceana* **14**, 39–44.

Warburg, M.R. (1987). Haemolymph osmolality, ion concentration and the distribution of water in body compartments of terrestrial isopods under different ambient conditions. *Comparative Biochemistry and Physiology* **86A**, 433–437.

Wells, K.D. (1977). The social behaviour of anuran amphibians. *Animal Behaviour* **25**, 666–693.

Wieser, W. (1966). Copper and the role of isopods in the degradation of organic matter. *Science* **153**, 67–69.

Wieser, W., Schweizer, G. and Hartenstein, R. (1969). Patterns in the release of gaseous ammonia by terrestrial isopods. *Oecologia* **3**, 390–400.

Williams, J.A. (1980). The light response rhythm and seasonal entrainment of the endogenous circadian locomotor rhythm of *Talitrus saltator* (Crustacea: Amphipoda). *Journal of the Marine Biological Association of the United Kingdom* **60**, 773–785.

Williams, J.A. (1983a). Environmental regulation of the burrow depth distribution of the sand-beach amphipod *Talitrus saltator*. *Estuarine, Coastal and Shelf Science* **16**, 291–298.

Williams, J.A. (1983b). The endogenous locomotor activity rhythm of four supralittoral peracarid crustaceans. *Journal of the Marine Biological Association of the United Kingdom* **63**, 481–492.

Williamson, D.I. (1951). Studies on the biology of Talitridae (Crustacea, Amphipoda): effects of atmospheric humidity in *Talitrus saltator*. *Journal of the Marine Biological Association of the United Kingdom* **30**, 73–90.

Yuasa, Y. (1973). Beach cleaner, *Tylos granulatus*. *Yama no Ue no Sakanatachi, Himeji City Aquarium*, No. 12. (In Japanese.)

Social Aggregation in Pelagic Invertebrates

D.A. Ritz

Zoology Department, University of Tasmania, Box 252C, GPO, Hobart, Tasmania 7001, Australia

ADVANCES IN MARINE BIOLOGY VOL 30
ISBN 0–12–026130–8

1. INTRODUCTION

1.1. Definitions

Social aggregations are a familiar sight among aquatic vertebrates: one has only to think of schools of dolphins, pods of whales, flocks of seagulls and schools of fish. But invertebrates too have evolved social aggregative behaviour though examples may not spring so readily to mind. Best known perhaps is the Antarctic krill (*Euphausia superba*) whose aggregations, in terms of numbers or biomass, rival or exceed those of any other animal on earth. Superswarms of *E. superba* have been estimated to contain up to 1×10^9 individuals (Nicol and de la Mare, 1993) and 2×10^6 tonnes (Macaulay *et al.*, 1984). Some examples of invertebrate aggregations are shown in Figure 1.

The paragraph above contains the terms social, swarm, school and aggregation in reference to groups of both vertebrate and invertebrate animals. The literature is replete with definitions of such terms and it is important to make clear from the start the sense in which I shall use them. Many authors have proposed schemes for categorizing invertebrate aggregations (e.g. Clutter, 1969; Mauchline, 1971, 1980a; Wittmann, 1977; Kalinowski and Witek, 1985; O'Brien, 1988a; Sauer *et al.*, 1992), and for the most part these have borrowed terms from the literature on fish aggregations though the meanings may not always be identical. Efforts to classify aggregations of Antarctic krill (*E. superba*) have spawned the greatest variety of terms (see Miller and Hampton, 1989). Some of these terms, especially those emphasizing dimensions and density of individuals, seem to apply only to *E. superba* and may not have relevance to aggregations of other species. Omori and Hamner (1982) caution against introducing new terms before the underlying behaviour is well understood, arguing that there is a danger of us searching for behaviour to fit the term rather than the other way round.

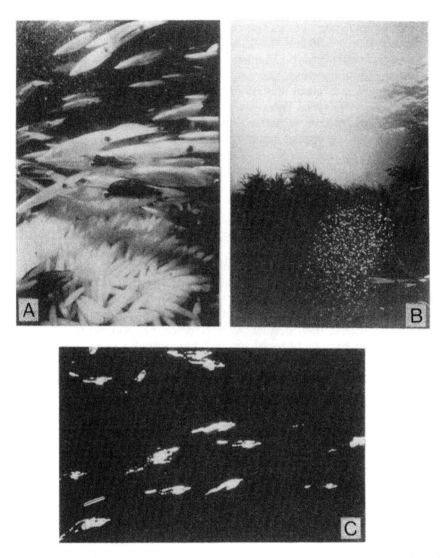

Figure 1 Examples of invertebrate aggregations. (A) School of spawning squid, *Loligo opalescens*. (B) Swarm of mysids in shallow coastal water, south east Tasmania, Australia. (C) School of mysids, *Paramesopodopsis rufa*. (A, courtesy of Planet Earth Pictures, photograph by N. Wu; B and C, courtesy of Jon Bryan.)

I follow most of the authors above in adopting the definitions below:

Aggregation: taken to mean a grouping of conspecific individuals without any connotation of mutual attraction. For example, individuals attracted to a common food source may form a feeding aggregation (= coincidental groups (Hamner, 1988)). I use this as a general term for any grouping where no particular social connotation is intended.

Social: taken to mean mutual attraction of conspecific animals leading to co-operative group behaviour (Wilson, 1975).

Swarm: taken here to mean a discrete integrated social group with members evenly spaced but not polarized.

School: is the same as above but members polarized and usually facing the same direction. (Note that because social squid can swim both backwards and forwards, a school need not imply that all individuals are facing the same direction.)

Shoal: is used to describe a (usually) larger grouping within which are contained subgroups conforming to the definitions of swarm and school.

I deliberately omit a range of additional terms introduced at intervals by various authors on the grounds that either the aggregations described are adequately covered by existing terms or that the behaviour appears to be restricted to one particular group of animals. Notwithstanding this, it is becoming evident that, despite many common behavioural traits in the aggregative behaviour of social species, there are also many instances of traits which appear to be unique to a particular group and for which new terms will have to be coined. Examples are clustering of scyphozoans (Zavodnik, 1987) (can they be regarded as social?); swirls of hyperiid amphipods (Lobel and Randall, 1986); groups of rock lobsters and hermit crabs in which physical contact between members is maintained (McKoy and Leachman, 1982; Gherardi and Vannini, 1989). It should be noted here that not all authors use the basic terms defined above to mean the same thing.

I follow O'Brien (1988a, 1989) in adopting Breder's (1967) terms for fish schools: *facultative* describes those species that form groups on an irregular basis using either intrinsic or extrinsic influences, or perhaps both, as their cues. *Obligate* describes those species that spend all of their lives in aggregations.

O'Brien (1988a) proposed a classification of mysid aggregations based on mechanisms responsible for their formation and maintenance (Figure 2). Application of such a scheme necessarily requires a detailed know-ledge of the behavioural ecology of the species concerned. O'Brien stresses the importance of understanding mechanisms of formation of aggregations before the introduction of new descriptive terms. While

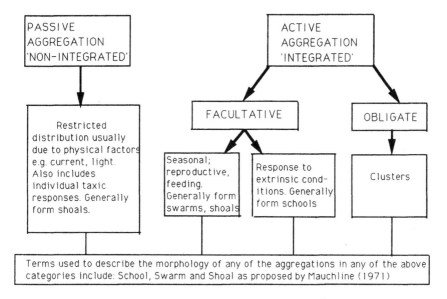

Figure 2 Classification of mysid aggregations in relation to mechanisms influencing their formation and maintenance. Terms describing the morphology of aggregations are as defined by Mauchline (1971). (After O'Brien, 1988a).

mostly using the terminology from Mauchline (1971), O'Brien proposed a new term, "cluster", to describe integrated aggregations formed in response to internal biological mechanisms. I am unconvinced of the need for this term and prefer to use the term "shoal" for such an aggregation, more in the sense that Clutter (1969) used it, or as adopted by Mauchline (1971). Unfortunately this conflicts to some extent with the definition of shoal in the fish literature. For example, Pitcher (1986a) uses the term to mean any social grouping of fish without implications of polarization or co-ordination.

1.2. Evidence for Natural Gregarious Behaviour

Evidence that many aquatic invertebrate aggregations form and are maintained because of a social attraction between members rather than through the agency of hydrographical forces is now overwhelming. Even so some authors still seem unwilling to accept this possibility. Factors which seem to support the idea of a natural gregarious response are abundantly evident within the crustacean groups. For example, reports by Hardy (1936) and Hampton (1981, *E. superba*); Kawamura (1974, copepods); Modlin (1990, mysids) and others present clear evidence of the resistance to disruption by severe water turbulence and/or predators of such groups. Hahn and Itzkowitz (1986) demonstrate the dogged fidelity of mysid swarms to their selected sites despite repeated disturbance by divers. Aggregations can be found at all times of year, and in some cases remain cohesive through whole diel cycles (Section 3). No single function yet advanced adequately explains all aggregations. In some species all ontogenetic stages form aggregations and it is rare to find solitary individuals of these animals. It is now widely agreed that intrinsic factors are the primary stimulus for the formation of such groups although, once formed, they may be modified according to a range of both biological and physicochemical stimuli. Hamner (1988) observed that patchiness can be generated by almost any animal behaviour but more sophisticated behaviour is required for social aggregation.

At this point it is worth making a distinction between proximate and ultimate causes of aggregations (Nicol, 1984; Miller and Hampton, 1989). *Proximate* causes are intrinsic (biological) or extrinsic (environmental) changes which induce or trigger aggregation. *Ultimate* causes are those that constitute the selective advantages gained by being in an aggregation.

1.3. Aggregations and Theoretical Behavioural Ecology

Much of the recent work on aggregative behaviour of "zooplanktonic" species is characterized by data on Antarctic krill, *Euphausia superba*. These have been reviewed recently and comprehensively by Miller and Hampton (1989) and suggest that the work has proceeded almost independently of parallel theoretical developments in behavioural ecology, notably optimal foraging theory (though see Antezana and Ray, 1983, 1984). The reasons for this, I believe, are largely because of the negative connotations of the term "plankton" which for too long have implied passivity, limited swimming power and limited possibilities for sophisticated behaviour because of small size (Hamner, 1988). This seems to have resulted in an unwillingness to accept the active rather than passive nature of such aggregations and perhaps explains why such words as school are less commonly applied than the alternative (and possibly functionally different) swarm. Nevertheless, many authors have emphasized their impressive cohesion, manoeuvring ability and broad behavioural repertoires (e.g. Zelikman, 1974; Hamner, 1985; Ritz, 1991). For example, all levels of escape responses (O'Brien and Ritz, 1988) echo those described for fish aggregations. It is becoming increasingly clear that, as Hamner (1985 and pers. commun.) has argued, krill are best viewed as "small sardines" since their group behaviour is so strongly reminiscent of that of fish.

Miller and Hampton (1989) draw three main conclusions about krill aggregations.

1. Krill feed both when aggregated in swarms and when dispersed, though feeding may be less efficient in more tightly organized faster moving or migrating swarms (a migrating swarm effectively must be a school (Pitcher, 1986a)).
2. No diel cycles of feeding are apparent and feeding appears to be unrelated to cycles of aggregation and dispersal.
3. Swarms probably arise and are maintained by a combination of abiotic and biotic factors including social interactions.

At the same time these authors acknowledge that "aggregation has not been conclusively linked to any one particular activity or environmental influence". This is not surprising in the light of theoretical development of conceptual models such as optimal foraging theory (see e.g. Antezana and Ray, 1983, 1984), and group foraging models which take account of predation pressure (see e.g. Daly and Macaulay, 1991). Early conceptual models relating feeding and swarming in euphausiids rested on the assumption that these are not co-occurring events because food would rapidly become limiting particularly in dense swarms. This view is not supported by more recent studies. Interpretation of data on swarms, including size, density, feeding success, etc., requires detailed knowledge of ethology, physiological state, threat from predators, etc. (O'Brien, 1987a).

Correlation between appearance of swarms and environmental variables alone will not, I believe, produce any new insights. On the other hand, analysis of the costs and benefits to individuals of remaining in the swarm, and behaviour when objectives are conflicting, is much more revealing. For example, Ritz (1994) showed that swarm size determined feeding success in a social mysid. When under threat, mysids in large swarms collected significantly more food than those in small swarms when food was patchy. In a conflict situation of limiting food and threat from a predator, survival demanded that swarms become more tightly cohesive even though food capture suffered as a result. I believe that swarms and schools must be viewed as groups of social but potentially selfish individuals which continuously monitor the costs and benefits of remaining within the group. The parameter under constant surveillance is presumably success at finding and capturing food. When food is patchily distributed, a group is more successful than individual animals at finding new patches (Pitcher et al., 1982) and large groups are better than small ones. But if food is limiting and the swarm is so large that the returns per individual fall below some threshold level, it would pay some individuals to leave and for the swarm to split. If, however, the swarm is threatened by a predator, the optimal strategy might be to ensure the swarm remains as large as possible even if this means that per capita food returns decline. Viewed in these terms it is not surprising that attempts to correlate appearance of swarms and superswarms with environmental variables have been distinctly unsuccessful.

1.4. General Features of Aggregations

Plankton patchiness has been noted by numerous authors and its origins frequently speculated upon. Hamner (1988) proposes that in open water, the most important cause of patchiness is the thermocline. But he stresses the importance of behaviour; for example "pelagic patchiness is often synonymous with behavioural aggregation". It is becoming clear that, despite many common stereotyped traits in the aggregative behaviour of social species, there are also many instances of traits that are restricted to a particular group. In common with fish schools, invertebrate aggregations adopt distinct shapes (see Section 2), have clearly defined boundaries, have no permanent leaders and members do not occupy fixed positions. Nearest neighbour distances (NND) appear to be species-specific and certain positions, for example directly above or below a neighbour, seem to be avoided (O'Brien and Ritz, 1988).

However, the growing number of reports of apparently unique behaviour displayed by aggregations (e.g. synchronous moulting by *E. superba* apparently to serve as a decoy to visual predators, Hamner *et al.*, 1983) highlights the flexibility of behavioural repertoires. So too does the realization that solitary, to facultative, to obligate aggregation is a behavioural continuum and that a given species might switch from one to another at different stages of its life cycle or under different environmental conditions. Thus, for example, Antarctic krill might form large dense shoals in open water in spring as a response to food availability and reduced predation pressure, but disperse under sea-ice in winter (Daly and Macaulay, 1991). This mirrors the situation found in facultatively schooling fish, for example rosyside dace (Freeman and Grossman, 1992). A further example of phenotypic plasticity is illustrated by inter-individual behaviour. Aggregations of higher vertebrates are often characterized by dominance and territoriality. The decision whether to defend a food source or territory may depend on an individual's evaluation of the cost–benefit balance sheet (Krebs and Davies, 1981). If no greater fitness results then it may be advantageous to join an aggregation rather than disperse. Group size may then become an issue because, if too small, it may be vulnerable to predatory attack, and if too large, members may suffer from intraspecific competition for food resources. Thus individual decisions to leave or join or actively deter new recruits from joining may become critical survival decisions. But are such considerations relevant to invertebrate aggregations? For the most part, defence of territory is not an issue for pelagic organisms. However, maintenance of position in the environment appears to be very important for inshore swarming mysids (Clutter, 1969; O'Brien, 1988a). Agonistic encounters between social squid at spawning sites seem to suggest a seasonal defence of resources (Hurley, 1978; Sauer et al., 1992). Hochberg and Couch (1971) observed signalling by some members of a school of *Sepioteuthis sepioidea* which may have served to prevent other squid from joining.

1.5. Origin and Maintenance of Social Systems

Social behaviour has been grouped into three major categories by Slobodchikoff and Shields (1988). These are: *genetic* hypotheses in which it is assumed that high levels of genetic relatedness among group members are necessary and sufficient to explain every kind of society; *phylogenetic* hypotheses which rest on the assumption that the evolution of a social system occurred earlier in the evolutionary history of a particular lineage and does not necessarily represent an adaptation to current conditions; and *ecological* hypotheses which, in turn, rest on an assumption that non-genetic environmental factors are necessary and sufficient to explain social systems.

In formulating ecological hypotheses it is assumed that cost–benefit functions have been major factors driving the evolution of sociality. Group behaviour is seen as a means of exploiting resources (e.g. food, space, mates, non-mate social partners, refugia, nest sites, climate, etc.) in ways not available, or at least much less efficiently so, to solitary individuals (Slobodchikoff and Shields, 1988). In this review I examine the data on invertebrate aggregations largely in the light of ecological hypotheses, not because I do not believe the other two are important, but because there is much less evidence in their support. I agree with the general conclusion of Slobodchikoff and Shields (1988) that "... genetic, ecological and phylogenetic factors have all contributed to the origin and maintenance of sociality in specific animal groups".

2. SHAPE, STRUCTURE AND PACKING RULES

2.1. Shape and Density of Aggregations

Invertebrate aggregations have been described as adopting shapes ranging from long ribbons to compact spheres (Omori and Hamner, 1982; Table

Table 1 Attributes of aggregation of pelagic invertebrates. (Values given in the table are crude approximations in most instances, based on data from available references. See text for further information.) Wave form type A, slow formation followed by rapid dispersal; B, rapid formation followed by slow dispersal; C, continuous presence fluctuating around some mean value; D, regular formation and regular dispersal, often fluctuating around 24 h; E, slow formation followed by slow dispersal. (Modified with permission from Omori and Hamner, 1982.)

Taxonomic group/clumping patterns	Temporal attributes			Spatial attributes		
	Duration/period length	Frequency	Wave form	Distance	Area	Volume
Noctiluca feeding aggregation	Months	1–10 years	Type A	1–100 m	10^{2-3} cm^2	10^{2-3} cm^2
Noctiluca dispersal swarms	Days	1–10 years	Type B	1–100 km	1–10^4 m^2	
Acoel flatworm swarms	Years	100s years	Type C	10–20 m	100–400 m^2	100–400 m^2
Scyphomedusan swarms (*Mastigias*)	Hours	24 h	Type D	100–1000 m	10^4 m^2	10^{4-5} m^3
Copepod swarms						
(*Calanus finmarchicus*)	12 h–days?	Seasonal	Type C?	1–3 m	1–3 m^2	1–5 m^3
(*Calanus tonsus*)	12 h–days?	Seasonal	Type C?	100–1000 m	10^{3-5} m^2	—
(*Oithona/Acartia*)	12 h	24 h	Type D	—	—	1–60 m^3
Copepod aggregations						
(*Centropages*)	Min–hours	Various	Type E	Various	Various	Various
(*Calanus pacificus*)	Hours–days?	Various	Type E	Various	Various	Various
Copepod schools (*Labidocera pavo*)	12 h	24h	Type D	1–5 m	5–10 m^2	5–10 m^3
Mysid schools	12 h	Seasonal	Type D	1–10 m	1–15 m^2	1–15 m^3
Euphausiid schools (*Euphausia*)	Days	Seasonal	Type C (D?)	100–1000 m	10^{2-4} m^2	10^{3-5} m^3
(*Meganyctiphanes norvegica*)	Days	Seasonal	Type C	1 m	1 m^2	1 m^3
Sergestid schools (*Acetes japonicus*)	Days	Seasonal	Type D	10–500 m	10^{2-4} m^2	10^{2-4} m^3
Sergestid swarms (*Sergia lucens*)	Days	Seasonal	Type D	1–100 m	1–10^3 m^2	$10^{-3} \times 10^5$ m^2

Table 1 cont.

Spatial attributes Shape	Density attributes Number of individuals	Spacing	Other attributes Presumed adaptive value	Reference
Thin parabolic strings	10^{3-4} m^{-3}	1–2 m	Feeding	Omori and Hamner (1982)
Linear surface swarms	10^{4-6} m^{-3}	0.5–2 cm	Dispersal	Omori and Hamner (1982)
Lens shaped	1.3×10^6 m^{-3}	1–2 cm	Feeding	Omori and Hamner (1982)
Irregular surface swarms	10^{2-3} m^3	10–100 cm	Reproductive/others?	Hamner and Hauri (1981)
Irregular balls	10^7 to 5×10^7 m^{-3}	—	—	Wiborg (176)
Irregular surface swarms	$3 \times 10^{3-4}$ m^3		Antipredator?	Kawamura (1974)
Round or irregular balls	10^{5-6} m^{-3}	1 cm	Antipredator	Hamner and Carleton (1979)
Linear	Various	Various	—	Omori and Hamner (1982)
Linear	10^{4-5} m^{-3}	Various	—	Omori and Hamner (1982)
Surface lens	10^4 m^{-3}	5 cm	Antipredator?	Omori and Hamner (1982)
Round/ribbon-shaped	10^3 to 3×10^3 m^{-3}	1–3 cm	Antipredator/feeding	Marr (1962), Komaki (1967) and others
Spherical/oval lens	$6 \times 10^{3-4}$ m^{-3}	1–3 cm	Antipredator/ Reproductive	
Spherical/surface swarms	10^3 m^{-3}	5–9 cm	Passive transportation by vertical turbulence	Brown et al. (1979)
Spherical/linear	10^4 to 5×10^4 m^{-3}	3–4 cm	?	Omori and Hamner (1982)
Spherical	1–10^2 m^{-3}	5–20 cm	Reproductive/others	Omori and Hamner (1982)

1). Table 1 incorporates a broad range of animal groupings, some clearly incidental, non-social aggregations, others clearly social. However, some defy categorization, for example acoel flatworm and scyphozoan "swarms". The surface expressions of aggregations, particularly euphausiids, have been described by many authors with many similarities apparent between different species (e.g. Marr, 1962; Cram and Malan, 1977; Hampton, 1981; Nemoto, 1983; Nicol, 1984, 1986; Hanamura *et al.*, 1984). Aggregations can vary in dimensions from dense fist-sized "balls" to immense shoals covering many hectares and extending 20 km or more in the longest dimension (Nicol, 1984; Shulenberger *et al.*, 1984; see Figures 3, 4 and 5). They can be found in the surface of ponds (Johnson and Chia, 1972), lakes (Byron *et al.*, 1983), tidal rivers (Xiao and Greenwood, 1992); quiet inshore waters (Ambler *et al.*, 1991); along the surf zone of sandy beaches (Wooldridge, 1983); in the open ocean (Hamner, 1984); and the deep sea (Van Dover *et al.*, 1992). Many authors note the clearly defined boundaries of aggregations though shape is not clearly correlated with a particular function. However, aggregations may adopt an elongated (cigar) shape when travelling (e.g. O'Brien, 1988b), and a more spherical or globular shape when stationary (Clutter, 1969; Nicol, 1984). This is not necessarily so, because more or less stationary aggregations of mysids (Wittmann, 1977) and copepods (Tanaka *et al.*, 1987) can be found in elongated ribbons or layers in shallow coastal water. Several authors describe diel and seasonal shape changes in aggregations (see below).

Information about the three-dimensional shapes of aggregations has

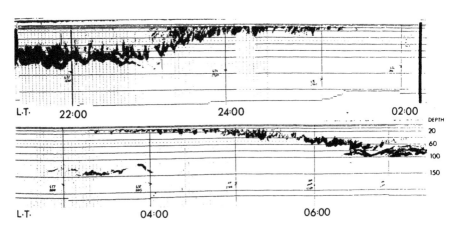

Figure 3 Migration of euphausiid scattering layer in the Antarctic, mostly *E. crystallorophias*. (From O'Brien, 1987b).

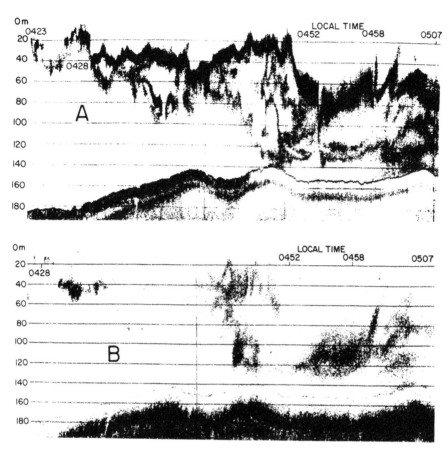

Figure 4 Echograms of large swarm of Antarctic krill, *Euphausia superba*. (A) Echo from 120 kHz transmission. (B) Echoes of the same shoal from 50 kHz transmission. (From Shulenberger *et al*., 1984.)

come from acoustic techniques, direct observations of SCUBA divers or remote stereophotographic methods. Ragulin (1969) described *E. superba* patches as truncated cones with the apex directed downwards, based on diver observations. Hamner (1984), while diving on aggregations of *E. superba*, noted that they were generally thin in at least one dimension. What appeared as a huge swarm at the surface would actually be found to be a sheet, hollow dome or flat ribbon. This may be a strategy to avoid severe depletion of food or oxygen, or a serious build up of excretory wastes within the aggregation (see below).

Densities within these aggregations can vary enormously (see Tables 1 and 2 and Clutter, 1969; Ragulin, 1969; Hamner and Carleton, 1979;

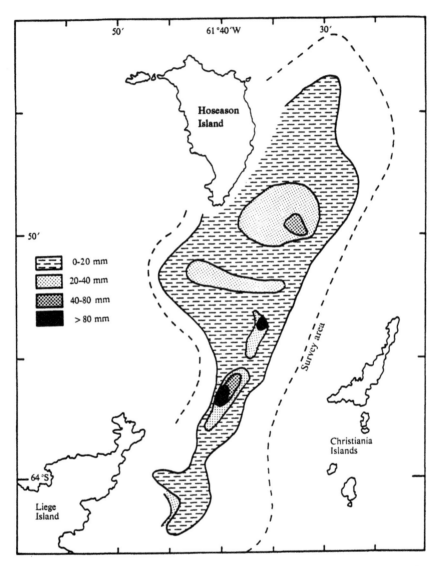

Figure 5 Shape of a krill swarm in the Gerlache Strait (approx. 64°30'S, 62°"W). The contour densities correspond to millimetres of integrator output per nautical mile. (From Cram *et al.*, 1979.)

Terazaki, 1980; Omori and Hamner, 1982; Byron *et al.*, 1983; Nemoto, 1983; Hamner, 1984; Alldredge *et al.*, 1984; Shulenberger *et al.*, 1984; Nicol, 1984, 1986; Hanamura *et al.*, 1984; Jillett and Zeldis, 1985; Kalinowsi and Witek, 1985; O'Brien *et al.*, 1986a; Nicol *et al.*, 1987; Tanaka *et al.*, 1987; O'Brien, 1987b, 1988b; Ambler *et al.*, 1991; Van Dover *et al.*, 1992). Methods used for calculating densities have varied greatly from visual estimates of NND, plankton net samples to sophisticated photogrammetric techniques, so direct comparisons could be misleading.

2.2. Internal Structure

The most detailed analysis of the internal structure of aquatic invertebrate aggregations is that of O'Brien (1989) using photogrammetric techniques described in an earlier paper (O'Brien *et al.*, 1986b). The indices used to describe positions of individuals within crustacean aggregations are the same as those used earlier for fish schools (Partridge *et al.*, 1980), and are shown in diagrammatic form in Figure 6. O'Brien found an overall similarity in internal structure of mysid and euphausiid aggregations, with a striking tendency for individuals to avoid positions directly above or below their neighbours. Indices of aggregation structure showed some consistent variations between mysids and euphausiids and between substratum-specialized and non-substratum-specialized mysids. Whereas euphausiid aggregations were less structured and tended toward facultative swarming or schooling, mysids showed a greater degree of structure and tended towards obligate social groups. Substratum-specialist mysids had smaller NNDs than non-substratum associated species. Importantly, behavioural stimuli were shown to have a more dramatic effect on aggregation structure than alterations to physical parameters of the environment such as current speed, light or substratum. These results indicate a remarkable convergence with fish shoals and suggest that similar selective pressures have shaped their evolution.

2.3. Density within Aggregations

As mentioned above, estimates of density within aggregations have been made using a wide variety of methods. Visual estimates and those from plankton net catches are notoriously unreliable. Those derived from echotraces require careful calibration and integration of target strength which is a function of echosounder frequency, density of organisms, their size distribution and also attitude in the water (see Miller and Hampton,

D.A. RITZ

Table 2 Densities of euphausiid swarms as calculated from estimations of nearest neighbour distance (NND) by direct observation or photography. (Modified with permission from O'Brien, 1988b.)

Species	Density (nos m^{-3})	Length (mm)	Source
Euphausia crystallorophias	100–30 000	16-20	O'Brien, 1987b
Euphausia lucens	4000–5 000 000	10–20	Nicol *et al.*, 1987
Euphausia pacifica	10–72 000	12.3–21.5	Hanamura *et al.*, 1984
Euphausia superba	20 000–60 000	>30?	Ragulin, 1969; Hamner *et al.*, 1983
Euphausia vallentini	70 000–80 000	35	Ragulin, 1969
Meganyctiphanes norvegica	9000–770 000	15–18	Nicol, 1986
Nyctiphanes australis	3000–480 000		O'Brien, 1988b

1989). Density estimates from photographic recording of aggregations underwater require the application of some kind of packing model. Several such models have been proposed. Close and cubic packing (Nicol, 1986; O'Brien, 1988b) is based on a cylindrical shape of the animal plus a clear volume separating it from its nearest neighbour. Densities are calculated from the following equations:

$$\text{Closest packing: } n = 1.155 \frac{1}{d^2 L}$$

$$\text{Cubic packing: } n = \frac{1}{d^2 L}$$

where n is number of animals m^{-3}; L is length of cylinder occupied in metres; and d is distance between centres of animals (i.e. NND + body width) in metres (Nicol, 1986). Using these equations to compare densities of the euphausiid *Nyctiphanes australis* from NND measurements and from volume estimated from stereophotographs, O'Brien (1989) found the close and cubic packing models overestimated density by 20 and 10% respectively.

Hamner and Carleton (1979) and Alldredge *et al.* (1984) used an isahedronic packing model to estimate densities of copepods. The equation is:

$$\text{No. of copepods m}^{-3} = \frac{1\,000\,000 \text{ cm}^3}{0.589 \times (\text{ave NN distance cm})^3}$$

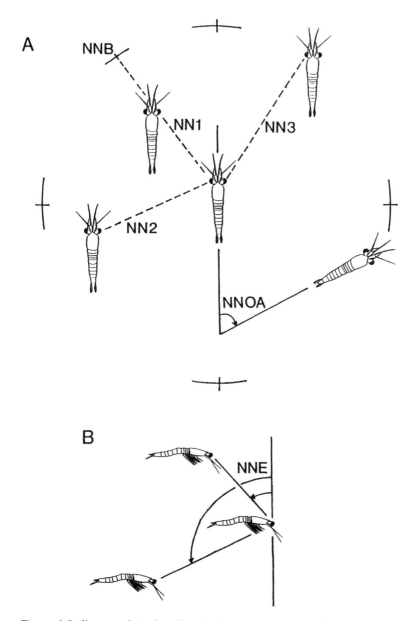

Figure 6 Indices used to describe the internal structure of crustacean swarms and schools. NN1, NN2, NN3: distances to 1st, 2nd and 3rd nearest neighbour from reference animal. NNOA, nearest neighbour orientation angle; NNE, nearest neighbour elevation; A, top view; B, side view.

Using these models, Alldredge *et al.* (1984) reported maximum densities of $26 \times 10^6 \, \text{m}^{-3}$ *Calanus pacificus californicus*. Nicol (1986) found up to $77 \times 10^4 \, \text{m}^{-3}$ *Meganyctiphanes norvegica* (compared with $6 \, \text{m}^{-3}$ from a plankton net catch!). Hamner and Carleton (1979) reported up to $1.5 \times 10^6 \, \text{m}^{-3}$ individuals of a range of copepod species.

Biomasses calculated from density estimates can be as high as $15 \, \text{kg} \, \text{m}^{-3}$ for *M. norvegica* and $11 \, \text{kg} \, \text{m}^{-3}$ for *Euphausia lucens* with total swarm biomass estimated to be around $5000 \, \text{t}$ for the latter species. The superswarm of *E. superba* sampled by Macaulay *et al.* (1984) may have contained $2.1 \times 10^6 \, \text{t}$.

2.4. Sorting within Aggregations

A characteristic feature of many invertebrate aggregations, and fish and tadpole shoals for that matter, is the uniformity of size of the members. Data for *E. superba* are the most extensive (see Miller and Hampton, 1989 for review). Hamner *et al.* (1989) found that krill larvae first exhibited aggregative behaviour as furcilia stages IV–V and were segregated by size in immediately adjacent schools (Figure 7). It appears that size-sorting mechanisms operate within a given aggregation, though not always perfectly (Watkins, 1986). However, adjacent aggregations are as unlikely to resemble each other in mean individual size as in a range of other biological characteristics (Watkins *et al.*, 1985, 1986, 1992; Watkins, 1986; Priddle *et al.*, 1990; Ricketts *et al.*, 1992). A selection pressure favouring segregation of sizes and stages within aggregations (ultimate cause) might be to minimize competition and/or cannibalism.

Length-dependent differences in swimming speeds are often invoked as a proximate cause to account for size-sorting within aggregations of euphausiids (e.g. Mauchline, 1980a; Hamner *et al.*, 1983; Nicol, 1984). O'Brien (1988b) found a greater range of sizes within swarms of juvenile *N. australis* than in adult swarms. He proposed that either this indicated a differential growth rate of immature krill in swarms or, more likely, a greater interchange of individuals between aggregations during their immature stage. Size segregation might then take place gradually over time, especially if groups tend to disperse and reform continually over a diel or seasonal cycle as is suggested to be the case (Everson, 1982). The apparently facultative nature of aggregation in euphausiids (O'Brien, 1989) lends some support to this idea.

Kils (1981) offered another suggestion to account for size-sorting in krill based on differential sinking rates. However, Watkins (1986) found no evidence of consistent variation in size of krill with depth to support this possibility.

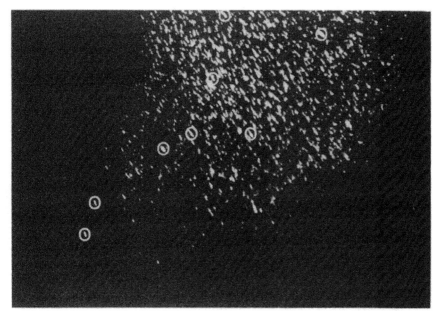

Figure 7 School of *Euphausia superba* in Croker passage, off the Antarctic Peninsula. The swimming direction is obliquely downward from left to right. There are no individuals outside the densely packed school. Unhealthy, whitish, individuals are distinguished (circles). (From Hamner, 1984.)

A further possibility advanced to account for size-sorting is that each swarm/school may represent a single cohort within which there is a higher degree of genetic relatedness than between such groups. This seems most unlikely for the more facultative krill aggregations (see above) but would be worth investigating in obligate mysid aggregations which show clear growth stage segregation (e.g. Modlin, 1990). Wittmann (1977) had earlier shown that mysid swarms remained faithful to an exact locality even after disturbance and marked individuals could be relocated in the field for up to 12 days (limit of intermoult period). Over much longer periods (years) swarms could still be found in many of the same locations though apparently aggregations were more site-faithful than were individual members. The brooding habit of mysids might contribute to cohort integrity over long periods. Genetic relatedness was also suggested as a possible explanation for non-mixing between tadpole schools of *Hyla geographica* by Caldwell (1989). This would require sibling recognition for which no mechanism has yet been advanced.

There is some evidence to suggest that social animals preferentially select neighbours of a similar size to themselves. For example, Ranta *et al.* (1992) showed that three-spined sticklebacks preferred the company of fish of matching body size. They believed this to be linked to an antipredatory adaptation since any detectable "oddity" among individuals in schools led to them being preyed on far more than inconspicuous fish. This was also demonstrated by Landeau and Terborgh (1986) for fish. Ranta and Lindstrom (1990) argued that, though a large fish might compete more effectively for food in a school of small fish, it did so at a cost of "oddity" when the aggregation was under threat from a predator. Pitcher (1986a, b) noted that multispecies fish shoals split up into species groups when under attack, presumably because the cost of being conspicuous outweighed any benefit gained from remaining in the mixed group. Hurley (1978) tested the idea of preferential selection of similarly sized neighbours in social squid, which also form schools of uniformly sized individuals, but did not obtain conclusive results.

2.5. Multispecies Aggregations

Multispecies aggregations among aquatic invertebrates are occasionally reported and members usually bear a close resemblance to each other. Such associations have been noted among copepod species (Kimoto *et al.*, 1988), euphausiid species (Nemoto *et al.*, 1981), euphausiids and juvenile fish (Slosarczyk and Rembiszewski, 1982), and mysids and fish (Wittmann, 1977). The brief association between mysids and postlarval grunts (McFarland and Kotchian, 1982) occurs at a stage of development of the fish at which it bears a remarkable similarity to mysids. As the fish grow and the distinction becomes more apparent, the association ends having lasted approximately 5 days, only to be renewed as more postlarvae recruit from the plankton.

2.6. Exchange between Aggregations

The permanence or otherwise of aggregations has long been a subject of speculation. One of the first to attempt to study this in the field was Wittmann (1977). He marked and released mysids in the vicinity of shallow, nearshore aggregations. The marked specimens were subsequently relocated for up to 12 days after release (maximum intermoult period). They were found in the same locations between 0 and 93% of occasions. On alternate days they were often found in neighbouring swarms suggesting regular interchange of members. There was none the less great site-fidelity of individual swarms as Hahn and Itzkowitz (1986) also showed for mysids. Hilborn (1991) showed that individual skipjack tuna exchanged rapidly between schools. Apparently between 16 and 63% of individuals leave a school each day and join another. On the other hand, Caldwell (1989) found that tadpole schools remained discrete and did not exchange members. She suggested genetic relatedness as a possible explanation since sibling tadpole schools are known in other species of frogs and toads (Waldman and Adler, 1979; Waldman, 1982). Sibling schools have also been reported for some fish (e.g. Ferguson and Noakes, 1981).

2.7. Mathematical Models

Mathematical models of aggregations have usually been based on the interaction of forces tending to attract and tending to repel individual organisms (Dill *et al.*, 1993). Examples are given in Okubo and Anderson (1984), Seno (1990) and Reynolds (1987), but they generally lack biological reality. Use of modelling techniques to propose and subsequently test hypotheses about spatial positioning in aggregations has been notably lacking in the literature (Dill *et al.*, 1993). This is despite a large and growing body of detailed descriptive studies of aggregation structure. Dill and coworkers hypothesized that, within an aggregation, there exist optimal positions from which to monitor changes in a neighbour's behaviour. They tested this using data from fish schools and bird flocks and found some support for the proposition that individuals monitor the optical specification of time to collision with their neighbour in front (tau). Clearly the field has advanced to the stage where we should be testing real predictions about spatial relationships. One example is the supposed hydrodynamic advantage to be gained by individuals adopting an appropriate position in order to benefit from vortices shed by a neighbour swimming in front. Fish and krill use different swimming mechanisms. As a result the wake is displaced rearwards at a different angle which should select for different preferred positions adopted by organisms in their respective schools if they are to gain some energetic advantage. O'Brien (1989) found little evidence to support this notion in crustacean aggregations.

3. DIEL AND SEASONAL PATTERNS

3.1. Diel Cycles

Diel cycles of vertical migration and aggregation/dispersal in social aquatic invertebrates present no clear conclusive picture. For example, after surveying a very large literature on *E. superba*, Miller and Hampton (1989) found conflicting data on migration in different areas. They also concluded that ". . . dispersal at night or at any other time is by no means a regular feature of krill's aggregating behaviour". One very characteristic feature of many species of euphausiid is daytime surface aggregation (Figure 8). Such aggregation has been reported for *Euphausia lucens* (Nicol *et al.*, 1987); *E. pacifica* (Komaki, 1967; Terazaki, 1980; Hanamura *et al.*, 1984); many instances for *E. superba* (e.g. Hardy, 1936); *Meganyctiphanes norvegica* (Brown *et al.*, 1979; Nicol, 1984); *Nyctiphanes australis* (O'Brien, 1988b); *N. capensis* (Thomas, 1980); *N. simplex* (Gendron, 1992); *Thysanoessa longicaudata* (Forsyth and Jones, 1966); *T. inermis* (Hanamura *et al.*, 1989); *T. spinifera* (Smith and Adams, 1988). Nicol *et al.* (1987) wrote that daytime swarming in euphausiids was a difficult phenomenon to study because of its sporadic nature. On the other hand, O'Brien (1988b) found evidence for surface swarms of *N. australis* throughout the year, and surface swarms of *E. pacifica* are regular enough off the coast of Japan to support a commercial fishery (Endo, 1984). The possible reasons for daytime surface aggregation have been speculated to be spawning or maturation, feeding, passive concentration by vertical turbulence or tidal streams, and that the euphausiids had been driven to the surface by predators (Komaki, 1967). In most cases no single cause seems to explain this phenomenon satisfactorily. Moreover, it is not only euphausiids that periodically appear in dense aggregations in the surface during the day. Many other crustaceans exhibit similar behaviour, for example mysids (Mauchline, 1980b); galatheids (Jillett and Zeldis, 1985); amphipods (Lobel and Randall, 1986); calanoid copepods (Kawamura, 1974; Byron *et al.*, 1983); cyclopoid copepods (Ambler *et al.*, 1991); brachyuran larvae (Rice and Kristensen, 1982); cladocerans (Johnson and Chia, 1972); and hoplocarid larvae (Komai, 1932).

There are many reports of aggregations undergoing diel cycles of vertical migration and/or dispersal and re-aggregation. For example, Wittmann (1977) observed mysid swarms over day–night cycles. At sunset, structure became looser as inter-individual distances increased and swarms expanded to form a layer 10–30 cm above the bottom in shallow water. As light level decreased, monospecific swarms that had occupied distinct bathymetric zones in the daytime, merged into a polyspecific night aggregation near the bottom. At sunrise, swarms rose in the water column and segregated once again into monospecific groups. Clear bathymetric segregation of monospecific mysid swarms in nearshore environments has been noted by other authors, for example Fenton (1992). The fact that mysid swarms did not entirely disperse at night was also noted by O'Brien (1988a) and Emery (1968).

Figure 8 Shoals of Antarctic krill, *Euphausia superba*. (A) Echogram showing dense shoals in mid-water and less dense shoals extending up to the surface. (From Nicol and de la Mare, 1993.) The frequency of the echosounder transmission was 120 kHz. The echoes have been quantified from weak (the palest areas) to dense (the darkest). By assuming that noise is insignificant, only krill are present, and target strength of krill is $-73\,dB$, the darkest areas correspond to 2000 krill m^{-2}. (B) A photograph of a similar shoal breaking the surface. (Also from Nicol and de la Mare, 1993.)

In contrast, McFarland and Kotchian (1982) describe a night-time dispersal by diurnally aggregated mysids which reform at dawn in association with postlarval fish (French grunts). Diel cycles of aggregation/dispersal appear to be rather similar in copepods. Hamner and Carleton (1979) reported that swarms of calanoid copepods formed in relation to topographic features such as coral outcrops during the day over tropical reefs, but dispersed at night. A similar picture was found for *Acartia* swarms by Kimoto *et al.* (1988) who described an increasing density near the bottom during the day through vertical migration. Here too, copepods aggregated in the lee of topographic features. After sunset, individuals swam up through the water column causing dispersal of the swarm. Ambler *et al.* (1991) also describe a dispersal of cyclopoid copepod swarms at night.

Many social euphausiid species undergo diel vertical migrations, not necessarily reaching the surface during these cycles. For example, *E. superba* (references in Miller and Hampton, 1989); *E. crystallorophias* (O'Brien, 1987b); *Thysanoessa inermis* and *T. raschii* (Zelikman *et al.*, 1979); and *M. norvegica* (Greene *et al.*, 1992). Such migrations in zooplanktonic organisms are commonly thought to be the result of a trade-off between foraging requirements and protection from predators (e.g. Gliwicz, 1986). If aggregative behaviour itself subserves both these needs, it might be expected that vertical movements of social aquatic animals would differ from non-social ones. Additionally, in facultative schoolers, it might be expected that vertical migration cycles would differ at times when populations were dispersed compared to times when they were aggregated. Such arguments might help to explain the sometimes conflicting data on vertical movements of *E. superba* (see Miller and Hampton, 1989).

In social animals that segregate by ontogenetic stage and/or sex, each category may adopt a different diel cycle. For example Omori (1974) showed that the males and females of some shoaling pelagic shrimps migrated differentially at night. Adult females migrated to a shallower layer than males. Wooldridge (1983) described dense communities of mysids along the surf zone of sandy beaches. The degree of pelagic activity varied with age, sex and within the females, with breeding status. During the day, males and immatures of *Gastrosaccus psammodytes* were farthest offshore. Inshore the females became progressively less active as they moved nearer to the beach, spending more time buried in the sand. Wittmann (1977) had earlier suggested a basic behavioural difference between substratum associated mysids and their more pelagic relatives that was reflected in their aggregative behaviour. O'Brien (1989) confirmed that degree of substratum-attraction in mysids did influence the characteristics of swarming: this was manifested in differences in NND between the two groups. Modin (1990) found that *Mysidium columbiae* segregated by life stage, forming vertically stratified shoals with smallest juveniles on top and mature males and gravid females at the bottom. During the night, the smallest juveniles maintained compact swarms while larger individuals, for example gravid females, became solitary. A day/night difference in aggregative behaviour was also noted in *Acetes sibogae australis* in saltwater lakes by Ross *et al.* (1988). During the day, schools remained in shallow water close to rocks or outcroppings. At dusk, these dispersed to deeper water to feed. Little interest in food was apparent during the daytime schooling phase. Ross *et al.* speculate that this aggregation/dispersal rhythm has evolved in response to conflicting pressures of the need to feed, avoid visual predators in deeper water (fish), shallow water predators (birds), and to perform social functions, for example reproduction. To satisfy these demands, *Acetes* schools in shallow shaded areas in daytime and disperses to deeper water at night.

Some hermit crabs exhibit endogenous circadian rhythms in the tendency to join groups or disperse (Pulliam and Caraco, 1984). Sea hares (*Aplysia* spp.) also show this tendency; in this case it is clearly associated with mating (Pennings, 1991).

3.2. Seasonal Cycles

In addition to diel cycles, many authors report seasonal cycles of aggregation and dispersal, sometimes in association with migration. Recently, conceptual models linking feeding, aggregation and seasonal distribution patterns in Antarctic krill have been suggested in an attempt to reconcile what hitherto has been a confused and conflicting picture (see Miller and Hampton (1989) for review of earlier data). Daly and Macaulay (1991) were the first to include seasonal variation in predation intensity as a proximate factor influencing swarming behaviour. In their view aggregation of *E. superba* in winter, despite low food abundance, is a response to predation pressures greatly increased compared to spring and autumn. When conditions of low food (or low food demand) coincide with low predation pressure, krill would be expected to disperse; conversely, in conditions of abundant food (or high demand) and high predation pressure, krill would be expected to aggregate into swarms. As simple as this model is, it appears to fit the observed facts more closely than earlier models which focused on feeding as the chief proximate cause of swarming behaviour (Miller and Hampton, 1989). Other relevant observations on migration of swarms of krill provide additional support for this general picture. For example, Bergstrom et al. (1990) describe more or less oriented swarms of E. superba in the western Weddell Sea forming among decaying sea ice. They speculate that this might indicate a transitional state from winter life as a grazer on ice algae to summer life as pelagic phytoplankton filterer. Ichii (1990) also describes high density concentrations at the pack ice edge in early summer in the Indian, Pacific and Atlantic sectors of the Southern Ocean. These concentrations moved further north in late summer. In addition, Sprong and Schalk (1992), working in the northern Weddell Sea, used continuous echosounder records to describe krill migrations. From late November to early January (spring to summer), swarms migrated northwards away from the ice edge towards greater depths, simultaneously growing in size. Whereas average number of swarms per 10 nautical miles remained fairly constant, average (and total) swarm sizes increased. These authors too believed that availability of food in open water in spring was an important factor in this migration. Growth in swarm size strongly suggests a response to increased predation pressure.

The concept of aggregation size changing seasonally in response to both food and predation pressure is well established in the literature on fish. For example, Blaxter and Hunter (1982) proposed that herring and anchovy schools changed seasonally to form large protective schools in autumn and winter when foraging was less prevalent. The occurrence of such group size changes in conflict situations was experimentally demonstrated by Pitcher et al. (1983). More recently, benefits of different group sizes of mysids in terms of food capture success were shown by Ritz (1994).

In one of the few experimental studies of seasonal change in aggregations, Koltes (1984) showed that Atlantic silversides exhibited strong rhythms of NND, depth of school and swimming speed. Mean direction of travel showed peaks in spring and autumn. These changes corresponded with shifts from active synchronized schools in spring and autumn, to inactive "mills" in summer and inactive, non-schooling groups in winter seen in wild fish. Patterns observed were believed to correlate with natural cycles of reproduction and migration and may have represented an antipredator strategy. The fact that they persisted in laboratory tanks suggests an endogenous component to the rhythms.

Modlin (1990) described seasonal changes in aggregation structure in the mangrove mysid Mysidium columbiae. Whereas in summer mysids aggregated in cylindrical schools >2 m in length and ≥0.5 m diameter, in winter they formed ovoid schools <2 m in length or diameter. Neighbouring schools did not mix during winter providing they were separated by a shaded area.

3.3. Other Cycles

Occasional reports in the literature describe aggregation changes apparently in response to tidal cycles. For example, Dadswell (1975) reported a regular concentration of a population of *Mysis gaspensis* at low tide along one bank of a small estuary. Apparently the entire population of mysids was to be found in this restricted location in a zone of medium current speed. At high tide the shoal broke up and became distributed more or less evenly throughout the estuary. Dadswell believed that this regular event was mainly in response to environmental stimuli, principally current speed and salinity. Aggregations of *Acetes sibogae australis* are common in tidal estuaries of south east Australia and are strongly seasonal (Omori, 1975). Xiao and Greenwood (1992) found such aggregations to undergo tidal and vertical movements that were believed to facilitate population maintenance in estuarine and coastal waters.

4. FORAGING AND FEEDING

4.1. Benefits of Aggregation

By far the great majority of aquatic invertebrates that form swarms or schools are feeders on small particles in the water column using some form of filtering. However, most if not all can switch to a raptorial mode in order to capture larger particles and many carnivorous species form aggregations. Debate in the literature is centred firstly on whether aggregation is a benefit in terms of foraging, and secondly whether it might be a cost in terms of feeding. There is little doubt that animals can and do feed whilst aggregated. The link between improved foraging and feeding and aggregative social behaviour of fish in schools has been conclusively demonstrated by many authors (see Pitcher, 1986a, b for reviews). However, this link, though suspected, has rarely been proved for social aquatic invertebrates. Many ways in which aggregations might benefit from foraging and feeding in groups have been suggested. Groups could locate new or richer food patches faster than individuals (Pitcher, 1986a). Larger groups might reduce per capita search time compared to smaller ones (Pulliam and Caraco, 1984). Schooling behaviour might itself promote long-distance horizontal or vertical search patterns (Hamner et al., 1983). The "information centre" concept could assist group members to gain a share of any new resource located (Krebs and Davies, 1981). Once a rich food patch has been located, a feeding swarm might have the effect of a physical filter capable of grazing down a parcel of water to much lower concentrations than otherwise possible (Antezana and Ray, 1984). Fish in schools have been shown to be less timid, to spend more time feeding, to switch more efficiently to more profitable patches, and passive information transfer between individuals is more effective in larger groups than in smaller ones (Pitcher, 1986a). In a patchy environment, increasing group size might decrease the variance in total amount of food captured per unit foraging time (Pulliam and Caraco, 1984). In mixed species groups one benefit might be the greater diversity of food items available (Pitcher, 1986b). Postlarval grunts that aggregate temporarily with similarly sized mysids appear to benefit both by gaining protection from predators and later, when they have increased in size, they begin to prey on the mysids (McFarland and Kotchian, 1982).

Schooling in fish has been suggested to increase the responsiveness of individual fish to food odours in the environment (Steele et al., 1991). In this context Poulet and Ouellet (1982) found that aggregation could be induced in copepods by dissolved amino acids and they proposed that optimization of food yield is one of the significant benefits. Ryer and Olla (1992) found that exploitation of patchy ephemeral food was more successful when juvenile walleye pollock foraged in groups. They attributed this to two mechanisms: *local enhancement*—that is, individuals finding a food patch and attracting others to it; and *social facilitation*—that is, increased frequency of a particular behaviour when in the presence of other individuals behaving in a like manner. Gotmark et al. (1986) showed that gulls foraged more successfully as flock size increased probably for the same reasons.

But it is important to remember that foraging and feeding decisions are not taken independently of other considerations and animals constantly assess risks and benefits. Strictly controlled laboratory conditions may not always apply to field situations. For example, Freeman and Grossman (1992) found little evidence of foraging benefits in groups of rosyside dace in a field study. Group size did not affect individual feeding rates. However, a stronger social attraction in autumn may have been a response to increased threat from a seasonal predator. Similarly, if it is profitable to defend a point source of food, fish may abandon schooling behaviour and show aggression to subordinate conspecifics and monopolize the resource. Once the food is eaten or removed, schooling behaviour may be resumed. This tendency to alternate facultatively between schooling and solitary behaviour seemed to increase in frequency and intensity with age in juvenile chum salmon (Ryer and Olla, 1991).

4.2. Costs of Aggregating

The chief cost of group foraging and feeding is generally held to be intraspecific competition for resources. The most obvious of these resources is food, but others might equally well include mates, spawning sites or territory. Competition (or interference) for food can take several forms (Milinksi and Parker, 1991). These include *exploitation*, occurring when an individual forages in an area that has already been exploited by a competitor thus reducing its success rate; *scramble*, occurring when several non-competitive individuals simultaneously attempt to eat the same item when only one can be successful; and *contest*, occurring when competitors interact aggressively for a given resource. In the case of contest competition, if the energetic costs of chasing away competitors or even fighting them is less than the benefit gained by monopolizing the resource, the latter becomes an economically defendable resource.

Many authors note the possible costs of group foraging/feeding but there has been little direct experimental testing of these concepts in aquatic invertebrates, Jakobsen and Johnsen (1988) and Tessier (1983) report that levels of food locally within swarms of cladocerans were drastically reduced compared to ambient levels. In laboratory experiments, Jakobsen and Johnsen found that swarming in *Bosmina* only occurred in the light and with abundant food. The authors interpreted their results as an adaptive diel rhythm in trading off maximal feeding for safety. When starvation risk is high (i.e. low ambient food levels) maximal feeding seems to have priority over swarming. Thus this might be an example of risk-sensitive facultative switching between aggregated and dispersed behaviour. In support of this, Jakobsen and Johnsen (1987) showed that *Daphnia*, when exposed to a gradient of food concentration, distributed themselves according to patch profitabilities, but only when food levels were low and limiting. Sih (1980) had earlier shown that *Notonecta* was capable of balancing two conflicting demands, foraging and safety, and behaving adaptively. Folt (1987) measured cost of aggregation as a reduction of feeding rate of zooplankton (both herbivores and omnivores). Reductions were thought to be due both to exploitative and interference mechanisms but not due directly to predation. Folt concluded that, overall, aggregative behaviour was a cost which was only compensated for by protection from predation. Therefore aggregations of zooplankton might only be advantageous in predator-prone environments. However, the possible advantages of foraging and feeding in groups are very numerous (see above) and extrapolations from controlled laboratory experiments in which food is distributed homogeneously rather than patchily can be misleading. Ritz (1993) showed that, in some circumstances, even in the absence of a predatory threat, mysids in large swarms captured more patchily distributed food than those in small ones. When under threat from a fish, the difference was magnified. In swarms of different sizes, highest food capture rate was attained by intermediate-sized swarms. This was interpreted as indicating that foraging or feeding was inefficient in the smallest swarms while competition was increasingly important in the larger ones. Hence, in terms of food capture success, aggregation favoured larger swarms when under predatory threat, but smaller ones when the threat was absent.

The idea of an ecological trade-off between foraging success and protection from predators thus seems to be well entrenched. Many authors hold the view that some feeding success will be sacrificed for safety (life before dinner principle) but that foraging/feeding benefits alone are insufficient to account for the evolution of aggregative behaviour (Ranta and Kaitala, 1991; Magurran, 1990).

4.3. Foraging Group Size Models

It is tempting to think that foraging groups should adopt some optimal size that maximizes fitness for the members. However, attempts to match naturally occurring groups with theoretically or experimentally determined optima have usually failed. Natural groups are almost always larger than optimal. Giraldeau (1988) proposes that optimal group size might not be a realistic expectation. The reasoning is that optimal group sizes are unstable because such groups would attract new recruits from suboptimal aggregations, each trying to maximize its own fitness. These kinds of arguments rest on certain assumptions of the "ideal free" model proposed by Fretwell (1972). These are that individuals are of equal competitive ability and that each is free to choose to feed where it can maximize its individual fitness. These assumptions are probably robust in some cases (Milinski and Parker, 1991), but are almost certainly violated in others. For example, the equality of individuals in squid schools may be an untenable concept (Hurley, 1978). The aggressive tendencies of squid in some situations could mean that new recruits to schools could be actively prevented from joining (e.g. Hochberg and Couch, 1971).

None-the-less it does appear that most assemblages so far studied are supra-optimal. Giraldeau (1988) calls this size the stable group, which will be of a size making it unattractive for solitary individuals to join. Figure 9 depicts the expected relationship between net benefit and group size

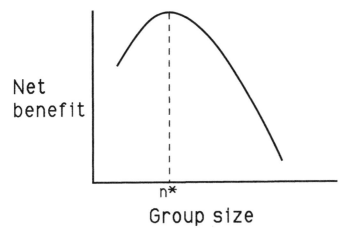

Figure 9 Relationship between foraging benefit per individual and group size. n^*, optimal group size. (Modified after Giraldeau, 1988.)

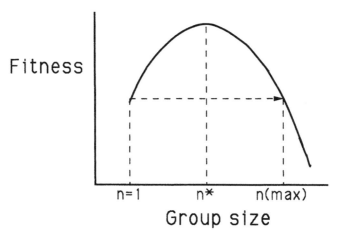

Figure 10 An hypothetical fitness function and its relationship with group size. The maximum fitness is obtained in the optimum group size (n^*). The stable group size (n) corresponds to a group size that provides its members and a potential recruit with as much or less fitness than remaining alone. (After Giraldeau, 1988.)

(Giraldeau, 1988). This is derived from the widely held view that benefits of group foraging increase at a decreasing rate while foraging requirements increase linearly. The stable group size model (Figure 10) then predicts that fitness is maximized at the optimal group size (n^*) while the stable group size (n) is one that provides members and a potential recruit with as much or less fitness than remaining alone. Thus, paradoxically, the model suggests that gregarious individuals, far from increasing their fitness by living in groups, end up with no more fitness than isolates. If groups could reach and maintain n it would be unlikely that sociality would have evolved. However, many factors serve to prevent groups from reaching or maintaining a size of n. Pitcher (1986b) reached a similar conclusion about size of fish shoals. He called his model zero benefit shoaling which was seen as an alternative to the non-achieved optimal group size and as a bet hedged against predator encounters. Facultative swarming in Antarctic krill can perhaps be viewed in this way (Daly and Macaulay, 1991).

4.4. Genetic Relatedness in Animal Aggregations

There is evidence for a few species of fish and tadpoles that individuals in schools are more closely genetically related than conspecifics in other schools (Ferguson and Noakes, 1981; Waldman, 1982). No data are available for social aquatic invertebrates. The effect of genetic relatedness on foraging group size has been modelled by Giraldeau (1988) and Giraldeau and Caraco (1993). Predicted group sizes differ according to the level of genetic relatedness among members. The outcome also differs according to whether there is free entry or group-controlled entry to new recruits. In the former case, which presumably applies to most invertebrates, increasing relatedness can reduce but never increase equilibrium group size. In the latter case, which may hold for social squid, increasing relatedness can increase but never decrease equilibrium group size.

5. REPRODUCTIVE FACILITATION

5.1. Advantages of the Social Habit

The commonly reported daytime surface aggregation in euphausiids is usually attributed to either feeding or reproductive functions. For example, Endo (1984), Nicol (1984), Naito et al. (1986) Nicol et al. (1987), O'Brien (1988b), Smith and Adams (1988), Hanamura et al. (1989) and Gendron (1992) all reported a preponderance of mature individuals in breeding condition. Often these aggregations have highly skewed sex ratios, with either males or females dominating, but usually the latter (e.g. Nicol, 1984; Hanamura et al., 1989). Swarms of entirely one sex are not uncommon, for example see Watkins et al. (1992). Such sexually segregated aggregations are frequently encountered among arthropods, for example in the Cladocera and Copepoda (Nicol, 1984), and are apparently the norm in flying dipteran swarms (Downes, 1969) in which swarming is almost invariably associated with mating (Cooter, 1989).

Many advantages of the social habit in reproduction have been suggested. The proximity of neighbours may facilitate mate location (Nicol, 1984); promote physical contact (Clutter, 1969); assist individuals to locate suitable spawning areas (Sauer *et al.*, 1992); increase the probability of fertilization (Byron *et al.*, 1983); and the possibility of grouping births in time might swamp predator appetites (Pulliam and Caraco, 1984). The timing of egg release can be critical in euphausiids and spawning schools of *E. superba* may need to migrate to deep water in order to prevent their rapidly sinking eggs from being deposited on the seabed (Hamner, 1984). Presumably timing of critical events in the reproductive cycle can be accomplished more effectively in a social group.

Despite the fact that surface aggregation of euphausiids is common in daylight, mating apparently only occurs at night (Naito *et al.*, 1986; O'Brien, 1988b). In considering the breeding-related adaptive value of surface swarming in *Thysanoessa spinifera*, Smith and Adams (1988) suggest timing of sexual maturation, egg release and larval dispersal in a changeable upwelling environment as likely factors.

Hamilton (1971) considered the ways in which selection would be likely to favour individuals in a nuptial aggregation compared with isolates. In nematocerous fly swarms, which usually consist wholly of males, the groups hover in a fixed spot. Females approach the swarm and are each seized by a male. A solitary male that does not join the swarm thus has little chance of mating. An optimal position for a male is probably on the periphery and not, as is argued for protection from predators, in the centre. The continual dance-like motions of swarms then could be due to attempts by its male membership to spend maximum time in favourable positions. This behaviour might have parallels in crustacean swarms. For example, Ritz (1994) suggested that the reason individual mysids made excursions throughout an aggregation was to gain a sense of its size. However, this behaviour could be a means of resolving conflicting requirements of feeding, safety and mating.

Flying dipteran swarms tend to hover in one position by reference to a "marker", that is, a fixed object or visual feature in their environment (Cooter, 1989). This too seems to have parallels in the social mysids, swarms of which often remain stationary with respect to a topographic feature in shallow nearshore habitats. In open water pelagic species, for example euphausiids, the swarm itself may act as the marker. Mature female flies, attracted to a male swarm, drop out once they have mated

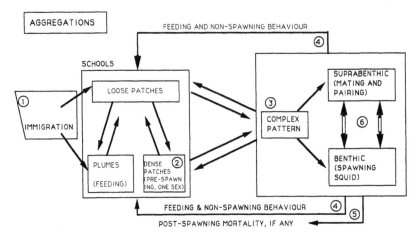

Figure 11 Proposed classification of behaviour of all schooling squid. Aggregations can be divided into non-spawning (schools) and spawning forms (concentrations). (1) Immigration: migration to spawning or feeding grounds. (2) Dense patches found intermittently throughout the year. (3) Usually form on a daily basis—often repeated for a few days up to as long as 3 months. Complex pattern may comprise different individuals each day. (4) Feeding mainly at night. These groups may move offshore to feed, if spawning is onshore. (5) Note post-spawning mortality on the spawning grounds which was not observed. (6) Probably applies to all loliginids but not oceanic squid. (After Sauer *et al.*, 1992.)

(Downes, 1969). Mated or gravid females of social crustaceans are known to show distinct behavioural changes (e.g. Terazaki, 1981 cited in Endo *et al.*, 1986; Modlin, 1990), and thus single sex schools may arise through active reproductive behaviour (Watkins *et al.*, 1992). Sauer *et al.* (1992) found that squid (*Loligo vulgaris reynaudii*) migrated as single sex schools which mixed during spawning periods (see Figure 1(A)). Their scheme for classifying aggregative behaviour of social squid during spawning is reproduced in Figure 11.

5.2. Sex Pheromones

Many female crustaceans release a sex pheromone at the time of a prenuptial moult. This is believed to be the case in mysids (e.g. Clutter, 1969) and might be expected in euphausiids, though I could find no evidence in the literature. According to Hardege *et al.* (1990) swarming (as a breeding-related behaviour) in marine invertebrates (they referred particularly to polychaetes) is controlled by daylength, tide, weather conditions and sex pheromones. In a similar way, adult male *Tribolium* beetles secrete an aggregating pheromone that is attractive to both sexes (Obeng-Ofori and Coaker, 1990). Little information is available for planktonic crustaceans but Katona (1973) and Griffiths and Frost (1976) provide some evidence for sex pheromones in copepods. Furthermore, Poulet and Ouellet (1982) demonstrated the importance of chemosensory detection of dissolved amino acids in stimulating aggregation and feeding in copepods. Though the latter authors referred to the resulting copepod behaviour as swarming, it was apparently an incidental aggregation (see Section 1) caused by group attraction to a chemical stimulus. It might be expected that any attractive dissolved organic molecule, either released during feeding activities or secreted by mature individuals of either sex, would be rapidly dispersed and diluted in water reducing the potential value of this signalling system to pelagic or planktonic animals. One would expect selection to favour either extreme sensitivity to very low concentrations (e.g. male gypsy moths) or the aggregative habit, either on a permanent (obligate) or as temporary reproductive groups (facultative). Aggregation of single sex groups has the advantage of reinforcement of signal providing that all individuals release pheromones simultaneously by entrainment. Thus, for example, swarming males would have more chance of attracting a female than any solitary male. This scenario seems to hold for the luminescent marine ostracods in which groups of swarming males produce synchronous bioluminescent displays presumed to be directed at sexually receptive benthic females (Morin, 1986). In this extraordinary system, there appears to be male to male competition both *within* and *between* groups. These displays of luminescent signalling show remarkable convergence with those of terrestrial fireflies. Whether individual swarms of other pelagic crustaceans compete for attention of the opposite sex in similar ways is unknown. The whole field of the importance of chemosense in sexual selection in relation to aggregative behaviour is in urgent need of further work.

Several authors have noted the relationship between aggregation and reproduction in cladocerans, for example Brandl and Fernando (1971), Ratzlaff (1974) and Young (1978). These authors imply that the chief stimulus for swarming in these animals is reproduction. However, Crease and Hebert (1983) suggest that the reverse may be true. In their view sexual reproduction and production of ephippia occur in response to high population densities. It is in such periods that swarming would be most apparent. They also searched but could find no evidence for sex pheromones in cladocerans. If true, this would be surprising and further work is warranted.

5.3. Reproductive Aggregations

Many aquatic animals form groups expressly for the purpose of breeding and remain solitary for most of their lives. These groups of individuals interact socially, but they are less than swarms because the group is not a cohesive whole in the sense that it could perform escape reactions if attacked, and individuals need not be evenly spaced. Classical examples of such groups are the "swarming" heteronereid stage of some polychaete worms. These aggregations are accurately cued by temperature often in conjunction with the lunar cycle (Goerke, 1984; Hardege et al., 1990). Zeeck et al. (1988) noted that sex pheromones were also a key controlling factor.

Sea hares of the genus *Aplysia* have been shown to aggregate for mating on a diel basis (Pennings, 1991). Benefits of such synchronized cycles are said to be the possibility of maximizing sexual encounters with potential mates; this is particularly enhanced because *Aplysia* spp. are simultaneous hermaphrodites. Mixed species aggregations formed suggesting a shared pheromone, but interbreeding was not observed.

Even jellyfish apparently actively "swarm" in order to improve the chances of successful fertilization (Larson, 1991). In the case of thimble jellyfish, the adults aggregate for their entire adult lives (4–5 months) becoming entrained in Langmuir circulation and swimming in tight little circles. Zavodnik (1987) reported mostly passive aggregation in another scyphozoan, but also described an active coupling termed "clustering" in which hundreds of medusae became physically interlaced whilst the bells pulsated at normal frequency. No explanation was offered for this phenomenon but some reproductive benefits are a possibility.

5.4. Costs of Reproduction in Groups

Pulliam and Caraco (1984) note that reproduction in groups could entail costs such as other group members killing and eating offspring, or the conspicuousness of groups attracting more predators to the young. These costs might provide an ultimate cause of the size and life stage sorting which seems to be nearly ubiquitous in aquatic aggregations (see Section 2). If costs of aggregating are unacceptably high, individuals might exercise an option to leave the group and defend a spawning territory. For example, social squid might facultatively alternate between schooling and seasonal competition for a mate and defence of a spawning site (Sauer et al., 1992). Costs of aggregating can be minimized by grouping births in time so that per capita predation on offspring is reduced by quickly satiating predators (Pulliam and Caraco, 1984). This seems to be the case for the highly synchronized larval release by nereid polychaetes (e.g. Hardege *et al.*, 1990) and may be true of individual swarms of Antarctic krill which Watkins *et al.* (1986, 1992) suggest are the basic unit of the population. Release of rapidly sinking eggs over deep water may be a means of separating them and the early larvae from the feeding adult population, as well as ensuring they do not sink onto the seabed (Hamner, 1984; Ritz, 1991).

6. PROTECTION FROM PREDATORS

6.1. Possible Benefits of Aggregation

One of the principal, if not the most important, functions of aggregation is thought to be that it confers protection from predators upon its members. Hobson (1978) compares the benefits of aggregating in water to those on land. On land, advance warning might assume greater import-ance since predators and prey can interact over considerable distances. In water, the interaction distance is often much shorter, and prey are commonly much smaller than their predators and consequently few can outrun them. Benefits of aggregation in the water then may be manifested chiefly after the prey has been seen (Godin, 1986), for example through the confusion effect. Note that such arguments assume that prey are located primarily through a visual sense.

Much of the theory on the adaptive value of shoaling in antipredation in aquatic habitats has been developed from work on fish. This has been reviewed relatively recently by Pitcher (1986a, b), Godin (1986) and Magurran (1990). I do not propose to include all of the arguments here but will consider the major theories in the context of what is known of invertebrate aggregations.

The range of possible antipredation benefits of aggregation include the following:

1. *Increased vigilance.* The fact that many individuals are alert means that possible danger can be detected earlier and concerted avoidance or evasion techniques can be initiated. This might be further improved as a consequence of inspection behaviour in which small groups break off from larger aggregations and closely approach the predator. This is believed to improve predator recognition and assessment of its attack readiness. It could also serve to show the predator that it had been discovered and might therefore divert it to less wary individuals (Magurran, 1990). Detection distance has been shown to increase with

increasing shoal size in fish (Godin *et al.*, 1988). There is little direct evidence that inspection behaviour occurs in invertebrate aggregations. It is possible that what O'Brien and Ritz (1988) described as "mobbing" behaviour by small groups of mysids approached by a model fish was, in fact, an example of inspection behaviour. Information transfer about the approaching predator is apparently transmitted by visual signals in fish (Magurran, 1990) and probably also in squid (Hochberg and Couch, 1971; Hurley, 1978). It is not known whether this is a feature of other invertebrate aggregations though maintenance of cohesion when under attack is equally as impressive as that in fish shoals.

2. *Attack inhibition*—may occur through sensory confusion of the predator or may be subserved by inspection behaviour of the prey. Solitary individuals or stragglers seem to provoke more attacks than groups (Milinski, 1977). Attacks on groups are often less successful, with capture rate decreasing as group size increases (Landeau and Terbogh, 1986). Many authors have noted that, providing groups stay intact in the field, predatory attacks are rarely seen (e.g. Modlin, 1990) or occur only around the margins (e.g. Hobson, 1968). On the other hand, Parrish *et al.* (1989) describe strikes by predators on a school of flat-iron herring as more common than those on stragglers, even though success rate was lower. Hobson (1968) also described tactics in which certain predatory fish deliberately lunged into the centre of prey schools apparently in order to break up the formation and facilitate capture of disorientated individuals. Just where a predator is prepared to risk striking at a school of prey seems to be the resultant of a complex and sometimes conflicting set of objectives in which hunger state and preservation of its own safety appear to play important roles.

Some of the earliest experimental evidence in support of the confusion effect induced in attacking predators by aggregated prey was provided by Radakov (1958) and Neill and Cullen (1974). The latter authors showed that hunting success of fish and cephalopod predators was increasingly impaired as shoal size of prey increased. Apparently the predator is unable to "lock on" to a single prey organism as it is continually distracted by other very similar individuals passing in and out of its visual range. The underlying mechanism is not yet clear but may involve sensory channel overload or cognitive confusion (Pitcher, 1986b).

Shoaling fish display behavioural tactics which tend to increase confusion in their predators (Pitcher, 1986b), and this is probably true also of invertebrate prey. Pitcher describes "skittering" behaviour, which consists of rapid startle accelerations, and flash expansion, as examples of a general increase in relative movement within the shoal after a predator has been detected or during an attack. These activities have their parallels within aggregations of mysids and euphausiids (O'Brien, 1987a; O'Brien and Ritz, 1988; Ritz, 1991).

One of the major advantages of the social habit is generally regarded as the possibility for *co-ordinated group avoidance and evasion tactics* when threatened. O'Brien and Ritz (1988) suggested a general classificatory scheme for such behaviour (Figure 12) which seems to apply widely within the social aquatic crustaceans. Key elements of this scheme are the three major levels of response. At the primary level, when the predator is first detected, the group polarizes, usually condenses and begins to swim in the opposite direction at increased speed. When attack cannot be avoided, the group resorts to secondary responses which are characterized by co-ordinated evasion tactics such as flash expansion and the fountain effect (Figure 13). Cohesion and synchrony of movement are maintained and collisions are rare. Aggregations sometimes split into smaller units during these escape manoeuvres but may reform after the threat has passed (O'Brien, 1987a, 1988b; O'Brien and Ritz, 1988; Ritz, 1991). At the point of attack, small subsets of larger aggregations of mysids and euphausiids may display a co-ordinated tail-flipping response. This response is short-lived and individuals quickly integrate their movements with the rest of the group. Co-ordinated tail-flipping may be the ecological analogue of "group jump"; that is, synchronized "skittering" behaviour in fish shoals (Pitcher, 1986b) which is believed to heighten the confusion in the predator.

Once a predator has been detected and a flight response is initiated among those individuals closest to the point of attack, the response is transmitted through the aggregation more rapidly than the approach speed of the predator. This phenomenon has been called the "Trafalgar effect" by Treherne and Foster (1981) who described it in the water skater *Halobates*. This effect enables individuals to initiate avoidance behaviour even before they are aware of the presence of the predator.

The integrity of the aggregation is only sacrificed as a last resort, in the tertiary level of response, highlighting its importance in protecting its members. In mysids and euphausiids the tertiary response consists of a series of tail flips by individuals in which the path of each is random with respect to its fellows. Thus cohesion of the group breaks down. Since the path of a particular individual is unpredictable to a predator, this behaviour is protean in nature (Humphries and Driver, 1967, 1970). It is a tactic used by such diverse animals as copepods, shore hoppers, cladocerans and squid as well as a range of terrestrial prey animals. Perfecting such a tactic could become critical since predators apparently rapidly learn how to exploit aggregated prey (Milinski, 1979). According

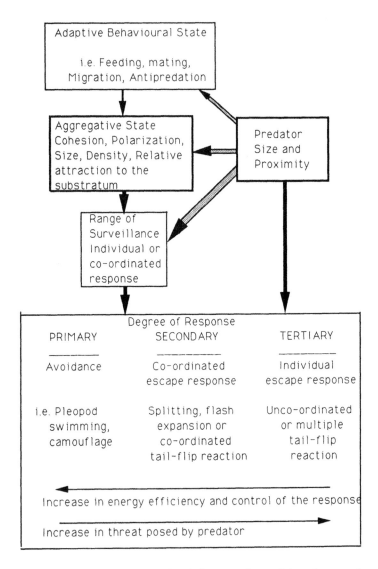

Figure 12 Scheme describing major influences determining degree of escape response in aggregated euphausiids, hyperbenthic mysids and possibly all gregarious crustaceans. Speckled arrows, feedback path; thickness of arrow, relative importance (subjective). (After O'Brien and Ritz, 1988.)

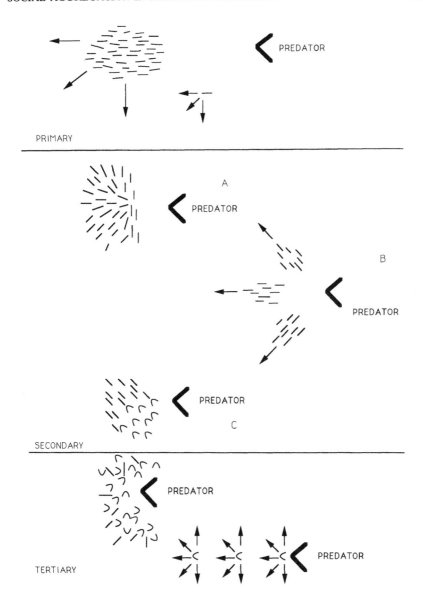

Figure 13 Diagrammatic representation of the three levels of escape response in euphausiids to a generalized predator, in this case a SCUBA diver. Predator–prey distances are not to scale but are relative between levels of escape response, assuming a constant size of predator. A, flash expansion; B, splitting; C, co-ordinated tail-flip. For krill, I = pleopod swimming; U = tail-flip. Arrows show general direction of movement of euphausiids; arrowheads show direction of movement of predators. (After O'Brien, 1987a.)

to O'Brien (1987a), which of three levels of escape response is manifested in an attack is a function of aggregative state of the prey, and size and proximity of the predator (see Figure 12).

6.2. Conflict between Antipredation and Other Activities

The question of animal decision-making where the objectives are conflicting has been referred to previously (Section 4). Most of the experimental work has addressed the ways in which problems of foraging/feeding are resolved while remaining alert against predatory attack. Many authors conclude that one must be traded off against the other. For example, Eggers (1976) developed a foraging model which predicts that schooling occurs at the expense of prey consumption because of overlap of perceptive fields of adjacent fish. Cost of schooling, therefore, increases as NNDs decrease or as school size increases. Blaxter and Hunter (1982) proposed that in fish, a school size which is optimal for foraging may be quite different to one which is optimal for protection. This may be why northern anchovy shoals continually break up into smaller feeding schools. The same appears to be true of mysid aggregations, although in some circumstances unthreatened individuals in larger swarms actually captured more food than those in smaller ones (Ritz, 1994). Folt (1987) suggested that aggregation in zooplankton may only be advantageous in predator-prone environments because of costs suffered through competitive feeding. Such costs can, of course, be minimized by regularly adjusting group size to get the "best of both worlds", as appears to happen in some social animals. Pulliam and Caraco (1984) had predicted that size of groups would be expected to increase as frequency of encounters with predators increases. Daly and Macaulay (1991) suggested that response to seasonal variation in food abundance and predation pressure could account for differences in level of aggregation and shoal size in Antarctic krill. Moreover, Pitcher *et al.* (1983) showed experimentally that median elective group size of minnows increased when they encountered a predator and declined only slowly after the threat had passed. Pitcher (1986b) sought to combine the ideas of attack dilution (apparent reduction of individual risk of predation by being in a group) along with predator search and encounter of different sized prey groups. His "attack abatement" model predicts that an individual fish will benefit from a reduction in risk by joining a group though this will not necessarily maximize benefit for all group members.

There may also be trade-offs between safety and other activities. For example, there is evidence that guppies show risk-sensitive courtship behaviour (Magurran and Seghers, 1990). It is not known whether such trade-offs are a feature of invertebrate aggregations but they might be expected, for example, in squid. As Magurran (1990) states, "each behaviour, such as schooling as a means of predator defence, cannot operate independently of other activities".

6.3. Aggregation as an Inappropriate Strategy

There are instances when it appears that aggregation is not an appropriate or successful defence strategy. One example of this is where predators themselves have formed foraging aggregations. Major (1978) found that grouped predators (jacks) were much more successful in capturing schooled prey (anchovy) than solitary individuals. He suggests that schooling in predators co-evolved with that of their prey as a means of breaking up large aggregations, thus isolating prey and facilitating their capture. That predators should need to disrupt prey groups in order to make successful captures serves to underline the protective nature of the group. "Mass" predators, such as baleen whales, seem to benefit positively from aggregation of their prey, and some have evolved capture tactics that encourage even tighter grouping before prey are engulfed. Humpback whales, for example, use elaborate "bubble pumping" and "herding" techniques to capture krill (Hamner, 1984). If certain predators appear to be ahead in the evolutionary "arms race" (Dawkins and Krebs, 1979), it must be remembered that perfecting escape tactics against the most common predator(s) might mean less than maximal success against less common ones (Hamilton, 1971). Notwithstanding this, aggregation by krill is unquestionably a successful strategy against the great majority of its predators.

Most of the antipredation strategies described above seem to be aimed at predators hunting visually. Predators using other senses to detect their prey may not, for example, suffer a confusion effect. Magurran (1990) describes predation on guppies by a freshwater prawn in Trinidadian streams. Schooling by the guppies as a defence against fish predation may actually be disadvantageous against prawns because the latter use olfactory and tactile cues to locate prey. This aspect of the defensive value of aggregative behaviour deserves further study.

Some squid forage in social groups but typically, when feeding on an aggregation of krill, they attacked alone with no co-ordination between individuals (Nicol and O'Dor, 1985). Squid were apparently able to select individual prey in the body of the swarm rather than peripheral prey suggesting they were not so prone to confusion. However, the results of Neill and Cullen (1974) indicated that squid, and (less obviously) cuttlefish, were equally as prone to confusion by their aggregated prey as fish predators. Attacking the densest region of a swarm (usually the centre) rather than the periphery may be a function of hunger state, as several authors have shown (e.g. Heller and Milinski, 1979; Morgan and Ritz, 1984).

The predatory cladoceran *Polyphemus* forms aggregations and behaviour was shown to be greatly influenced by group size. Large groups swam more slowly and covered less ground than small ones. Interestingly, swimming tracks were oriented orthogonal to those of their swarming prey *Bosmina* (also a cladoceran). Thus encounter rate, and presumably also capture rate, was maximized (Young and Taylor, 1990).

6.4. Bioluminescence

The function of bioluminescence in aquatic animals has long been debated and is still far from clear. Buskey and Swift (1985) compared the responses of luminescent and non-luminescent copepods to simulated bioluminescent flashes. The former were much more responsive and the authors concluded that bioluminescence may be a recognizable warning signal to conspecifics as well as a possible defence against predators. The former interpretation seems to be group selectionist and untenable unless it had evolved as a form of individual defence but proved useful for maintaining group cohesion at night. Oddly, Buskey and Swift found no behavioural response to luminescence in the social euphausiid *Meganyctiphanes*. There are records of predation on luminescing schools of krill (*E. superba*) at night (Tomo, 1983); O'Brien (1987a) suggests that, if bioluminescence is used as an antipredator strategy, it may be used to augment confusion in combination with other evasion tactics. Cypridinid ostracods apparently use bioluminescence as a highly successful predator deterrent (Morin, 1983, 1986). When attacked by planktivorous fishes or crustaceans, the ostracods produce a brilliant cloud of luminescence that probably startles and temporarily blinds the predator. Those that are engulfed during these attacks have been seen to be regurgitated unharmed. Morin speculates that the luminescent cloud might also attract a second order predator which serves to increase the risk for the initial predator. Attacks that trigger luminescent displays by these ostracods in the field are apparently rare suggesting that predators rapidly learn to avoid these animals. It is frequently suggested that bioluminescent signalling by krill might be maladaptive in that it could serve to attract predators, but this is apparently not the case in the luminescent ostracods which produce complex and vivid sexual signals lasting about an hour near the end of twilight on Caribbean reefs (Morin, 1986) (see Section 5).

Synchronous group moulting as a predator decoy has only been observed in *E. superba* (Hamner *et al.*, 1983). It does not involve the whole aggregation, only a subset closest to the point of attack. Presumably the process is under neurological control but further work on this unique behaviour is clearly warranted.

7. SENSES EMPLOYED

7.1. Vision

There is little doubt that vision plays an important, if not the most important, role in maintaining group cohesion in aquatic invertebrates just as it does in fish schools. Many observations lend support to this but there are few direct experimental tests. Clutter (1969) showed that visual cues between social mysids were transmitted across a transparent barrier. However, the cohesion between individuals thus separated was less precise than among those which had chemical and tactile contact in addition to visual. Because aggregations persisted at night, Clutter concluded that vision was the primary but not exclusive mode of establishing contact between individual mysids. Steven (1961) also showed that individual mysids (*Mysidium columbiae*), when separated from a shoal, would try to rejoin although a sheet of glass was placed between, provided they were within a critical visual distance. Squid (*Loligo opalescens*) responded to a transparent barrier in a similar fashion, apparently schooling normally (Hurley, 1978).

Hamner and Carleton (1979) reported that swarms of copepods responded rapidly to moving objects at a distance of about 20 cm. They state that swarm integrity is maintained by a combination of visually mediated behaviour, orientation to features of the habitat and mechanical winnowing by currents. In addition, apparently visually mediated reactions of swarms to divers at a distance of several metres were observed.

Strand and Hamner (1990) found that individual krill (*E. superba*) with damaged eyes, or those whose eyes had been coated with black paint, were unable to maintain position in a school. They also reported that schooling krill avoided visually contrasting stimuli and responded to abrupt changes in light intensity. Ragulin (1969) had reported that *E. superba* reacted to a diver at a distance of 1.5–2 m suggesting acute visual perception of movement. Differences in plankton net catches of krill during the night compared with daytime also suggest that animals are strongly reliant on visual cues (Everson and Bone, 1986). Schooling in the cladoceran *Moina* sp., which has only a simple compound eye, led Johnson and Chia (1972) to suggest that maintenance of cohesion could be based on contrast orientation. A mechanism had earlier been proposed by Ringelberg (1969). O'Brien (1988a) also noted the strong response to light intensity change in mysids, and speculated that NND may be visually mediated since it increased with increasing light (O'Brien, 1989). NND was also positively correlated with light intensity in tadpoles of *Xenopus laevis* (Katz *et al.*, 1981). In contrast, in field observations, Modlin (1990) found that inter-individual distances in aggregations of *Mysidium columbiae* increased and decreased as light intensity decreased and increased (i.e. negative correlation).

According to Partridge and Pitcher (1980), blinded fish can continue to maintain position in a school, but NND increases. Apparently schooling in fish depends largely on the interaction of vision, which provides the attractive stimulus, and the lateral line system, which provides the repulsive stimulus. Whereas vision is primarily responsible for maintenance of position and angle between fish, the lateral line system is responsible for monitoring swimming speed and direction of travel of neighbours. When the two systems are in conflict, for example when the school is turning, vision has priority (Partridge, 1982). If both vision and lateral line senses are lost a fish is unable to maintain schooling cohesion (Burgess and Shaw, 1979; Partridge and Pitcher, 1980). Katz *et al.* (1981) concluded that tadpoles of *Xenopus laevis* also use both visual and lateral line input in order to school.

Galle *et al.* (1991) suggest a novel visually mediated mechanism to account for swarming in planktonic organisms. They describe a spontaneous, ultra-weak photoemission (biophoton emission) which is produced by individuals and may be detected by neighbours. It is too weak to be classified as bioluminescence. Intensity of light from *Daphnia magna* is in the range of a few to hundreds of photons $s^{-1} cm^{-2}$ of surface and is suggested to be a factor in self-regulation of swarm density.

7.2. Rheotaxis and Response to Water Movement

Invertebrates have no lateral line system but do have an array of sensory receptors performing similar functions. Hamner *et al.* (1983) believed that schools of *Euphausia superba* were maintained by rheotactic cues with individuals orienting into currents produced by neighbours swimming above and in front. Certainly the evidence of mysids and euphausiids maintaining cohesion in light levels presumed to be too low for vision suggests that at least one other sensory mode is adequate for this purpose. However, I could find no experimental test of this hypothesis. There are many recorded instances of animals orienting into currents, even rhythmically reversing direction with incoming and outgoing waves (e.g. O'Brien, 1988a). Many authors have suggested that hydrodynamic features themselves, or a common response by the organisms to such features, are sufficient to explain aggregations of zooplankton (e.g. Rice and Kristensen, 1982; McClatchie *et al.*, 1989). The currents may serve to modify or organize aggregations but the primary stimulus in most cases remains a biological attraction between conspecifics. Zeldis and Jillett (1982) described aggregations of juvenile *Munida gregaria* which are ordered by hydrodynamic processes of convergence at coastal fronts and internal wave sites. But the aggregations persisted even when the influence of these ordering processes was lost.

7.3. Chemosensitivity

Chemical agents play an important role in swarming activity during the nuptial dance of heteronereid stages of some polychaete annelids which are promoted by the release of sex pheromones (Hardege *et al.*, 1991). There is no doubt that chemosense is also well developed in planktonic crustaceans and plays an important role in feeding, social and sexual behaviour (Poulet and Ouellet, 1982). These authors also describe a chemosensory-mediated locomotory and "swarming" response to dissolved amino acids. What they describe as swarming appears to be an aggregative response to a feeding stimulus; that is, the aggregation is incidental in otherwise non-social copepods. Interestingly, though, aggregations once formed in this way remained clumped and "sensory bonded" for >24 h even after the initial stimulus had presumably faded. They believe that such "sensory bonding" could explain longer-term persistence of some zooplankton patches in the sea, though they did not consider the other benefits accruing from aggregation. They noted also that some categories of amino acids are effective olfactory stimuli for both zooplankton and (potentially predatory) fish species (see also Hamner and Hamner, 1977). This could have been a potent selection pressure favouring aggregations which remained cohesive in order to benefit from protection (see Section 6). The fact that aggregations which have fed tend to become tighter (e.g. Jakobsen and Johnsen, 1988; Ritz, 1994) may be a reinforcement of the chemically mediated response.

Strand and Hamner (1990) tested the effect on schooling *E. superba* of addition of a range of chemical substances to the water. Introduction of food substances caused a reduction in swimming speed and often a cessation of schooling formation. Addition of nitrogenous compounds, for example ammonium chloride, caused the school to disperse. This might reflect the build up of excretory products in the denser region of the school. Moss and McFarland (1970) had shown that a slow reduction of dissolved oxygen or reduction in pH did not affect nearest neighbour orientation or swimming speed of anchovy schools, whereas rapid reductions by even very small amounts did. Although extrapolation of these results to the natural situation must be done with caution, Strand and Hamner suggest that continual readjustment of position of individuals within a school is necessary to permit regular contact with clean, oxygenated and food-rich water. Given the swimming capability of *E. superba*, and the active current regimes typical of its habitat, it is possible that conditions within aggregations rarely deteriorate sufficiently to limit swarm size in the natural environment (Johnson *et al.*, 1984).

Steele *et al.* (1991) suggest the possibility that schooling in fish actually facilitates foraging by individuals by enhancing the group's responsiveness to food odours. They investigated the responsiveness of zebrafish in different group sizes to L-alanine. Surprisingly there was a bimodal relationship with groups of four being most responsive. Presumably increasing the power of the analysis, for example with more replicates, would have resolved this apparent anomaly. However, the principle remains and could be explained simply on the basis that the group is as sensitive to a stimulus as its most sensitive member, given that information is readily transferred by social facilitation (Ryer and Olla, 1992).

Wittmann (1978) reported that female mysids would capture and introduce liberated embryos of other females in their brood pouches. He termed this behaviour adoption. Moreover, parts of embryos, for example single eyes or pairs of eyes with eyestalks, were treated in the same manner suggesting that chemosensory identification was involved. Since mysids will readily capture and eat small crustaceans such behaviour has presumably evolved to protect the young against cannibalism.

7.4. Tactile

In most cases individuals in aquatic aggregations do not make physical contact with their neighbours. But there are occasions where physical contact occurs, apparently by accident, and other instances in which it appears to be important for maintaining the cohesion of the group. Individuals in fish schools rarely make contact even when performing complex escape manoeuvres such as flash expansion or fountain effect (Partridge, 1982). Nevertheless, Breder (1959) described groups he called "pods" in which fish were touching. McKoy and Leachman (1982) reported that *Jasus edwardsii* formed tight groups of up to 200 individuals in close contact. Wasserug (1973) suggested two categories of tadpole group behaviour in one of which, the *Bufo* mode, tadpoles form dense mats in which individuals are in physical contact. This behaviour seems similar to "matting" behaviour of the euphausiid *Nyctiphanes australis* in which individuals lie in dense mats, dorsal side down on the substratum in shallow water (O'Brien *et al.*, 1986a). Caldwell (1989) also found groups of tadpoles of *Hyla geographica* which were always in physical contact with one another. Physical contact seems to be a feature of intertidal aggregations of hermit crabs (Gherardi and Vannini, 1989).

Clutter (1969) believed that maintenance of mysid aggregations, even parallel orientation, at night was achieved by body contact as well as perception of currents created by neighbours. Probably the latter would be sufficient since body contact does not seem to be a normal feature of mysid aggregations. O'Brien (pers. commun.) noted that collisions occurred during the unco-ordinated tail-flipping escape response of euphausiids. Additionally Hamner *et al.* (1983) reported that individuals of *E. superba* collided while tail flipping as two schools attempted to merge.

Evidently body contact is actively avoided in some aggregations but may be important for cohesion in others. It is not simply a question of sophistication of neurological equipment since body contact seems relatively unimportant in mysids and euphausiids but integral in some tadpole aggregations.

8. CONCLUSIONS

In this review I have tried to summarize the information regarding social aquatic invertebrates by reference to developing theoretical behavioural ecology and also by comparison with what is known of social aquatic vertebrates, chiefly fish. The ecological and behavioural analysis of invertebrate aggregations has lagged some way behind that of fish mainly, it appears, because being for the most part small, sophisticated behaviour was "unexpected" and perhaps overlooked. This is partly a legacy of the tag "plankton". Where appropriate, I have also drawn on examples from terrestrial social groups, for example insects. There are plenty of parallels apparent, but also there are many unique traits to be found amongst aquatic invertebrate social systems.

In the search for function in aggregative behaviour it must be remembered that present behaviour may not necessarily yield a present advantage (Vine, 1971). This possibility is incorporated in one of the three categories of hypotheses advanced to explain social behaviour—that is, phylogenetic (Slobodchikoff and Shields, 1988 and see Section 1). The other two categories are genetic and ecological hypotheses. Genetic hypotheses concerning the relatedness of group members cannot be evaluated for aquatic invertebrates without more data. However, Slobod-chikoff and Shields caution against regarding genetic and ecological arguments about sociality as alternative hypotheses. They may have operated independently or jointly during the evolution of a social system. For example, kinship may have influenced the kinds of behaviour displayed by animals once they had begun to live in groups that had formed because of favourable ecological circumstances.

In view of the weight of evidence, it seems reasonable to accord predators, food and reproduction the highest priority in assigning ecological function to the social habit. I do not imply a hierarchy of importance, nor that the same factors will necessarily apply with equal weighting in all social animals. Plainly from the discussions above, the potencies of selection pressures vary during the ontogeny of an animal and also seasonally during its life. Thus flexibility in social response may have been a crucial issue in its evolution. As Magurran (1990) states in the case of social fish, ". . . individual fish are skilled at constantly reassessing their environment and adjusting their behaviour accordingly".

In proposing ecological hypotheses about the functions of sociality, which applies to the great majority of papers referred to above, it is generally assumed that groups are maintained in order to exploit resources more efficiently than solitary animals (Slobodchikoff and Shields, 1988). Resources are taken to include a variety of factors that affect fitness, namely food, space, mates, spawning sites, other animals (including therefore group defence). Magurran (1990) and others believe that schooling in fish evolved as a defence against predation. There is no doubting the power of this argument: an individual must, after all, be alive to carry out any other activity! Other necessary activities can be subordinated to safety and some of the most revealing studies have been those dealing with ecological trade-offs (see e.g. Dill, 1987; Slobodchikoff and Shields, 1988). The most often studied are feeding and antipredation, but similar trade-offs between courtship, migration, timing of spawning and safety are to be expected. Investigation of behavioural trade-offs in invertebrates are uncommon (e.g. Sih, 1980), and those in social aquatic species even more so (Jakobsen and Johnsen, 1988; Ritz, 1993). The risk-sensitive nature of these trade-offs is shown by the following example. If, say, possibility of starvation is high, social animals might be inclined to disperse in order to feed maximally, whereas such risky behaviour would not be countenanced in the face of strong predation pressure selecting for tight grouping.

Although invertebrates do not appear to use active co-operative feeding (besides finding food sources faster than solitary animals), none the less the presence of other conspecifics may facilitate food capture. Even the behaviourally highly advanced squids, although hunting in schools, typically attack their prey alone once located. Nicol and O'Dor (1985) watched squid attack a swarm of krill and reported that movements of the predators were not co-ordinated spatially or temporally with each other. However, it may be that multiple attacks are more effective in breaking up the prey swarm than single ones, thus making subsequent capture of separated and disorientated prey more successful. Many social crustaceans feed by filtering small particles from the water. It has often been suggested that aggregations could filter a parcel of water much more efficiently than an individual. For example, swarms of krill might act like a physical filter grazing down phytoplankton to much lower levels than solitary animals could (Antezana et al., 1982). It is possible that suspended particles are detained longer when passing through an aggregation, making eventual capture more likely. It is even possible that the more turbulent flow through non-aligned animals in swarms favours particle retention and capture compared with the more laminar flow through polarized schools. However, these ideas have not been experimentally tested. The patchy nature of food resources in water has probably had a strong influence on the evolution and maintenance of sociality. Such conditions need to be reproduced in experimental studies in order to interpret fully the functional significance of social behaviour in foraging and feeding.

Social behaviour can potentially influence reproductive success in aquatic invertebrates in many ways (see Section 5) but experimental proof of the effectiveness of this strategy is often lacking. Animals may aggregate for reproductive purposes on a diel or seasonal basis or they may reside permanently in groups. The common occurrence of single-sex aggregations probably reflects the fact that the cost/benefit equation is often different for the two sexes. Such groups of a single sex might more effectively advertise their presence with pheromones as appears to be the case in flying dipteran swarms. Similarly the synchronously luminescent displays of groups of male marine ostracods seems to be a reinforcement of the sexual signalling to solitary females (Morin, 1986). Although the dictum "animals behave so as to maximize their reproductive success" guides much behavioural ecology (Krebs and Davies, 1991), it is rare that reproductive success is measured directly especially in experiments on social aquatic invertebrates. Since many aquatic invertebrates have short life cycles it should be possible to gain valuable information within a fairly short time span.

New insights into the behaviour of social aquatic invertebrates will probably come from studies of behavioural choices when objectives are conflicting—that is, trade-offs (Dill, 1987; Slobodchikoff and Shields, 1988). This approach has been productive in analysis of the benefits of fish schooling but has only just begun for aquatic invertebrate social systems. Now that techniques exist for the accurate description of the internal organization of swarms and schools, it is possible to test real hypotheses about preferred positions of individuals under different conditions. Modelling studies should begin to assume greater biological reality. What have been notably lacking in early models have been *a priori* predictions about structure (Dill *et al.*, 1993).

ACKNOWLEDGEMENTS

I express my sincere thanks to Professors L.M. Dill and L.-A. Giraldeau for allowing me to see their prepublication manuscripts. Also I am very grateful to Dr S. Nicol, Australian Antarctic Division, for allowing me to use his photographs.

REFERENCES

Alldredge, A.L., Robinson, B.H., Fleminger, A., Torres, J.J., King, J.M. and Hamner, W.N. (1984). Direct sampling and *in situ* observations of a persistent copepod aggregation in the mesopelagic zone of the Santa Barbara Basin. *Marine Biology* **80**, 75–81.

Ambler, J.W., Ferrari, D.D. and Fornshell, J.A. (1991). Population structure and swarm formation of the cyclopoid copepod *Dioithona oculata* near mangrove cays. *Journal of Plankton Research* **13**, 1257–1272.

Antezana, T. and Ray, K. (1983). Aggregation of *Euphausia superba* as an adaptive group strategy to the Antarctic ecosystem. *Berichte zur Polarforschung* **4**, 199–215.

Antezana, T. and Ray, K. (1984). Active feeding of *Euphausia superba* in a swarm north of Elephant Island. *Journal of Crustacean Biology* **4** (Spec. No. 1), 142–155.

Antezana, T., Ray, K. and Melo, C. (1982). Trophic behaviour of *Euphausia superba* Dana in laboratory conditions. *Polar Biology* **1**, 77–82.

Bergstrom, B.I., Hempel, G.G., Marschall, H.-P., North, A., Siegel, V. and Stromberg, J.-O. (1990). Spring distribution, size composition and behaviour of krill *Euphausia superba* in the western Weddell Sea. *Polar Record* **26**, 85–89.

Blaxter, J.H.S. and Hunter, J.R. (1982). The biology of clupeid fishes. *Advances in Marine Biology* **20**, 1–223.

Brandl, Z. and Fernando, C.H. (1971). Microaggregation of the cladoceran *Ceriodaphnia affinis* Lilljeborg with a possible reason for microaggregation of zooplankton. *Canadian Journal of Zoology* **49**, 775.

Breder, C.M. Jr (1959). Studies on social groupings in fish. *Bulletin of the American Museum of Natural History* **117**, 397–481.

Breder, C.M. Jr (1967). On the survival value of fish schools. *Zoologica* **52**, 25–40.

Brown, R.G.B., Barker, S.P. and Gaskin, D.E. (1979). Daytime surface swarming by *Meganyctiphanes norvegica* (M. Sars) (Crustacea Euphausiacea) off Brier Island, Bay of Fundy. *Canadian Journal of Zoology* **57**, 2285–2291.

Burgess, J.W. and Shaw, E. (1979). Development and ecology of fish schooling. *Oceanus* **22**, 11–17.

Buskey, E.J. and Swift, E. (1985). Behavioural responses of oceanic zooplankton to simulated bioluminescence. *Biological Bulletin, Woods Hole* **168**, 263–275.

Byron, E.R., Whitman, P.T. and Goldman, C.R. (1983). Observations of copepod swarms in Lake Tahoe. *Limnology and Oceanography* **28**, 378–382.

Caldwell, J.P. (1989). Structure and behaviour of *Hyla geographica* tadpole schools, with comments on classifications of group behaviour in tadpoles. *Copeia* **4**, 938–950.

Clutter, R.I. (1969). The microdistribution and social behaviour of some pelagic mysid shrimps. *Journal of Experimental Marine Biology and Ecology* **3**, 125–155.

Cooter, R.J. (1989). Swarm flight behavior in flies and locusts. In "Insect Flight" (G.J. Goldsworthy and C.H. Wheeler, eds), pp. 165–203. CRC Press, Boca Raton, Florida.

Cram, D.L. and Malan, O.G. (1977). On the possibility of surveying krill (*Euphausia superba* Dana) in the Southern Ocean by remote sensing. *South Africa Journal of Antarctic Research* **7**, 14–19.

Cram, D.L., Agenbag, J.J., Hampton, I. and Robertson, A.A. (1979). SAS Protea Cruise, 1978; The general results of the acoustics and remote sensing study, with recommendations for estimating the abundance of krill (*Euphausia superba* Dana). *South African Journal of Antarctic Research* **9**, 3–14.

Crease, T.J. and Herbert, P.D.N. (1983). A test for the production of sexual pheromones by *Daphnia magna* (Crustacea: Cladocera). *Freshwater Biology* **13**, 491–496.

Dadswell, M.J. (1975). Some notes on shoaling behaviour and growth of *Mysis gaspensis* (Mysidacea) in a small Newfoundland estuary. *Canadian Journal of Zoology* **53**, 374–377.

Daly, K.L. and Macaulay, M.C. (1991). Influence of physical and biological mesoscale dynamics on the seasonal distribution and behaviour of *Euphausia superba* in the antarctic marginal ice zone. *Marine Ecology Progress Series* **79**, 37–66.

Dawkins, R. and Krebs, J.R. (1979). Arms races between and within species. *Proceedings of the Royal Society of London* B **205**, 489–511.

Dill, L.M. (1987). Animal decision making and its ecological consequences: the future of aquatic ecology and behaviour. *Canadian Journal of Zoology* **65**, 803–811.

Dill, L.M., Holling, C.S. and Palmer, L.H. (1994). Predicting the 3-dimensional structure of animal aggregations from functional considerations: the role of information. In "Three Dimensional Animal Aggregations" (J.K. Parrish, W.M. Hamner and C.T. Prewitt, eds). Cambridge University Press, Cambridge, in press.

Downes, J.A. (1969). The swarming and mating flight of Diptera. *Annual Review of Entomology* **14**, 271–293.

Eggers, D.M. (1976). Theoretical effect of schooling by planktivorous fish predators on rate of prey consumption. *Journal of the Fisheries Research Board of Canada* **33**, 1964–1971.

Emery, A.R. (1968). Preliminary observations on coral reef plankton. *Limnology and Oceanography* **13**, 293–303.

Endo, Y. (1984). Daytime surface swarming of *Euphausia pacifica* (Crustacea: Euphausiacea) in the Sanriku coastal waters off northeastern Japan. *Marine Biology* **79**, 269–276.

Endo, Y., Imaseki, T. and Komaki, Y. (1986). Biomass and population structure of Antarctic krill (*Euphausia superba* Dana) collected during SIBEX II cruise of R.V. Kaiyo Maru. *Memoirs of the National Institute of Polar Research, Tokyo* (Spec. issue) **44**, 107–117.

Everson. I. (1982). Diurnal variation in mean volume backscattering strength of an Antarctic krill (*Euphausia superba*) patch. *Journal of Plankton Research* **4**, 155–161.

Everson, I. and Bone, D.G. (1986). Effectiveness of the RMT8 system for sampling krill (*Euphausia superba*) swarms. *Polar Biology* **6**, 83–90.

Fenton, G.E. (1992). Population dynamics of *Tenagomysis tasmaniae* Fenton, *Anisomysis mixta-australis* (Zimmer) and *Paramesopodopsis rufa* Fenton from south-eastern Tasmania (Crustacea: Mysidacea). *Hydrobiologia* **246**, 173–193.

Ferguson, M.M. and Noakes, D.L.G. (1981). Social grouping and genetic variation in common shiners, *Notropis cornutus* (Pisces, Cyprinidae). *Environmental Biology of Fishes* **6**, 357–360.

Folt, C.L. (1987). An experimental analysis of costs and benefits of zooplankton aggregation. *In* "Predation: Direct and Indirect Impacts on Aquatic Communities" (W.C. Kerfoot and A. Sih, eds), pp. 300–314. University Press of New England, Hanover.

Forsythe, D.C.T. and Jones, L.T. (1966). Swarming of *Thysanoessa longicaudata* (Kroyer) (Crustacea, Euphausiacea) in the Shetland Islands. *Nature, London* **212**, 1467–1468.

Freeman, M.C. and Grossmann, G.D. (1992). Group foraging by a stream minnow: shoals or aggregations? *Animal Behavior* **44**, 393–403.

Fretwell, S.D. (1972). "Populations in a Seasonal Environment." Princeton University Press, Princeton, New Jersey.

Galle, M., Neurohr, R., Altmann, G., Popp. F.A. and Nagl, W. (1991). Biophoton emission from *Daphnia magna*: a possible factor in the self-regulation of swarming. *Experientia* **47**, 457–460.

Gendron, D. (1992). Population structure of daytime surface swarms of *Nyctiphanes simplex* (Crustacea: Euphausiacea) in the Gulf of California, Mexico. *Marine Ecology Progress Series* **87**, 1–6.

Gherardi, F. and Vannini, M. (1989). Field observations on activity and clustering in two intertidal hermit crabs. *Clibanarius virescens* and *Calcinus laevimanus* (Decapoda, Anomura). *Marine Behaviour and Physiology* **14**, 145–159.

Giraldeau, L.-A. (1988). The stable group and the determinants of foraging group size. *In* "The Ecology of Social Behaviour" (C.N. Slobodchikoff, ed.), pp. 33–53. Academic Press, San Diego.

Giraldeau, L.-A. and Caraco, T. (1993). Genetic relatedness and group size in an aggregation economy. *Evolutionary Ecology*, **7**, 429–438.

Gliwicz, M.Z. (1986). Predation and the evolution of vertical migration in zooplankton. *Nature, London* **320**, 746–748.

Godin, J.-G. J. (1986). Antipredator function of shoaling in teleost fishes: a selective review. *Le Naturaliste Canadien* **113**, 241–251.

Godin, J.-G. J., Classon, L.J. and Abrahams, M.V. (1988). Group vigilance and shoal size in a small characin fish. *Behaviour* **104**, 29–40.

Goerke, H. (1984). Temperature-dependence of swarming in North Sea Nereidae. *Fortschritte der Zoologie* Band **29**, 39–43.

Gotmark, F., Winkler, D.W. and Andersson, M. (1986). Flock feeding on fish schools increases individual success in gulls. *Nature, London* **319**, 589–591.

Greene, C.H., Widder, E.A., Youngbluth, M.J., Tamse, A. and Johnson, G.E. (1992). The migration behavior, fine structure, and bioluminescent activity of krill sound-scattering layers. *Limnology and Oceanography* **37**, 650–658.

Griffiths, A.M. and Frost, B.W. (1976). Chemical communication in the marine planktonic copepods *Calanus pacificus and Pseudocalanus* sp. *Crustaceana* **30**, 1–8.

Hahn, P. and Itzkovitz, M. (1986). Site preference and homing behaviour in the mysid shrimp *Mysidium gracile* (Dana). *Crustaceana* **51**, 215–219.

Hamilton, W.D. (1971). Geometry for the selfish herd. *Journal of Theoretical Biology* **31**, 295–311.

Hamner, P.P. and Hamner, W.M. (1977). Chemosensory tracking of scent trails by the planktonic shrimp *Acetes sibogae australis*. *Science* **195**, 886–888.

Hamner, W.M. (1984). Aspects of schooling in *Euphausia superba*. *Journal of Crustacean Biology* **4** (Spec. No. 1), 67–74.

Hamner, W.M. (1985). The importance of ethology for investigations of marine plankton. *Bulletin of Marine Science* **37**, 414–424.

Hamner, W.M. (1988). Behaviour of plankton and patch formation in pelagic ecosystems. *Bulletin of Marine Science* **43**, 752–757.

Hamner, W.M. and Carleton, J.H. (1979). Copepod swarms: attributes and role in coral reef ecosystems. *Limnology and Oceanography* **24**, 1–14.

Hamner, W.M., Hamner, P.P., Strand, S.W. and Gilmer, R.W. (1983). Behaviour of Antarctic krill *Euphausia superba*: chemoreception, feeding, schooling and molting. *Science* **220**, 433–435.

Hamner, W.M., Hamner, P.P., Obst, B.S. and Carleton, J.H. (1989). Field observations of the ontogeny of schooling of *Euphausia superba* furciliae and its relationship to ice in Antarctic waters. *Limnology and Oceanography* **34**, 451–456.

Hampton, I. (1981). Suggested methods for observation and measurement of visible swarms of Antarctic krill. *Fishery Bulletin of South Africa* **15**, 99–108.

Hanamura, Y., Endo, Y. and Taniguchi, A. (1984). Underwater observations on the surface swarm of a euphausiid, *Euphausia pacifica* in Sendai Bay, northeastern Japan. *La Mer* **22**, 63–68.

Hanamura, Y., Kotori, M. and Hamaoka, S. (1989). Daytime surface swarms of the euphausiid *Thysanoessa inermis* off the west coast of Hokkaido, northern Japan. *Marine Biology* **102**, 369–376.

Hardege, J.D., Bartels-Hardege, H.D., Zeeck, E. and Grimm, F.T. (1990). Induction of swarming of *Nereis succinea*. *Marine Biology* **104**, 291–295.

Hardege, J.D., Bartels-Hardege, H. and Zeeck, E. (1991). Volatile compound from the coelomic fluid of *Nereis succinea*: biological activity as sex pheromones. *Invertebrate Reproduction and Development* **19**, 83–85.

Hardy, A.C. (1936). Observations on the uneven distribution of oceanic plankton. *Discovery Reports* **11**, 511–538.

Heller, R. and Milinski, M. (1979). Optimal foraging of sticklebacks on swarming prey. *Animal Behaviour* **27**, 1127–1141.

Hilborn, R. (1991). Modeling the stability of fish schools: exchange of individual fish between schools of skipjack tuna (*Katsuwonus pelamis*). *Canadian Journal of Fisheries and Aquatic Science* **48**, 1081–1091.

Hobson, E.S. (1968). Predatory behavior of some shore fishes in the Gulf of California. *United States Fish and Wildlife Service Research Reports* **73**, 1–92.

Hobson, E.S. (1978). Aggregating as a defense against predators in aquatic and terrestrial environments. *In* "Contrasts in Behaviour: Adaptations in the Aquatic and Terrestrial Environments" (E.S. Reese and F.J. Lighter, eds), pp. 219–234. John Wiley, Toronto.

Hochberg, F.G. and Couch, J.A. (1971). Biology of cephalopods. *In* "Tektite 2. Scientists in the Sea" (J.W. Miller, J.G. VanDerwalker and R.A. Waller, eds), pp. 221–228. U.S. Dept Int., Washington DC.

Humphries, D.A. and Driver, P.M. (1967). Erratic display as a device against predators. *Science* **156**, 1767–1768.

Humphries, D.A. and Driver, P.M. (1970). Protean defence by prey animals. *Oecologia* **5**, 285–302.

Hurley, A.C. (1978). School structure of the squid *Loligo opalescens. Fishery Bulletin U.S.* **76**, 433–442.

Ichii, T. (1990). Distribution of Antarctic krill concentrations exploited by Japanese krill trawlers and Minke whales. *Proceedings of the NIPR Symposium on Polar Biology* **3**, 36–56.

Jakobsen, P.J. and Johnsen, G.H. (1987). Behavioural response of the water flea *Daphnia pulex* to a gradient in food concentration. *Animal Behaviour* **35**, 1891–1895.

Jakobsen, P.J. and Johnsen, G.H. (1988). The influence of food limitation on swarming behaviour in the waterflea *Bosmina longispina. Animal Behaviour* **36**, 991–995.

Jillett, J.B. and Zeldis, J.R. (1985). Aerial observations of surface patchiness of a planktonic crustacean. *Bulletin of Marine Science* **37**, 609–619.

Johnson, D.S. and Chia, T.E. (1972). Remarkable schooling behaviour of a water flea, *Moina* sp. (Cladocera). *Crustaceana* **24**, 332–333.

Johnson, M.A., Macaulay, M.C. and Biggs, D.C. (1984). Respiration and execretion within a mass aggregation of *Euphausia superba*: implications for krill distribution. *Journal of Crustacean Biology* **4**, (Spec. No. 1), 174–184.

Kalinowski, J. and Witek, Z. (1985). "Scheme for Classifying Aggregations of Antarctic Krill." *BIOMASS Handbook* **27**, 9 pp. SCAR SCOR, Scott Polar Institute, Cambridge.

Katona, S.K. (1973). Evidence for sex pheromones in planktonic copepods. *Limnology and Oceanography* **18**, 574–583.

Katz, L.C., Potel, M.J. and Wasserug, R.J. (1981). Structure and mechanisms of schooling in tadpoles of the clawed frog, *Xenopus laevis. Animal Behaviour* **29**, 20–33.

Kawamura, A. (1974). Food and feeding ecology in the Southern Sei whale. *Science Reports of the Whales Research Institute* **26**, 25–143.

Kils, U. (1981). Size dissociation in krill swarms. *Kieler Meeresforschungen* **5**, 262–263.

Kimoto, K., Nakashima, J. and Morioka, Y. (1988). Direct observations of copepod swarm in a small inlet of Kyushu, Japan. *Bulletin of the Seikai Regional Fisheries Research Laboratory* **66**, 41–55.

Koltes, K.H. (1984). Temporal patterns in three-dimensional structure and activity of schools of the Atlantic silverside *Menidia menidia. Marine Biology* **78**, 113–122.

Komai, T. (1932). An enormous swarm of stomatopod larvae. *Annotationes zoologicae japonenses* **13**, 351–354.

Komaki, Y. (1967). On the surface swarming of euphausiid crustaceans. *Pacific Science* **21**, 433–448.

Krebs, J.R. and Davies, N.B. (1981). "An Introduction to Behavioural Ecology." Blackwell, Oxford.

Krebs, J.R. and Davies, N.B. (1991). "Behavioural Ecology: An Evolutionary Approach", 3rd edn. Blackwell, Oxford.

Landeau, L. and Terborgh, J. (1986). Oddity and the "confusion effect" in predation. *Animal Behaviour* **34**, 1372–1380.

Larson, R. (1991). Why jellyfish stick together. *Natural History* **10**, 66–71.

Lobel, P.S. and Randall, J.E. (1986). Swarming behavior of the hyperiid amphipod *Anchylomera blossevilli*. *Journal of Plankton Research* **8**, 253–262.

Macaulay, M.C., English, T.S. and Mathisen, O.A. (1984). Acoustic characterisation of swarms of Antarctic krill (*Euphausia superba*) from Elephant Island and Bransfield Strait. *Journal of Crustacean Biology* **4**, (Spec. No. 1), 16–44.

Magurran, A.E. (1990). The adaptive significance of schooling as an anti-predator defence in fish. *Annales Zoologici Fennici* **27**, 51–66.

Magurran, A.E. and Seghers, B.H. (1990). Risk-sensitive courtship in the guppy (*Poecilia reticulata*). *Behaviour* **112**, 194–201.

Major, P.F. (1978). Predator–prey interactions in two schooling fishes, *Caranx ignobilis* and *Stolephorus purpureus*. *Animal Behaviour* **26**, 760–777.

Marr, J.W.S. (1962). The natural history and geography of the Antarctic krill (*Euphausia superba* Dana). *Discovery Reports* **32**, 33–464.

Mauchline, J. (1971). Seasonal occurrence of mysids (Crustacea) and evidence of social behaviour. *Journal of the Marine Biological Association of the United Kingdom* **51**, 809–825.

Mauchline, J. (1980a). "The Biology of Mysids and Euphausiids." *Advances in Marine Biology* **18**, 1–681.

Mauchline, J. (1980b). "Studies on Patches of Krill, *Euphausia superba* Dana." *BIOMASS Handbook* **16**, 36 pp.

McClatchie, S., Hutchinson, D. and Nordin, K. (1989). Aggregation of avian predators and zooplankton prey in Otago shelf waters, New Zealand. *Journal of Plankton Research* **11**, 361–374.

McFarland, W.N. and Kotchian, N.M. (1982). Interaction between schools of fish and mysids. *Behavioural Ecology and Sociobiology* **11**, 71–76.

McKoy, J.L. and Leachman, A. (1982). Aggregations of ovigerous female rock lobsters, *Jasus edwardsii* (Decapoda: Palinuridae). *New Zealand Journal of Marine and Freshwater Research* **16**, 141–146.

Milinski, M. (1977). Do all members of a swarm suffer the same predation? *Zeitschrift für Tierpsychologie* **45**, 373–388.

Milinski, M. (1979). Can an experienced predator overcome the confusion of swarming prey easily? *Animal Behaviour* **27**, 1122–1126.

Milinski, M. and Parker, G.A. (1991). Competition for resources. *In* "Behavioural Ecology: An Evolutionary Approach" (J.R. Krebs, and N.B. Davies, eds), pp. 137–168. Blackwell, Oxford.

Miller, D.G.M. and Hampton I. (1989). "Biology and Ecology of the Antarctic Krill." *BIOMASS Scientific Series* **9**, SCAR SCOR, Scott Polar Institute, Cambridge, England.

Modlin, R.F. (1990). Observations on the aggregative behavior of *Mysidium columbiae*, the mangrove mysid. *P.S.Z.N.I. Marine Ecology* **11**, 263–275.

Morgan, W.L. and Ritz, D.A. (1984). Effect of prey density and hunger state on capture of krill *Nyctiphanes australis* Sars, by Australian salmon, *Arripis trutta* (Bloch and Schneider). *Journal of Fish Biology* **24**, 51–58.

Morin, J.G. (1983). Coastal bioluminescence: patterns and functions. *Bulletin of Marine Science* **33**, 787–817.

Morin, J.G. (1986). "Firefleas" of the sea: luminescent signaling in marine ostracode crustaceans. *Florida Entomologist* **69**, 105–121.

Moss, S.A. and McFarland, W.N. (1970). The influence of dissolved oxygen and carbon dioxide on fish schooling behavior. *Marine Biology* **5** 100–107.

Naito, Y., Taniguchi, A. and Hamada, E. (1986). Some observations on swarms and mating behavior of Antarctic krill (*Euphausia superba* Dana). *Memoirs of the National Institute of Polar Research* Special Issue **40**, 178–182.

Neill, S.R.St.J. and Cullen, J.M. (1974). Experiments on whether schooling by their prey affects hunting behaviour of cephalopods and fish predators. *Journal of the Zoological Society of London* **172**, 549–569.

Nemoto, T. (1983). Net sampling and abundance assessment of euphausiids. *Biological Oceanography* **2**, 211–226.

Nemoto, T., Ishimaru, K. and Shirai, T. (1981). Swarm of euphausiids in the ocean. *Antarctic Record* **73**, 103–112.

Nicol, S. (1984). Population structure of daytime surface swarms of the euphausiid *Meganyctiphanes norvegica* in the Bay of Fundy. *Marine Ecology Progress Series* **18**, 241–251.

Nicol, S. (1986). Shape, size and density of daytime surface swarms of the euphausiid *Meganyctiphanes norvegica* in the Bay of Fundy. *Journal of Plankton Research* **8**, 29–39.

Nicol, S. and de la Mare, W. (1993). Ecosystem management and the Antarctic krill. *American Scientist* **81**, 36–47.

Nicol, S. and O'Dor, R.K. (1985). Predatory behaviour of squid (*Illex illecebrosus*) feeding on surface swarms of euphausiids. *Canadian Journal of Zoology* **63**, 15–17.

Nicol, S., James, A. and Pitcher, G. (1987). A first record of daytime surface swarming by *Euphausia lucens* in the Southern Benguela region. *Animal Biology* **94**, 7–10.

Obeng-Ofori, D. and Coaker, T.H. (1990). *Tribolium* aggregation pheromone: monitoring, range of attraction and orientation behaviour of *T. castaneum* (Coleoptera: Tenebrionidae). *Bulletin of Entomological Research* **80**, 443–451.

O'Brien, D.P. (1987a). Description of escape responses of Krill (Crustacea: Euphausiacea) with particular reference to swarming behaviour and the size and proximity of the predator. *Journal of Crustacean Biology* **7**, 449–457.

O'Brien, D.P. (1987b). Direct observations of the behaviour of *Euphausia superba* and *Euphausia crystallorophias* (Crustacea: Euphausiacea) under pack ice during the Antarctic spring of 1985. *Journal of Crustacean Biology* **7**, 437–448.

O'Brien, D.P. (1988a). Direct observations of clustering (schooling and swarming) behaviour in mysids (Crustacea: Mysidacea). *Marine Ecology Progress Series* **42**, 235–246.

O'Brien, D.P. (1988b). Surface schooling behaviour of the coastal krill *Nyctiphanes australis* (Crustacea: Euphausiacea) off Tasmania, Australia. *Marine Ecology Progress Series* **42**, 219–233.

O'Brien, D.P. (1989). Analysis of the internal arrangement of individuals within crustacean aggregations (Euphausiacea, Mysidacea). *Journal of Experimental Marine Biology and Ecology* **128**, 1–30.

O'Brien, D.P. and Ritz, D.A. (1988). The escape responses of gregarious mysids (Crustacea: Mysidacea): towards a general classification of escape responses in aggregated crustaceans. *Journal of Experimental Marine Biology and Ecology* **116**, 257–272.

O'Brien, D.P., Ritz, D.A. and Kirkwood, R.J. (1986a). Stranding and mating behaviour in *Nyctiphanes australis* (Euphausiacea: Crustacea). *Marine Biology* **93**, 465–473.

O'Brien, D.P., Tay, D., and Zwart, P.W. (1986b). Laboratory method of analysis of swarming behaviour in macroplankton: combination of a modified flume tank and stereophotographic techniques. *Marine Biology* **90**, 517–527.

Okubo, A. and Anderson, J.J. (1984). Mathematical models for zooplankton swarms: their formation and maintenance. *Eos* **65**, 731–732.

Omori, M. (1974). The biology of pelagic shrimps in the ocean. *Advances in Marine Biology* **12**, 233–324.

Omori, M. (1975). The systematics, biogeography and fishery of epipelagic shrimps of the genus *Acetes* (Crustacea, Decapoda, Sergestidae). *Bulletin of the Ocean Research Institute, University of Tokyo* **7**, 1–91.

Omori, M. and Hamner, W.M. (1982). Patchy distribution of zooplankton: behaviour, population assessment and sampling problems. *Marine Biology* **72**, 193–200.

Parrish, J.K. Strand, S.W. and Lott, J.L. (1989). Predation on a school of flat-iron herring, *Harengula thrissina*. *Copeia* **4**, 1089–1091.

Partridge, B.L. (1982). The structure and function of fish schools. *Scientific American* **246**, 90–99.

Partridge, B.L. and Pitcher, T.J. (1980). The sensory basis of fish schools: relative roles of lateral line and vision. *Journal of Comparative Physiology* **135**, 315–325.

Partridge, B.L., Pitcher, T.J., Cullen, J.M. and Wilson, J. (1980). The three-dimensional structure of fish schools. *Behavioural Ecology and Sociobiology* **6**, 277–288.

Pennings, S.C. (1991). Reproductive behavior of *Aplysia californica* Cooper: diel patterns, sexual roles and mating aggregations. *Journal of Experimental Marine Biology and Ecology* **149**, 249–266.

Pitcher, T.J. (1986a). Functions of shoaling behaviour in teleosts. *In* "The Behaviour of Teleost Fishes", pp. 294–336. The Johns Hopkins University Press. Baltimore, MD.

Pitcher, T.J. (1986b). Predators and food are the keys to understanding fish shoals: a review of recent experiments. *Naturaliste Canadienne* **113**, 225–233.

Pitcher, T.J., Magurran, A.E. and Winfield, I.J. (1982). Fish in larger shoals find food faster. *Behavioural Ecology and Sociobiology* **10**, 149–151.

Pitcher, T.J., Magurran, A.E. and Allan, J.R. (1983). Shifts of behaviour with shoal size in cyprinids. Proceedings of the 3rd British Freshwater Fisheries Conference, University of Liverpool, pp. 220–228.

Poulet, S.A. and Ouellet, G. (1982). The role of amino acids in the chemosensory swarming and feeding of marine copepods. *Journal of Plankton Research* **4**, 341–359.

Priddle, J., Watkins, J., Morris, D.J., Ricketts, C. and Bucholz, F. (1990). Variation of feeding by krill in swarms. *Journal of Plankton Research* **12**, 1189–1205.

Pulliam, H.R. and Caraco, T. (1984). Living in groups: is there an optimal group size? *In* "Behavioural Ecology: An Evolutionary Approach" (J.R. Krebs and N.B. Davies, eds), pp. 122–147. Blackwell, Oxford.

Radakov, D.V. (1958). On the adaptive significance of shoaling in young coalfish (*Pollachia virens* L.). *Voprosy Ikhtiologii* **11**, 69–74.

Ragulin, A.G. (1969). Underwater observations of krill. *Trudy VNIRO*: **66**, 231–234.

Ranta, E. and Kaitala, V. (1991). School size affects individual feeding success in three-spined sticklebacks (*Gasterosteus aculeatus* L.). *Journal of Fish Biology* **39**, 733–737.

Ranta, E. and Lindstrom, K. (1990). Assortative schooling in three-spined sticklebacks? *Annales Zoolologici Fennici* **27**, 67–75.

Ranta, E., Lindstrom, K. and Peukhuri, N. (1992). Size matters when three-spined sticklebacks go to school. *Animal Behaviour* **43**, 160–162.

Ratzlaff, W. (1974). Swarming in *Moina affinis*. *Limnology and Oceanography* **19**, 993–995.

Reynolds, C.W. (1987). Flocks, herds and schools: a distributed behavioural model. *Computer Graphics* **21**, 25–34.

Rice, A.L. and Kristensen, I. (1982). Surface swarms of swimming crab Megalopae at Curacao (Decapoda, Brachyura). *Crustaceana* **42**, 233–240.

Ricketts, C., Watkins, J.L., Priddle, J. Morris, D.J. and Bucholz, F. (1992). An assessment of the biological and acoustic characteristics of swarms of Antarctic krill. *Deep-Sea Research* **39**, 359–371.

Ringelberg, J. (1969). Spatial orientation of planktonic crustaceans. *Verhandlungen internationale Verein Limnologie* **17**, 841–847.

Ritz, D.A. (1991). Benefits of a good school. *New Scientist* **1761**, 41–43.

Ritz, D.A. (1994). Costs and benefits of aggregation in aquatic crustaceans: experiments on a swarming mysid *Paramesopodopsis rufa* Fenton. In "Three-Dimensional Animal Aggregations: Mechanisms and Functions" (J.K. Parrish, W.M. Hamner and C.T. Prewitt, eds). Cambridge University Press, Cambridge, in press.

Ross, R.M., Quetin, L.B. and Ball, E.E. (1988). *In situ* observation of the school behaviour of the sergestid shrimp *Acetes sibogae australis*. *Bulletin of Marine Science* **43**, 849.

Ryer, C.H. and Olla, B.L. (1991). Agonistic behavior in a schooling fish: form, function and ontogeny. *Environmental Biology of Fishes* **31**, 355–363.

Ryer, C.H. and Olla, B.L. (1992). Social mechanisms facilitating exploitation of spatially variable ephemeral food patches in a pelagic marine fish. *Animal Behaviour* **44**, 69–74.

Sauer, W.H.H., Smale, M.J. and Lipinksi, M.R. (1992). The location of spawning grounds, spawning and schooling behaviour of the squid *Loligo vulgaris* (Cephalopoda: Myopsida) off the Eastern Cape Coast, South Africa. *Marine Biology* **114**, 97–107.

Seno, H. (1990). A density-dependent diffusion model of shoaling of nesting fish. *Ecological Modelling* **51**, 217–226.

Shulenberger, E., Wormuth, J.H. and Loeb, V.J. (1984). A large swarm of *Euphausia superba*: overview of patch structure and composition. *Journal of Crustacean Biology* **4** (Spec. No. 1), 75–95.

Sih, A. (1980). Optimal behaviour: can foragers balance two conflicting demands? *Science* **210**, 1041–1043.

Slobodchikoff, C.N. and Shields, W.M. (1988). Ecological trade-offs and social behaviour. In "The Ecology of Social Behaviour" (C.N. Slobodchikoff, ed.), pp. 3–10. Academic Press, San Diego.

Slosarczyk, W. and Rembiszewski, J.M. (1982). The occurrence of juvenile Notothenioidei (Pisces) within krill concentrations in the region of the Bransfield Strait and the southern Drake Passage. *Polish Polar Research* **3**, 299–312.

Smith, S.E. and Adams, P.B. (1988). Daytime surface swarms of *Thysanoessa spinifera* (Euphausiacea) in the Gulf of Farallones, California. *Bulletin of Marine Science* **42**, 76–84.

Sprong, I. and Schalk, P.H. (1992). Acoustic observations on krill spring-summer migration and patchiness in the northern Weddell Sea. *Polar Biology* **12**, 261–268.

Steele, C.W., Scarfe, A.D. and Owens, D.W. (1991). Effects of group size on the responsiveness of zebrafish, *Brachydanio rerio* (Hamilton Buchanan), to alanine, a chemical attractant. *Journal of Fish Biology* **38**, 553–564.

Steven, D.M. (1961). Shoaling behaviour in a mysid. *Nature, London* **192**, 280–281.

Strand, S.W. and Hamner, W.M. (1990). Schooling behaviour of Antarctic krill (*Euphausia superba*) in laboratory aquaria: reactions to chemical and visual stimuli. *Marine Biology* **106**, 355–360.

Tanaka, M., Ueda, H. and Azeta, M. (1987). Near-bottom copepod aggregations around the nursery ground of the juvenile red sea bream in Shijiki Bay. *Nippon Suisan Gakkaishi* **53**, 1537–1544.

Terazaki, M. (1980). Surface swarms of a euphausiid *Euphausia pacifica* in Otsuchi Bay, northern Japan. *Bulletin of the Plankton Society of Japan* **27**, 19–25.

Tessier, A.J. (1983). Coherence and horizontal movements of patches of *Holopedium gibberum* (Cladocera). *Oecologia* **60**, 71–75.

Thomas, R.M. (1980). A note on swarms of *Nyctiphanes capensis* (Hansen) larvae off South West Africa. *Fisheries Bulletin of South Africa* **13**, 21–23.

Tomo, A. (1983). Observations on krill shoal luminescence at the sea surface and the accompanying fauna. *Berichte zur Polarforschung* **4**, 196–198.

Treherne, J.E. and Foster, W.A. (1981). Group transmission of predator avoidance behavior in a marine insect: The Trafalgar Effect. *Animal Behaviour* **29**, 911–917.

Van Dover, C.L., Kaartvedt, S., Bollens, S.M., Wiebe, P.H., Martin, J.W. and France, S.C. (1992). Deep-sea amphipod swarms. *Nature, London* **358**, 25–26.

Vine, I. (1971). Risk of visual detection and pursuit by a predator and the selective advantage of flocking behaviour. *Journal of Theoretical Biology* **30**, 405–422.

Waldman, B. (1982). Sibling association among schooling toad tadpoles: field evidence and implications. *Animal Behaviour* **30**, 700–713.

Waldman, B. and Adler, K. (1979). Toad tadpoles associate preferentially with siblings. *Nature, London* **282**, 611–613.

Wasserug, R.J. (1973). Aspects of social behaviour in anuran larvae. In "Evolutionary Biology of the Anurans" (J.L. Vial, ed.), pp. 273–297. University of Missouri Press, Columbia.

Watkins, J.L. (1986). Variations in the size of Antarctic krill, *Euphausia superba* Dana, in small swarms. *Marine Ecology Progress Series* **31**, 67–73.

Watkins, J.L., Morris, D.J. and Ricketts, C. (1985). Nocturnal changes in the mean length of a euphausiid population: vertical migration, net avoidance, or experimental error? *Marine Biology* **86**, 123–127.

Watkins, J.L., Morris, D.J., Ricketts, C. and Priddle, J. (1986). Differences between swarms of Antarctic krill and some implications for sampling krill populations. *Marine Biology* **93**, 137–146.

Watkins, J.L., Buchholz, F., Priddle, J., Morris, D.J. and Ricketts, C. (1992). Variation in reproductive status of Antarctic krill swarms: evidence for a size-related sorting mechanism? *Marine Ecology Progress Series* **82**, 163–174.

Wiborg, K.F. (1976). Fishery and commercial exploitation of *Calanus finmarchicus* in Norway. *Journal du Conseil Permanent International pour l'Exploration de la Mer* **36**, 251–258.

Wilson, E.O. (1975). "Sociobiology." Harvard University Press, Boston, MA.

Wittmann, K.J. (1977). Modification of association and swarming in north Adriatic Mysidacea in relation to habitat and interacting species. *In* "Biology of Benthic Organisms" (P. O'Ceidigh and P.J.S. Boaden, eds), pp. 605–612. 11th European Marine Biology Symposium, Pergamon Press, London.

Wittmann, K.J. (1978). Adoption, replacement, and identification of young in marine Mysidacea (Crustacea). *Journal of Experimental Marine Biology and Ecology* **32**, 259–274.

Wooldridge, T.H. (1983). Ecology of beach and surf-zone mysid shrimps in the eastern Cape, South Africa. *In* "Sandy Beaches as Ecosystems" (A. McLachlan and T. Erasmus, eds), pp. 449–460. W. Junk, The Hague.

Xiao, Y. and Greenwood, J.G. (1992). Distribution and behaviour of *Acetes sibogae* Hansen (Decapoda, Crustacean) in an estuary in relation to tidal and diel environmental changes. *Journal of Plankton Research* **14**, 393–407.

Young, J.P.W. (1978). Sexual swarms in *Daphnia magna*, a cyclic parthenogen. *Freshwater Biology* **8**, 279–281.

Young, S. and Taylor, V.A. (1990). Swimming tracks in swarms of two cladoceran species. *Animal Behaviour* **39**, 10–16.

Zavodnik, D. (1987). Spatial aggregations of the swarming jellyfish *Pelagia noctiluca* (Scyphozoa). *Marine Biology* **94**, 265–269.

Zeeck, E., Hardege, J., Bartels–Hardege, H. and Wesselmann, G. (1988). Sex pheromone in a marine polychaete: determination of the chemical structure. *Journal of Experimental Zoology* **246**, 285–292.

Zeldis, J.R. and Jillett, J.B. (1982). Aggregation of pelagic *Munida gregaria* (Fabricius) (Decapoda, Anomura) by coastal fronts and internal waves. *Journal of Plankton Research* **4**, 839–857.

Zelikman, E.A. (1974). Group orientation in *Neomysis mirabilis* (Mysidacea: Crustacea). *Marine Biology* **24**, 251–258.

Zelikman, E.A., Lukashevich, I.P. and Drobysheva, S.S. (1979). Year-round diurnal vertical migrations of the euphausiids *Thysanoessa inermis* and *T. raschii* in the Barents Sea. *Oceanology* **19**, 82–85.

An Appraisal of Condition Measures for Marine Fish Larvae*

A. Ferron and W.C. Leggett

*Department of Biology, McGill University, Montréal, Québec, Canada,
H3A 1B1*

1. INTRODUCTION

Recruitment, usually defined as the number of young fish from a single year-class that enter the fishery, might be expected to relate to the total

*Contribution to the research programs of GIROQ (Groupe Interuniversitaire de Recherches Océanographiques du Québec) and OPEN (Ocean Production Enhancement Network).

ADVANCES IN MARINE BIOLOGY VOL 30
ISBN 0–12–026130–8

biomass of the spawners comprising a stock. However, the typical dome-shaped or asymptotic stock-recruitment relationships reported for most marine fish stocks are characterized by high variance, with "outliers" reflecting the presence of exceptionally large or small year-classes (Sissenwine, 1984; Rothschild, 1986; Fogarty et al., 1991; Koslow, 1992). The low predictive power of such relationships, and the difficulties associated with predicting recruitment with accuracy, has been a central problem in fisheries management (Sissenwine, 1984; Rothschild, 1986; Koslow, 1992). Since very few of the many eggs produced by marine fish species eventually give rise to adults (typically less than 0.01%), it is believed that small changes in the mortality rates of the young stages can lead to the observed high variability in recruitment (Underwood and Fairweather, 1989; Sale, 1990; Fogarty et al., 1991; Koslow, 1992; Fogarty, 1993). Although there has been substantial controversy over the ontogenetic stage at which recruitment is determined (Peterman et al., 1988; Bradford, 1992), the larval stage is still considered important in this respect (Smith, 1985; Sundby et al., 1989; Bailey and Spring, 1992).

Many biological and physical factors can interact to affect the growth and mortality of young fish during their first year (Fogarty et al., 1991). Starvation and predation, acting independently or together with the modifying impact of abiotic factors (physical processes), have often been suggested as the main sources of mortality (Hjort, 1914; Cushing, 1972; Hunter, 1976, 1984; Lasker, 1981; Rothschild and Rooth, 1982; Sissenwine, 1984; Hewitt et al., 1985; Rothschild, 1986; Rice et al., 1987; Sinclair, 1988; Owen et al., 1989). Differential growth may also be important because it determines larval stage duration and thus the length of time the larvae are susceptible to any mortality source (Buckley and Lough, 1987; Buckley et al., 1987). The potential for growth rate and stage duration influences on survival has been illustrated by Chambers and Leggett (1987), Houde (1987), Pepin (1991), and Rice et al. (1993). The extent to which these potential effects are realized is, however, the subject of debate (Litvak and Leggett, 1992; Pepin et al., 1992). While predation may be the ultimate cause of mortality, the measurement of its direct impact in the field is exceptionally difficult and has yet to be conclusively demonstrated (Hunter, 1984; Bailey and Houde, 1989). Potential or realized starvation mortality, in contrast, can be estimated directly from the measurement of the nutritional condition of the larval fish, provided that the threshold for irreversible starvation (often referred to as the point-of-no-return or PNR) can be reliably established (O'Connell, 1980; Theilacker, 1986; Owen et al., 1989; Theilacker and Watanabe, 1989).

Hunter (1984) proposed to quantify predation mortality by measuring the incidence of starvation at sea and subtracting it from the total mortality. This approach was used by Hewitt *et al.* (1985), and Theilacker (1986), who showed that starvation mortality could be equal to total mortality during the early larval stages of jack mackerel (*Trachurus symmetricus*). While other approaches have been adopted to measure predation (Bailey and Houde, 1989; Purcell and Grover, 1990), the assessment of the physiological condition of larval fishes at sea, in order to evaluate their vulnerability to death due to food limitation and predation, continues to be a major focus of research (Buckley and Lough, 1987; Robinson and Ware, 1988; Frank and McRuer, 1989; Håkanson, 1989b; Owen *et al.*, 1989; Hovenkamp, 1990; Powell *et al.*, 1990; Canino *et al.*, 1991; Hovenkamp and Witte, 1991; McGurk *et al.*, 1992; Sieg, 1992a; Suthers *et al.*, 1992; Margulies, 1993; McGurk *et al.*, 1993).

Larval condition can be assessed from measurements taken at three organizational levels: organismal, tissue, and cellular. At the organism level, interest focuses on detecting changes in external body shape which relate to condition. These morphological changes of shape are usually depicted as ratios or multivariate vectors of body measurements. At the tissue level, changes in condition are detected through analyses of modifications in the appearance of the cells and their arrangement in different tissues of the animal. This is made possible because the tissue histology of starved fish larvae often differs from that of well-fed specimens. Finally, biochemical condition can be assessed by quantifying chemical constituents used as energy substrates or by measuring physiological rate indicators that are known to vary at the cellular level in relation to the nutritional status of the animal.

Notwithstanding the growing number and use of condition indices in the study of fish growth and survival, no comprehensive review of the topic is available. In this review, we define important considerations in the use of condition indices as applied to marine fish larvae, and describe the present state of knowledge concerning each index. An appraisal of their relative merits is then attempted by comparing the strengths and limitations of each category. Finally we provide suggestions for the use of condition indices in different situations and identify topics requiring further research.

2. RELIABILITY OF CONDITION INDICES

2.1. Morphometric Indices

The first person to use the term "condition" to describe the nutritional status of larval fish was Shelbourne (1957) whose qualitative classification of condition in larval plaice (*Pleuronectes platessa*) was based on an impression that larvae in poor condition were typically leaner for a given length than were healthier larvae. He established five categories of robustness based on a subjective assessment of the proportion of soft parts (gut and muscular axis) to hard parts (head structure). Such shape differences were later quantified using dry weight/length^3 (also known as the Fulton-K index and previously applied to adult fish by LeCren (1951)). This isometric index was applied to larvae of herring (*Clupea harengus*) sampled at sea (Hempel and Blaxter, 1963; Blaxter, 1971; Vilela and Zijlstra, 1971), and to fed and starved grunion (*Leuresthes tenuis*) larvae reared in the laboratory (May, 1971). The reliability of the Fulton-K index as applied to larval fishes was soon challenged by Ehrlich (1974a), who noted problems associated with the concurrent loss of both length and weight during starvation of larvae of plaice and the allometric growth and ossification of the skeleton in older individuals.

Several other morphological ratios have since been proposed. Wyatt (1972) obtained significant correlations between length, height (at pectorals), and height/length ratios of plaice larvae and the food densities offered to these larvae in the laboratory. The correlation between ration and the height/length ratio was greater than that for either length or height taken separately. Ehrlich *et al.* (1976) showed that the pectoral angle (angle made by the ventral body contour and the pectoral girdle), the body height of herring and plaice larvae, and the eye height/head height ratio of herring larvae, were all affected by starvation. However, problems associated with size and species-specificity, and with between-individual variability, restricted the broad application of these ratios. Yin and Blaxter (1986), working with larvae of cod (*Gadus morhua*) and flounder (*Platichthys flesus*), reported that the ratio of gut height/myotome height was useful for discriminating fed and starved specimens shortly before the point of irreversible starvation. They also reported the total length/body height and eye height/head height ratios to be effective in categorizing nutritional status.

Theilacker (1978) first suggested the use of ratios in which the numerator alone was sensitive to starvation to allow the detection of starving specimens. However, the finding that few of these morphological measurements proved to be indicative of starvation in laboratory-reared jack mackerel larvae led her to propose the application of multivariate stepwise discriminant analysis (SWDA). SWDA conducted with eleven different morphometric variables (including some ratios) showed five to have significant discriminating power. These were, in order, (1) anal body depth/standard length, (2) anal body depth/head length, (3) pectoral body depth/head length, (4) eye diameter/standard length, and (5) pectoral body depth. When applied to the same data-set, 87% of the fed and 94%

Table 1 Condition of larvae of *Trachurus symmetricus* fed or starved for 1, 2 or 3 d (column 1). Each larva was classified using morphometric (column 2) and histological (column 3) methods. The number of larvae correctly classified by each method, and by both methods, is indicated. The numbers in brackets are percentages. Asterisks* refer to the number of larvae classified similarly by the two methods. (From Theilacker, 1978.)

(1) Experimental conditions		(2) Morphometric analysis	(3) Histological analysis		
			Healthy	Inter-mediate	Starved
Fed	Fed	53 (82.8)	*44 (68.8)	8	1
	Starved 1–2 d	10 (15.6)	8	2	0
	Starved 3 d	1 (1.6)	1	0	0
	Total	64	53 (82.8)	10 (15.6)	1 (1.6)
Starved 1–2 d	Fed	1 (3.7)	0	1	0
	Starved 1–2 d	26 (96.3)	5	*11 (40.7)	10
	Starved 3 d	0	0	0	0
	Total	27	5 (18.5)	12 (44.5)	10 (37.0)
Starved 3 d	Fed	0	0	0	0
	Starved 1–2 d	4 (22.2)	1	1	2
	Starved 3 d	14 (77.8)	1	0	*13 (92.9)
	Total	18	2 (11.1)	1 (5.6)	15 (83.3)

of the starved specimens could be correctly assigned to their feeding treatment (Table 1). Powell and Chester (1985) reanalysed Theilacker's (1978) measurements for laboratory-reared spot (*Leiostomus xanthurus*) larvae and noted that anal body depth and pectoral body depth declined faster than any other measurements during starvation. However, a multivariate analysis of variance (MANOVA) showed all measurements to be significantly different between fed and starved larvae. Canonical discriminant functions calculated from these data showed pectoral and anal body depths to have the highest discriminating power, leading to 84% of the fed and 83% of the starved larvae being correctly classified.

Martin and Wright (1987) applied Theilacker's measurements to delayed-starved and delayed-fed striped bass (*Morone saxatilis*) larvae in the laboratory. A stepwise discriminant analysis retained eight ratios with F-values higher than 4.0, from which head length/eye diameter, notochord length/eye diameter, head length/anal body depth and notochord length/anal body depth had the highest discriminating power. However, overlapping canonical centroids prevented the distinction between larvae of different delayed-feeding and starved treatments. This failure was attributed to the fact that most computed ratios shared size information (r values between 0.86 and 0.98 for regressions of ratios against standard length). Koslow *et al.* (1985) computed linear regressions relating body measures of cod larvae believed to be sensitive to starvation (body height at the pectoral fins and at the anus, interorbital distance and total dry weight) with body measures believed to be insensitive to starvation (total length, eye diameter and head length), and used the residuals as inputs to a principal component analysis. They concluded that the multivariate approach offered no apparent advantage over a bivariate analysis.

McGurk (1985b) recorded six morphological measurements (head width, pectoral and anal body depth, eye diameter and standard and total length) from laboratory starved and fed herring larvae. By contrasting the results obtained from a stepwise discriminant analysis (SWDA) of ratios, residuals, and principal components, he found that several ratios and residuals were correlated, but that principal components based on these measures were not. He determined that ratios and residuals were correlated because they shared size information, which lead to biased estimates of classification. However, estimates based on the principal components were not biased, since shape differences between starved and fed animals could be obtained from the second principal component after size effects had been removed with the first principal component. A SWDA using the second principal component of the six variables measured successfully classified 95.8% of the starved larvae. McGurk's (1985a, b) approach was justified on the basis of arguments that it conformed to the three essential elements of the ideal morphometric condition factor, (1) size-independence, (2) biological meaning, and (3) orthogonality. These conditions appear to be met rarely in bivariate analyses.

An important confounding factor associated with the use of morphometric condition indices is the presence of rapid morphological changes which are more related to important developmental events such as the transition to exogenous feeding (Theilacker, 1978; Neilson *et al.*, 1986) than to starvation. Farbridge and Leatherland (1987) also suggest that growth in length and mass may be out of phase in young fish, such that an increase in mass is preceded by an increase in length. This, too, can potentially influence the accuracy of morphometric indices.

2.2. Histological Indices

Histological changes in the digestive tract, liver, pancreas, and intermuscular tissues associated with starvation in laboratory culture were first described for larvae of yellowtail (*Seriola quinqueradiata*) by Umeda and Ochiai (1975) and northern anchovy (*Engraulis mordax*) by O'Connell (1976). O'Connell (1976) employed a 10 variables SWDA to demonstrate that pancreas condition, notochord shrinkage, muscle fibre separation, intermuscular tissue and liver cytoplasm, in that order, were effective for distinguishing robust, severely and moderately starved larvae. Over 90% of the larvae surveyed were correctly classified using these variables. The number and size of hindgut inclusions had a high discriminating power, but this variable was later discarded because it was believed to be more related to feeding status than condition. Ehrlich *et al.* (1976) examined the histology of herring and plaice larvae following starvation. They reported that head shrinkage in herring, in response to starvation, was demonstrated in the histological sections by a reduction of the interorbital distance. A contraction of the gut (in length and diameter) and a reduction in the area of the liver in plaice larvae was also observed. In starving black bullhead (*Ictalurus melas*) adults, Kayes (1978) measured the largest weight loss in the liver, followed by intestine, stomach and eviscerated carcass. The liver was more responsive to starvation in the older fish because of its greater energy storage efficiency as compared with fish larvae; the atrophy of the gut was, however, as responsive to starvation as in larval fish.

Theilacker (1978) recorded 12 different histological characteristics associated with the brain, liver, pancreas, digestive tract and musculature of jack mackerel larvae starved in the laboratory. SWDA revealed that two of these variables (acinar arrangement of the pancreatic cells and the presence of midgut mucosal cells sloughed in the lumen) allowed correct classification of 83% of the fed animals into a healthy class, and 83% of larvae starved for 3 d into a starved class. However, only 44% of the larvae starved for 1–2 d were correctly classified into an intermediate class (Table 1). Inclusion of the 10 remaining variables yielded no improvement in group classification. Cousin et al. (1986) reported that the first histological anomalies observed in turbot (Scophthalmus maximus) larvae after 6 d of starvation in the laboratory were atrophy of the digestive tract and/or a desquamation of the intestinal mucosa. Concurrent with these changes these authors also observed hyperplasia (high rate of cell division) of the cardiac muscle and of the gut epithelium, especially in the hindgut area. Some degeneration of the muscle tissue, but not of the pancreas, was also observed. Several older moribund individuals (10–14 d posthatching), collected on the bottom of the tanks, showed guts full of prey while some areas of the gut epithelium were degenerated. There is reason to believe that these larvae had passed the point of irreversible starvation and were unable to digest the food ingested.

Oozeki et al. (1989) studied the effect of starvation on the histology of sea-caught and laboratory-reared larvae of stone flounder (Kareius bicoloratus). They found that the epithelial cell heights of the anterior, mid, and posterior parts of the gut, and the heights of the liver and pancreatic cells decreased during starvation. They noted that these quantitative measurements were easy to obtain and did not require interpretation by experienced workers. Theilacker and Watanabe (1989) evaluated the utility of the size of the midgut mucosal cells as an index of starvation in northern anchovy larvae. These cells were selected for study because they were not affected, as were other tissue cells, by autolysis which occurred 3–4 min after death. The size of the midgut mucosal cells was strongly correlated with other histological scores, but was a better predictor of recent feeding history in larvae for which first-feeding had been delayed. This index correctly classified 95% of the fed, 77% of the severely starved, and 74% of the moderately starved larvae.

The process of scoring histological changes related to starvation is typically qualitative. Hence, the quality of the data is strongly coupled to the experience of the observer. O'Connell and Paloma (1981), who conducted a histochemical study of liver glycogen content in larvae of northern anchovy, reported that two experienced readers disagreed on 12% of the samples in three categories (low, medium and high) based on the degree of PAS (Periodic Acid-Schiff) coloration. These differences were reconciled only after the sections had been re-examined and the different ratings discussed. (Similar calibration problems are faced by those who employ otoliths to determine fish ages. Here, strict protocols and interlaboratory exchanges have been instituted to overcome the bias that could result. We are not aware of similar protocols having been initiated by those using histological methods to assess condition in larval fish, and feel that such procedures should be instituted.)

The results of the histologically based studies reported to date suggest that severely emaciated larvae would die of starvation even if food became available because of the permanent loss of digestive abilities following the degeneration of the digestive tract and its associated organs. For this reason, histological examination is likely to be the most reliable method now available with which to identify correctly the point of irreversible starvation or point-of-no-return (PNR) in starving larvae.

2.3. Biochemical Indices

Condition is linked to the quantity of energy reserves available to an organism. However, the storage of energy after feeding, or its mobilization following starvation, is very complex when measured at the cellular level, because of the numerous energy substrates and metabolic pathways available. This reality was highlighted by Frolov and Pankov (1992), who studied the effect of starvation on the biochemical composition of rotifers (*Brachionus plicatilis*). They therefore subdivided major biomolecules into three main classes: (1) substances that are quickly and readily mobilized following starvation, (2) substances forming the bulk of energy reserves, and (3) molecules that are mobilized only during severe starvation and which elicit profound metabolic reorganizations. Frolov and Pankov (1992) suggested that the pattern of mobilization of biomolecules could differ between species. Given what is currently known about biochemical processes during starvation in fishes, small molecules such as glucose, free fatty acids and free amino acids are likely to be included in the first category; larger molecules such as glycogen, lipids, and small soluble proteins in the second; and some lipids and large non-soluble structural proteins in the third (Love, 1970, 1980).

The sequence of mobilization of primary energy stores in adult cod following starvation and refeeding was studied by Black and Love (1986). In these fishes which have less than 3% muscle fat, and which have the majority of their lipids stored in the liver, liver lipids, white muscle and liver glycogen were the first to be mobilized. Liver lipids were also the first to be exhausted about 10 weeks after the onset of starvation, just before liver and white muscle glycogen were completely depleted. Red and white muscle protein and red muscle glycogen begin to decrease 9 weeks after the onset of starvation. Red muscle glycogen was depleted after 22 weeks leaving red and white muscle proteins as the only energy source. When cod were re-fed after 11 weeks of starvation, liver glycogen, red and white muscle glycogen, and plasmatic fatty acids all increased rapidly to above prestarvation levels. The pattern of mobilization described by Black and Love (1986) was also reported for starved barred sand bass (*Paralabrax nebulifer*) by Lowery *et al.* (1987), and was paralleled by a decrease in the activity of glycolytic enzymes. In fish which have high lipid content (>20% muscle fat), such as herring or eels, lipids occur in abundance not only in the liver but also in the muscle tissues, and the pattern of utilization seems to differ slightly. Moon (1983) showed that in American eels (*Anguilla rostrata*) no changes in muscle glycogen occurred when they were starved for 6 months; there was, however, a decrease in muscle protein and an increase in muscle lipids over this interval.

No comprehensive study of the sequence of mobilization of the energy stores following starvation has yet been reported for larval fishes. Early attempts to correlate chemical composition with larval condition employed measurements of proximal components such as carbon, hydrogen, nitrogen and ash (May, 1971; Ehrlich, 1974a, b; Ehrlich, 1975). The carbon/nitrogen ratio, an index believed to reflect the quantity of lipid relative to the quantity of protein, was derived from these measurements. Proteins have a C/N ratio near 3.0, while lipids are carbon-rich molecules. Therefore, when lipids are present, the ratio increases above a value of 3.0 (Harris *et al.*, 1986). In grunion larvae (May, 1971), the C/N ratio decreased following starvation due to fat catabolism; well-fed larvae, however, also exhibited a decrease in C/N ratios with age as a consequence of protein deposition. Increasing trends in the Fulton-K index and decreasing trends of the C/N ratio with age led May (1971) to conclude that the C/N ratio was an unreliable index of condition. Ehrlich (1974a, b) reached a similar conclusion, noting no consistent pattern of change in the C/N ratio during starvation in plaice larvae. Both authors measured the quantity of nitrogen (as an index of proteins), carbohydrates and neutral fats, and found that these constituents exhibited a more or less gradual decline until the larvae died of starvation. Kiørboe *et al.* (1987) reported that in laboratory-reared larvae of herring, a higher C/N ratio (4.54) in yolk-sac larvae than in post-yolk-sac larvae (3.8) was a result of a lower carbon and a higher nitrogen content in the latter. There was also no significant difference in the C/N ratio of starved larvae and larvae fed at different food rations. Moreover, the C/N ratio measured in the larvae reflected the same value as in the copepods used as food. The authors therefore concluded that herring yolk-sac larvae metabolize lipids preferentially and utilize protein for growth. The post-yolk-sac larvae, on the other hand, metabolized body constituents or utilized ingested protein and lipids in proportion to their occurrence in the diet, and did not form significant lipid energy stores but rather channelled the ingested matter into protein and growth.

Several factors might explain the lack of sequential mobilization expected from nitrogen, carbon, and carbohydrates measurements during starvation. The most obvious (also problematic in the evolution of the C/N ratio during starvation) is that nitrogen is associated not only with protein but also with several other molecules (Pfeila and Luna, 1984) including lipoproteins and phospholipids. Hence, protein estimates derived from the nitrogen content are likely to be overestimated. Second, sequential mobilization may be masked by integration of tissue and individual differences resulting from groups of whole larvae being homogenized. Finally it is possible that a decrease in the quantity of one component such as protein (nitrogen) could be compensated for by *de novo* synthesis from non-protein precursors such as carbohydrates or lipids which contribute to the carbon value. Such a continuous flow of molecules through different metabolic pathways complicates, and may confound, analyses of the sequential mobilization of various metabolic compounds during starvation and may, in fact, require simultaneous measurement of all major substrates in order to assess condition accurately.

Ehrlich (1974a, b) computed the quantity of energy obtained from the catabolism of the main chemical components available to starving larvae of herring and plaice prior to the point-of-no-return. The majority of this energy was obtained from the catabolism of proteins (74.3%), followed by triglycerides (20.7%) and carbohydrates (5.0%). This corresponds with the order of importance attributed to food reserves by Smith (1957) and Lasker (1962). Ehrlich (1974a, b) also observed that the very limited lipid and carbohydrate reserves possessed by post-yolk-sac larvae caused protein catabolism to commence almost immediately following the onset of starvation. Nitrogen and carbohydrates were laid down faster than were neutral fat stores (triglycerides) in well-fed first-feeding plaice (Ehrlich, 1974a) and herring (Ehrlich, 1974b) larvae. This finding was interpreted as evidence that food was converted to growth (protein), rather than to energy stores in larval fish which are believed to operate on a tighter energy budget than adults.

Govoni (1980) observed that high carbohydrate storage (in the form of hepatic glycogen) in the larvae of spot had a sparing effect on protein catabolism. He also determined, from histological sections, that these carbohydrate stores were readily metabolized to meet short-term energy needs. O'Connell and Paloma (1981) also reported liver glycogen to be the first line energy reserves in northern anchovy larvae. However changes in liver glycogen were less effective in reflecting the degree of energy deprivation than were other histological characteristics. Margulies (1993) observed that liver glycogen was the first energy reserve to be mobilized at the onset of starvation in wild preflexion scombrids (*Scomberomorus sierra*, *Euthynnus lineatus*, *Auxis* spp.) larvae sampled in the Panama Bight.

Free amino acids are believed to be important energy substrates in developing embryos and in first-feeding larvae (Fyhn and Serigstad, 1987; Fyhn, 1989; Rønnestad *et al.*, 1992a, b). These amino acids are obtained from the egg yolk during embryonic development, and a small proportion of these available egg reserves are incorporated into the embryo body protein. When the yolk is exhausted after hatching, catabolism of body protein becomes the main source of amino acids prior to first-feeding (Fyhn and Serigstad, 1987; Fyhn, 1989). Species which possess oil globule(s) in their eggs are less dependent on amino acids immediately after hatching since they can rely on lipid reserves for a longer interval (Eldridge *et al.*, 1982; Vetter *et al.*, 1983; Rønnestad *et al.*, 1992a). Lipid reserves are, however, relatively small in larval fish, and appear to increase significantly only in the later juvenile stages (Marshall *et al.*, 1937; Balbontin *et al.*, 1973; Ehrlich, 1974a, b, 1975; Fukuda *et al.*, 1986; Suthers *et al.*, 1992), at which time they provide a good buffer against starvation (Love, 1980). This is true of species having high and low lipid stores.

The overall patterns of energy mobilization and storage in fish larvae seem to be similar to those documented for adult fish. Glycogen, free amino acids and fatty acids are utilized as short-term energy stores, and are rapidly depleted following starvation. Lipids yield some energy after glycogen reserves are consumed, but are also rapidly exhausted because of their low quantity in the liver which is not well developed as an efficient storage organ in fish larvae. Structural proteins thus become the ultimate energy substrate sustaining larval fishes during starvation. As in adult fish, it is expected that these proteins are the last to be utilized and the first to be replenished when feeding is resumed.

2.3.1. *Proteins*

Unlike higher vertebrates, in which muscle protein is protected from being metabolized during starvation (Stryer, 1981), fish are well adapted to mobilize these body constituents as fuel for survival (Love, 1970). Because fish contain more than 50% protein by weight, increases in fish biomass are accomplished primarily through protein synthesis (Love, 1970; Bulow, 1970, 1987; Lied and Roselund, 1984). It has been estimated that in adult fish under optimal feeding conditions, 20% of the protein synthesis is allocated to maintenance metabolism while the remaining 80% is incorporated into muscle growth (Roselund *et al.*, 1983). In fish larvae, these proportions are likely to differ because, as Kiørboe (1989) suggested, larvae have a lower rate of protein turnover than adults, and are able to retain a higher proportion of synthesized protein which is laid down as growth.

Lasker (1962), working with sardine (*Sardinops caerulea*) larvae, observed an energy deficit prior to complete yolk-sac resorption. This deficit was evident in sardine larvae (average dry weight of 40 μg) by a resorption of muscle tissue and a loss of dry weight at a rate approaching 2 μg d^{-1}. Ehrlich (1974a, b), Zeitoun *et al.* (1977), Buckley (1979, 1980), Clemmesen (1987) and Buckley *et al.* (1990) all concluded that protein is an important energy source during the period immediately following hatching. In rainbow trout (*Oncorhynchus mykiss*), for example, 19.4% of the protein present in unfertilized eggs had been metabolized or lost at the time of hatching and up to 50% by the time the yolk-sac was fully absorbed (Zeitoun *et al.* 1977). In starved cod larvae, absolute protein content decreased by 45% between 2 and 11 d posthatching although relative protein content, expressed as a percentage of dry weight, did not change (Buckley, 1979). A large part of the weight loss was therefore attributed to a net loss of protein. In winter flounder (*Pleuronectes americanus*) larvae, 45% of the protein present at hatching was lost prior to initiation of feeding 7 d later (Buckley, 1980). Powell and Chester (1985), employing some of Houde and Schekter's (1983) bioenergetic calculations to determine the amount of energy provided by the yolk and oil globule of spot larvae, concluded that these larvae could survive in a food-limited environment only if they catabolized existing tissue (mainly composed of protein). Pedersen *et al.* (1990) demonstrated a decrease in protein content between yolk absorption (4 d posthatch) and 14 d after hatching in groups of larval herring fed at low food levels (15 copepod nauplii/larva). Herring larvae fed high rations (80 nauplii/larva) showed no significant difference in their protein content during the same time interval.

The mobilization of muscle protein following months of starvation in adult cod is known to produce an increase in intercellular space and water content which leads to a negative relationship between protein content and the water content of muscle tissue (Love, 1970). Starved fish are able to undergo considerable degeneration in the cells of their body musculature and still survive, but their buoyancy changes significantly under such conditions. In larval fish, an increase in water content and buoyancy is often observed concurrent with muscle protein catabolism during the interval between hatching and the beginning of exogenous feeding. This has been reported both in laboratory (Blaxter and Ehrlich, 1974; Neilson *et al.*, 1986; Yin and Blaxter, 1987; Mathers *et al.*, 1993) and in wild larvae (Frank and McRuer, 1989). Blaxter and Ehrlich (1974), who examined the buoyancy forces acting on larvae of plaice and herring, determined that buoyancy resulted from the balance between lift forces provided by water and fat content and sinking forces exerted by proteins. Increased buoyancy following starvation was caused primarily by a significant decrease in protein content. Neilson et al. (1986) concluded that for larvae of a given age and size, buoyancy was a better predictor of condition than were morphometric indices.

2.3.2. Lipids

Total fat content was related to condition and growth in adult (Olivier et al., 1979; Wicker and Johnson, 1987) and juvenile fish (Keast and Eadie, 1985). However, no survival value could be assigned to lipid levels. The fat content of larval fish has been measured (Balbontin et al., 1973; Ehrlich, 1974a, b; Fukuda et al., 1986) but has rarely been related to condition because of its high rate of turnover. Although lipids are typically found in small quantities in first-feeding larvae (Ehrlich, 1974a, b), some lipid classes appear to be preferentially used or retained at given periods of fish development (Tocher et al., 1985a, b). Tocher et al. (1985a) and Fraser et al. (1987) reported a predominance of triacyl-glycerols (TAG) over other lipid classes in yolk-sac herring larvae. TAG were also preferentially stored in feeding larvae that were able to meet their basic lipid metabolic requirements from dietary sources. Håkanson (1989a) also reported that TAGs were used preferentially for short-term energy needs by starving northern anchovy larvae, and that cholesterol and polar lipids, which are cell membrane constituents, were conserved except in cases of severe starvation. This phenomenon has been observed in several species of bivalve larvae, in which high TAG levels have been correlated with high growth rates and higher rates of survival to metamorphosis (Gallager et al., 1986). Delauney et al. (1992) found no such correlation in scallops (Pecten maximus).

The level of TAG in larval fish was first proposed as an index of condition by Fraser et al. (1987). However, high variability in absolute TAG levels, which were unrelated to diet in first-feeding larvae, led these authors to modify the index to the ratio of TAG/sterol. This is because sterols, which are important membrane components, are believed to be correlated with the biomass of the animal, and therefore to be unaffected by nutritional status. They are also easily obtained from TLC/FID (thin layer chromatography–flame ionization detection) chromatograms using a non-polar solvent (Fraser et al., 1985; Fraser, 1989). However, Håkanson (1989a), who reared northern anchovy larvae in the laboratory under different starvation and feeding regimes, found that both polar (including sterols) and neutral lipids varied with starvation. Cholesterol levels, in contrast, were unaffected by starvation. Polar lipid–standard length relationships were also more variable than were cholesterol–standard length relationships. He concluded that a ratio of TAG/cholesterol would be superior to a TAG/sterol ratio as an index of condition. TAG/cholesterol ratios smaller than 0.2 were determined as being indicative of poor nutritional condition in anchovy larvae kept in the laboratory (Håkanson, 1989a). Using this criteria, Håkanson (1989b) estimated that from 8 to 27% of the anchovy larvae sampled over a large grid of stations in the Southern California Bight were in poor nutritional condition. He also reported important between-station differences in the proportion of starving larvae as indicated by their TAG/cholesterol ratio.

Relatively few studies have yet employed TAG/sterol or TAG/cholesterol ratios as indices of condition. Problems associated with their use have, however, been encountered due to the small quantities of lipids present in individual marine fish larvae and the difficulty of quantifying these levels. Levels of these constituents are often close to, or below, the detection limits of the chromatography techniques used. For example, Gatten et al. (1983) could detect TAG in herring larvae only 10 d after hatching, when they had begin to feed. McClatchie et al. (pers. commun.), who used TLC/FID techniques, reported no detectable quantities of TAG in 90% of cod larvae that had been feeding for 6 d in the laboratory. Fluorescence techniques, which are more sensitive and require less time and equipment (Gleeson and Maughan, 1986; Nemeth et al., 1986), and the availability of numerous fluorescent binding probes specific for different classes of lipids (Haugland, 1992), might help to resolve these detection difficulties.

Exogenous lipids are lacking in starving larvae and hence endogenous lipids become increasingly important. Under these conditions, it appears that amino acids released through protein breakdown provide most of the energy required to prolong life (Ehrlich, 1974a, b). For adult fish, Love (1970) commented that: "... while a steady depletion of lipid characterizes subsistence without food in almost all cases, the lipid content cannot necessarily be taken as a reliable index of the nutritional status of the fish". Because lipids, like carbohydrates, are often utilized first during starvation, but can be obtained from dietary sources, the TAG/cholesterol ratio of individual larvae can provide a reasonable indication of the time since the last ingestion of planktonic prey. However, it gives only a poor indication of the ability of the animal to sustain a longer period of starvation and to recover from it. It is possible, for instance, that a larvae could be judged in good condition on the basis of a high protein level, and yet exhibit a low TAG/cholesterol ratio simply because these two indices respond on different time-scales. We shall return to this question when comparing the sensitivity, and the time response, of different indices.

The composition of fatty acids in larval fish has also been proposed as an indicator of the quality and the quantity of food intake (Tocher *et al.*, 1985b; Fraser *et al.*, 1987; Klungsøyr *et al.*, 1989; Navarro and Sargent, 1992). Soivio *et al.* (1989) showed that for whitefish (*Coregonus muksun*) larvae, the fatty acid composition of the body lipids often resembles that of their food. Fatty acid composition may also reflect variation in the quantity of non-essential relative to essential fatty acids which are thought to be preferentially conserved during starvation. However, the variety of fatty acid types and the numerous transformations they undergo (desaturation and saturation) render their dynamics difficult to follow. Further, similar fatty acids can occur in different lipid classes, which, in turn, are differentially mobilized during starvation. For example, Frolov and Pankov (1992) measured the fatty acid composition of polar and neutral lipids in starving rotifers (*Brachionus plicatilis*). They observed that the composition of these fatty acids was rather stable, in spite of a decrease of polyunsaturated acids in neutral lipids and an increase in polar lipids. We conclude from the available data that the potential for the use of fatty acid composition as an index of condition in larval fish is low. However, the potential for the use of fatty acids as guides to the composition of the diet, or as biomarkers for studying food webs, may be real.

We now turn to indicators of the rate of synthesis or degradation of some of the energy substrates discussed above. These include nucleic acids as indicators of the rate of protein synthesis, and enzymes which have been suggested as indicators of aerobic and anaerobic metabolism. Digestive enzymes, which are released in the gut of fish larvae when food is ingested, have been used as indicators of the feeding status of fish larvae and will also be reviewed.

2.3.3. *Nucleic Acids*

The quantity of DNA per cell is believed to be a species-constant. In contrast, the quantity of RNA (primarily associated with ribosomes) varies with the rate of protein synthesis. Since growth in fish is mainly accomplished through protein synthesis, the RNA/DNA ratio has been proposed as a short-term (Bulow, 1970) and as a long-term (Haines, 1973) index of growth in adult fish.

Buckley (1979, 1980) found that in the larvae of cod and winter flounder reared in the laboratory, the RNA/DNA ratio was positively correlated to food densities offered and to growth rates achieved. He therefore proposed its use as an index of the nutritional condition of larval fishes. In the larvae of winter flounder, the RNA/DNA ratio more consistently tracked changes in food availability than did individual measures of RNA, DNA, protein content or dry weight (Buckley, 1980). Clemmesen (1987) confirmed these findings for larvae of herring and turbot. All four species (Buckley, 1979, 1980, 1981; Clemmesen, 1987) also exhibited a decrease in protein content following hatching, and a decrease in RNA and protein content at first-feeding, if food was not available. Although first used as an index of nutritional condition for fish larvae, it soon became evident that the RNA/DNA ratio was also an index of larval growth, since good condition is a prerequisite for sustained exponential growth. Significant correlations have now been reported between growth rate, expressed as the increase in protein quantity per unit time, and the RNA/DNA ratio of several species of phytoplankton, molluscs, crustaceans, echinoderms, fish larvae and juveniles (Frantzis *et al.*, 1993).

It is well known that growth of fish larvae is mainly affected by food levels and temperatures (Heath, 1992). The relationship between protein growth rate and the RNA/DNA ratio in larval winter flounder and sand lance (*Ammodytes* spp.) was evaluated at different temperatures by Buckley (1982), and Buckley *et al.* (1984). Under constant food levels, protein growth rates of larval winter flounder were higher at higher temperatures, even though RNA/DNA ratios remained constant (Buckley, 1982). In larval sand lance, food levels had a greater impact on growth than did temperature or the interaction of temperature and food levels. Here too, the RNA/DNA ratio was less variable than was growth in response to rearing temperatures. The addition of temperature as an independent variable in regressions relating RNA/DNA ratios to protein growth rates significantly reduced the unexplained variance in growth rates. Buckley (1984) reported that protein growth rate in larvae of eight different marine species was related to the RNA/DNA ratio and to temperature as follows:

$$G_p = 0.93\,T + 4.75\,RNA/DNA - 18.18\ (r^2 = 0.92) \qquad (1)$$

where G_p = protein growth rate, and T = temperature. Goolish *et al.* (1984) reported a similar relationship for adult carp (*Cyprinus carpio*):

$$G_p = (0.065\,T + RNA/DNA)/0.707\ (r^2 = 0.76) \qquad (2)$$

The terms of these two relationships differ substantially because fish larvae tend to have much higher RNA/DNA ratios at a given growth rate and temperature than do older fish (Mathers *et al.*, 1993). When all temperatures were analysed simultaneously for adult carp, the RNA/ DNA ratio was more highly correlated to growth rate than was RNA/protein or RNA/tissue. However, at a given temperature, all three correlation coefficients had the same magnitude (Goolish *et al.*, 1984). At lower temperatures a higher quantity of RNA was required to achieve a given growth rate (compensation mechanism), emphasizing the fact that RNA activity was lower and therefore temperature-dependent. This was also confirmed by Ferguson and Danzmann (1990) who worked on juveniles of rainbow trout and by Foster *et al.* (1992, 1993) who worked on juveniles of cod. Mathers *et al.* (1993) measured the RNA/DNA ratio and the protein growth rate of rainbow trout fry of different ages (21, 30, 58 and 65 d posthatch) maintained at different temperatures in the laboratory. They found no significant differences in RNA concentration between fry maintained at 5, 10 and 15°C. They explained the absence of RNA compensation mechanisms at lower temperatures by the fact that larvae already possess very high RNA levels relative to juveniles, and that compensation at lower temperature could only be achieved through increased RNA activity. They also computed the RNA/protein ratio, an index of RNA efficiency, and showed that the ratio doubled between 5 and 10°C indicating that the rate of protein synthesis does vary independently of RNA concentrations at different temperatures. Temperature effects must therefore be assessed, if estimates of protein growth rates are to be obtained from the RNA/DNA ratio of field-collected fish larvae.

If the protein growth rate of fish larvae is assigned a value of zero in Buckley's (1984) model, the critical RNA/DNA ratio can be calculated after rearranging Eq. (1):

$$RNA/DNA_{crit.} = (19.18 - 0.93T)/4.75 \qquad (3)$$

This critical ratio, which typically ranges from 1.0 to 3.0 depending on temperature, was used by Robinson and Ware (1988), Hovenkamp (1990) and McGurk *et al.* (1992) to calculate the proportion of animals showing negative growth rates. Robinson and Ware (1988) reported that the critical RNA/DNA ratio corresponded closely to RNA/DNA ratios exhibited by cultured larval Pacific herring (*Clupea pallasi*) at the point of irreversible starvation. They also reported a decrease in standard length and a net catabolism of protein (as indicated by predicted negative protein growth rates) soon after complete yolk absorption. Buckley and Lough (1987) and Robinson and Ware (1988) were the first to report RNA/DNA ratios for field-collected larvae. In both cases, the frequency distribution of the RNA/DNA ratio was log-normal and truncated on the left side, suggesting growth-selective (or condition-selective) mortality in larvae characterized by RNA/DNA values below the critical ratio.

Hovenkamp (1990) estimated the growth of larval plaice sampled in the North Sea using three different methods: (1) width of otolith growth increments, (2) size-at-age data, and (3) RNA/DNA ratios based on Buckley's (1984) model (Eq. (1)). The results obtained from the three methods were consistent. However, more recently, Bergeron and Boulhic (1994) evaluated Buckley's (1984) growth model (Eq. (1)) using laboratory-reared sole (*Solea solea*) and found no direct relationship between observed and predicted specific protein growth rates. They concluded that extreme caution should be used when employing Buckley's model to predict growth rates of larval fishes, for species other than those used to develop the model. Canino *et al.* (1991) also employed this model to calculate the protein growth rate of field-collected larvae of walleye pollock (*Theragra chalcogramma*), and found that most specimens sampled on two different cruises had negative growth rates despite normal ingestion rates. They suggested that intercalibration studies of RNA/DNA ratios should be conducted to reconcile the results obtained using different standards, buffers and extraction techniques.

One major limitation of the methods used by Buckley (1979, 1980, 1981, 1982, 1984), Buckley *et al.* (1984) and Buckley and Lough (1987) is the large quantity of material ($>800 \mu g$) required for accurate nucleic acid determination. Such quantities can only be obtained by pooling many individuals when marine fish larvae are assayed. To overcome this limitation, Clemmesen (1988) proposed new methods for determining nucleic acids in individual fish larvae. Individual nucleic acid determinations have, however, revealed a large and unexplained inter-individual variability in the RNA content and in the RNA/DNA ratios of fed larvae reared under identical conditions (Clemmesen, 1988, 1989, 1992; Robinson and Ware, 1988; Westerman and Holt, 1988; McGurk and Kusser, 1992; Ueberschär and Clemmesen, 1992; Clemmesen and Ueberschär, 1993; Mathers *et al.*, 1993). Westerman and Holt (1988) suggested that this variability between individuals might be caused by periods of cell proliferation and growth during the development of organs. These findings led Bergeron and Boulhic (1994) to question the reliability of RNA/DNA ratios as indices of the condition or growth rate of individual larval fish.

The rationale underlying the use of the RNA/DNA ratio dictates that DNA should increase during development since DNA content per cell is constant and the total number of cells increases as the animal grows. However, Clarke *et al.* (1989) found no significant relationship between DNA content and cuttlefish (*Sepia officinalis*) (Mollusca: Cephalopoda) size. It should be noted, however, that their reports of some negative RNA values suggest that their DNA values may have been overestimated. We have experienced, as did Clarke *et al.* (1989), the problems of negative RNA values when using Hoechst 33258 (the dye recommended in Clemmesen (1988) protocol), because this sensitive dye can yield high fluorescence in the presence of SDS (a strong detergent used to disrupt the cells) or other extraction solvents. We found it necessary to extract buffer blanks in a manner identical to that used for samples in order to correct for this effect (Ferron et al. pers. data). The rationale underlying the use of the RNA/DNA ratio also dictates that DNA should not vary with starvation. However, exceptionally high DNA values (μg/larvae) have been reported in starved post-yolk-sac larvae of cod reared in the laboratory and in an outdoor pond (Raae et al., 1988). This was argued as being the result of rapid unscheduled DNA synthesis following a collapse of cellular control mechanisms resulting from inadequate nutrition. While the possibility of a link between high DNA synthesis and hyperplasia of the digestive tract epithelium and liver in starving larvae, as proposed by Cousin et al. (1986), cannot be excluded, the possibility of non-rigorous methodological procedures should first be addressed. Several recent studies have confirmed that the observed variability in the RNA/DNA ratio results primarily from variation in RNA levels (Clemmesen, 1988, 1989, 1992; Ferron, 1991; Bergeron and Bouhlic, 1994). Recent work by Clemmesen (1988, 1993) and McGurk and Kusser (1992) has also clearly demonstrated that purification and extraction of nucleic acids is an essential step in the accurate determination of RNA/DNA ratios in larval fishes. This purification was not a component of the standard protocol used prior to Clemmesen's (1988) study, and conclusions based on these earlier analyses should therefore be interpreted with caution. When highly purified material and a DNA-specific dye such as Hoechst 33258 are employed, DNA variability (coefficient of variation) can be reduced to about 8.4% (Clemmesen, 1988). This contrasts with variability as high as 35.4% when RNAse digestion and non-specific dyes such as ethidium bromide are employed (Clemmesen, 1988). Most recently, Clemmesen (1993) has recommended the use of ethidium bromide rather than Hoechst 33258, because of quenching and self-fluorescence problems associated with the latter. We have been successful in overcoming self-fluorescence problems associated with Hoechst 33258 by washing the extracted homogenates with diethyl ether to remove traces of phenol and chloroform, and by reading the fluorescence at a low temperature (1°C) to reduce background noise (Ferron et al. pers. data).

Several studies in which RNA/DNA ratios were employed have reported an increase in DNA content (expressed as a percentage of dry weight) concurrent with the decrease in the RNA/DNA ratio associated with starvation in fish larvae (Buckley, 1979; Clemmesen, 1987; Raae et al., 1988; Richard et al., 1991). This appears to occur because starved animals lose weight while DNA is conserved. Hence, a ratio of DNA to dry weight (DNA/DW) has potential as an indicator of starvation. Berg-eron *et al.* (1991) and Richard *et al.* (1991) reported that fed sole larvae exhibited lower between-individual variability in DNA/DW ratios than in RNA/DW or RNA/DNA ratios. They therefore suggested the use of the DNA/DW ratio in studies of larval fish condition because it appears to reflect more accurately condition changes related to starvation. The approach has the added advantage that it is easier to obtain reliable DNA values in isolation than it is to assess DNA and RNA simultaneously. However, the DNA/DW ratio does not allow evaluation of the growth potential, and of the ability of the larvae to recover from starvation at the time of sampling, as does the RNA/DNA ratio. Moreover, reduction in individual variability should not be seen as an end in itself, because there is no reason to expect individual larvae to respond similarly in a common feeding environment. Given the reliability of protocols assessed to date, simultaneous evaluation of DNA, RNA, protein and total dry weights of individual larvae can easily be achieved. This allows various ratios (RNA/DNA, DNA/DW, RNA/DW, RNA/prot, prot/DNA) to be calcu-lated and used in order to investigate details of growth response and condition (Lone and Ince, 1983). Multivariate approaches to the analysis of such data, similar to those now applied to morphometric variables, should also be explored (Farber-Lorda, 1991). Buckley and McNamara (1993) have recently used nucleic acid probes to examine changes in 18S rRNA, and actin and myosin mRNA in fed and starved larvae of cod. The use of these molecular techniques to assess larval fish condition, however, remains in its infancy and it is premature to assess the significance of these results.

2.3.4. *Digestive Enzymes*

In an effort to assess the nutritional status of fish larvae, some workers have quantified the content and activity of various digestive enzymes. These enzymes, which are known to be produced and secreted when food is ingested, have activity levels that should reflect the extent to which fish larvae have been feeding. Because most early stage larval fish are stomachless and possess only a functional intestine, trypsin-like enzymes are the main proteolytic enzymes produced (Vu, 1983; Hjelmeland *et al.*, 1984; Govoni *et al.*, 1986). However, their activity can arise from two sources, endogenous (produced by the larvae) or exogenous (obtained from the guts of ingested food items). Repeated studies have shown that these two sources can contribute equally to the total enzyme activity measured (Dabrowski and Glogowski, 1977; Dabrowski, 1982; Munilla-Moran *et al.*, 1990). Dabrowski (1982) measured pepsin and trypsin-like activity in starving pollan (*Coregonus pollan*) and rainbow trout larvae and found that the enzymes were at their lowest level at the point of irreversible starvation.

Three problems confront the measurement of trypsin-like enzyme levels: (1) the presence of proteolytic enzymes other than trypsin can react with synthetic substrates, (2) the presence of trypsin inhibitors in homogenates of fish larvae can cause underestimation of enzyme activity, and (3) the presence of trypsinogen, the inactive proform of trypsin, can contribute to the determination. Hjelmeland and Jørgensen (1985) proposed a radioimmunoassay method to overcome these problems. Using these methods, Pedersen et al. (1987) reported that intestinal tryptic enzyme levels in larval herring were positively related to the number of planktonic prey in the gut, at values up to five prey. Beyond that prey level, an asymptote in enzyme activity was reached. They also estimated the enzyme/substrate ratio to be between 1 and 4. However, Hjelmeland et al. (1988) found no significant correlation between trypsin activity and the number of prey ingested as a result of large variations observed between individuals. Hjelmeland et al. (1984) observed that only growing cod larvae produced typsin and its precursor trypsinogen. They suggested that the level of these compounds could be used as a growth index and as an indicator of nutritional condition. Their results were, however, confounded by distinct phases of enzyme activity related to specific development stages.

Ueberschär (1988) developed a fluorometric method to measure the tryptic enzyme content of individual fish larvae. The method is equal in sensitivity to radioimmunoassays, but is less trypsin-specific because 7% of the proteolytic activity measured is attributable to proteolytic enzymes other than trypsin. Fluorescence determination (FL) and radioimmunoassay (RIA) do, however, give similar results and significant ($P < 0.05$) correlations have been obtained between (RIA) quantity and (FL) activity of trypsin in all tissues of larval herring (Ueberschär et al., 1992), except the pancreas in which trypsinogen leads to overestimated trypsin content when assessed by RIA methods.

Adult fish muscle is rich in proteolytic enzymes and this apparently facilitates the catabolism of body tissues when energy is required after long periods of starvation (Love, 1980). Larval fish, too, contain endogenous proteolytic enzymes which are probably produced under the same starvation situations. However, to date, the relative contribution of such endogenous sources to the total proteolytic enzyme activity in a larvae has not been adequately assessed. As is the case with triglycerides, proteolytic enzymes are greatly affected by exogenous sources of food, hence their activity is likely to be more indicative of short-term feeding status than longer-term condition and susceptibility to starvation.

2.3.5 Metabolic Enzymes

Quantification of the activity levels of key enzymes involved in different metabolic pathways has also been suggested as a means of assessing condition in fish. The activity of various metabolic enzymes has provided good information on the rate of synthesis or degradation of major energy substrates (Moon and Johnston, 1980; Moon, 1983) as well as the organs affected during starvation in juvenile and adult fish (Goolish and Adelman, 1987, 1988; Lowery et al., 1987; Lowery and Somero, 1990). Citrate synthase (a Krebs cycle enzyme) and cytochrome c oxidase (an enzyme of the oxidative phosphorylation) are two enzymes for which activity is believed to be related to feeding-induced increases in aerobic metabolism. In contrast, lactate dehydrogenase (an enzyme of the Cori cycle) activity is thought to vary with the rate of feeding, anaerobic metabolism and locomotory performance (Sullivan and Somero, 1980). Mathers et al. (1992) reported significant correlations between individual growth rate and the activity of citrate synthase, cytochrome oxidase, lactate dehydrogenase and the amount of RNA in the muscle of adult saithe (Pollachius virens), supporting the hypothesis of a link between growth rate and aerobic and anaerobic capacity.

For larval fish, the interest in metabolic enzymes as tools for assessing condition is relatively new, and has been restricted to the use of two major enzymes, citrate synthase (CS) and lactate dehydrogenase (LDH). Clarke et al. (1992) measured LDH and CS activity in fed and starved larvae of red drum (Sciaenops ocellatus) and lane snapper (Lutjanus synagris). Although LDH and CS activities changed markedly with age, independently of food densities, the LDH activities were significantly influenced by the quantity and the quality of food offered. The enzyme activities were also found to be affected by temperature, although this effect was not quantified. Mathers et al. (1993) reported no significant correlation between CS and LDH activity and RNA concentration and protein growth rate in fry of rainbow trout. Despite this, the activity of the two enzymes differed significantly between fed and starved individuals, but not between individuals fed at different rations. It seems from the limited data available that metabolic enzymes can be affected by the feeding status of larval fish. However, it is important to realize that metabolic enzyme activity provides no more than a "snapshot" of the current rate of metabolism of the animal. Such measures will need to be combined with some estimates of the quantity of various substrates involved if an assessment of condition in response to past food deprivation is to be attempted.

2.4. Sources of Variability other than Nutrition

The main source of variability evaluated in studies dealing with the measurement of condition in larval fish has been food availability, since starvation was long believed to be the predominant source of mortality for first-feeding larval fish (reviewed in May, 1974; Leggett, 1986). However, when studying the impact of food availability on the condition of larval fish, it is vital that other potential sources of variability are identified and controlled. Such sources have been identified and are briefly detailed below.

For morphological measurements, Blaxter and Hempel (1963) and Hempel and Blaxter (1963) demonstrated that length, height and weight of herring larvae are influenced by the spawning stock from which the specimens originated. Variability in the morphology of individual fish larvae, often originating from egg size variation, is also known to have maternal sources (Wilson and Millemann, 1969; Beacham and Murray, 1985; Knutsen and Tilseth, 1985; Marsh, 1986; Bengtson et al., 1987; Meffe, 1987, 1990; Hislop, 1988; Chambers et al., 1989; Zastrow et al., 1989; Buckley et al., 1991b; De March, 1991; McEvoy and McEvoy, 1991; Panagiotaki and Geffen, 1992). Martin and Wright (1987) also reported between-year differences in the morphometric condition of laboratory-reared larvae of striped bass (Morone saxatilis), which they assigned to genetic differences between broodstocks.

When grading histologically fed and starved anchovy larvae, O'Connell (1976) noticed the presence of several moderately and severely emaciated specimens in the fed group. He concluded that food availability may not always be the underlying cause of declining condition. When examining the condition of larvae collected at sea, O'Connell (1980, 1981) noted that three-quarters of the emaciated larvae were concentrated in four of the 64 net tows. However, numerous emaciated larvae were found to occur within samples characterized by larvae in good condition that were collected in areas offering good feeding conditions. O'Connell attributed this low condition to genetic factors, accident, or chance failure to capture the available food. Changes in the glycogen content of hepatocytes seems to be under the influence of rhythmical (diel) feeding periods related to time of the day (O'Connell and Paloma, 1981; Segner and Möller, 1984; Sieg, 1992b, 1993). Rather than accounting for this source of variability, it has been suggested that other histological criteria insensitive to diel changes be used. However, three additional histological traits (the number of inclusions present in hindgut mucosal cells, the number of midgut mucosal cells and the inflation of the swimbladder) were also found to vary on a diel basis in jack mackerel larvae sampled at sea (Theilacker, 1986). Increased hindgut inclusions, a lower rate of mitosis of midgut mucosal cells, and a deflation of the swimbladder during daytime were also reported to be associated with feeding activity. At night, the empty gut produced an inflation of the swimbladder, a disappearance of hindgut inclusions and a higher rate of midgut mucosal cell mitosis (Theilacker, 1986).

Among the different sources of environmental variability shown to influence condition in fish, the largest number has been associated with biochemical indices. For nucleic acids and protein measurements, these include biotic sources such as maternal origin (Buckley *et al.*, 1990, 1991a, b; Ferron *et al.*, pers. data), diel variations (Ferron, 1991; Mugiya and Oka, 1991) and parasitism (Steinhart and Eckmann, 1992), all of which can produce strong confounding effects. Several abiotic sources of variability have also been identified. These include water temperature (Buckley, 1982; Buckley *et al.*, 1984, Jürss *et al.*, 1987), salinity (Jürss *et al.*, 1986), toxicant levels (Rosenthal and Alderdice, 1976; Barron and Adelman, 1984, 1985), oxygen levels (Peterson and Brown-Peterson, 1992) and lunar cycles (Farbridge and Leatherland, 1987). Variability in lipid-based indices are suspected to result from diet composition (Gatten *et al.*, 1983; Klungsøyr *et al.*, 1989), diel periodicity in lipid metabolism and deposition (De Vlaming *et al.*, 1975; Farbridge and Leatherland, 1987; Boujard and Leatherland, 1992), maternal source (Tocher and Sargent, 1984; Fraser *et al.*, 1988; Zastrow *et al.*, 1989) and pollution stress (Fraser, 1989). Tryptic enzyme levels are affected by time of day (Pedersen *et al.*, 1987), higher values occurring in the afternoon relative to morning. Variability in trypsin activity levels in larval herring has also been attributed to interstock differences (Pedersen *et al.*, 1987). In the balance of this review, we focus on food availability as the main source of environmental variability, because it is the only source for which different aspects of condition have been documented.

3. SENSITIVITY

Variability in the ocean's physical forcing occurs throughout a spectrum of space and time scales ranging from seconds to years, and from centimetres to hundreds of kilometres (Harris, 1980; Denman and Powell, 1984; Horne and Platt, 1984). Plankton patchiness is produced through biological responses to such forcing, and as a consequence of interactions between the numerous species comprising the plankton (Platt and Denman, 1975; Steele, 1978; Mackas *et al.*, 1985). Marine fish larvae thus experience during their planktonic life a dynamic, diverse and patchy feeding environment. It is therefore necessary to consider the time

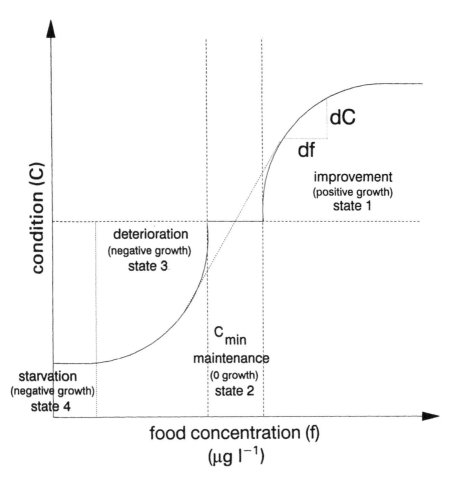

Figure 1   Expected relationship between condition (*C*) and food concentration
(*f*) based on bioenergetic models proposed by Kamler (1992). The sensitivity (*S*)
of a particular index of condition is defined by ($S = dC/df$), where d*C* = changes
in condition and d*f* = changes in prey concentration. Note that sensitivity is low
(small d*C*/d*f*) at or near high and low food concentrations and at food
concentrations yielding near maintenance rations.

response of larval condition to varying food levels, if the dynamic nature
of processes relating food availability to condition and survival are to be
studied effectively.

We define sensitivity as the minimum environmental change (e.g. food
ration) that can be detected by a particular index of condition (Figure 1).
The sensitivity of the condition index employed is of primary importance,
particularly in situations where small changes in environmental conditions

lead to significant short time-scale changes in growth rate and/or survival probabilities. The term "sensitivity" has often been used to express the time duration of a particular set of environmental conditions necessary to induce changes in a given condition index (which we define below as latency) or to express the rate of change in the index of condition after a response has been elicited (which we define as dynamics). In reality, sensitivity is likely to be a more useful measure than latency or dynamics when applied to field studies, since a total lack of food is likely to be improbable, and temporal and temporary changes in food availability are more likely to be commonly experienced by wild larvae (Vlymen, 1977; Lasker and Zweifel, 1978; Hunter, 1981; Rothschild and Rooth, 1982). To measure sensitivity effectively and to quantify the extent to which individual indices differ in their sensitivity, it is necessary that their response be evaluated against a common source of variability (e.g. food ration) and over a range of conditions that produce positive responses. Below, we propose a bioenergetic framework within which the sensitivity of condition indices to changes in the rate of food ingestion may be assessed.

Bioenergetic models initially developed for adult fish, and subsequently applied to younger life stages, indicate that all the energy assimilated from ingested food by larvae is used for growth, metabolism and excretion (Laurence, 1977; Houde and Schekter, 1983; Checkley, 1984; Kiørboe et al., 1987; Houde, 1989; Kiørboe, 1989; Yamashita and Bailey, 1989). A standard bioenergetic model (see Kamler, 1992) describing this relationship is:

$$C = R + P + F + U \qquad (4)$$

where C is consumption (energy assimilated from ingested food), R is respiration (energy losses through aerobic and anaerobic metabolism), P is production (energy used for growth), F is faeces (energy losses through faeces production), and U is non-faecal excretion (energy losses through osmoregulation and urine production). Kamler (1992) identified four different feeding states based on Eq. (4) which are directly relevant to the consideration of larval fish condition. In state 1, $C > 0$, $C > C_{min}$ (C_{min} is maintenance consumption), growth rate increases with consumption to an asymptote defined by the maximum rate at which food can be processed. In state 2, $C > 0$, $C = C_{min}$, the energy budget becomes $C = R + F + U$ because growth is zero across some finite range of consumption rates, the width of this range being a function of variable maintenance costs associated with the animal's ability to regulate metabolism through behavioural or physiological means. In state 3, $C > 0$, $C < C_{min}$, negative growth (catabolism) increases as consumption declines toward zero, the energy budget being then expressed by $C + P = R + F + U$. Finally, state 4 relates to total deprivation of food $C = 0$, $F = 0$, the energy budget is then given by $P = R + U$, negative growth reaches an asymptote defined by the maximum rate at which stored energy (protein/fat) can be catabolized at that time.

If, as a first approximation, we consider condition to be a reasonable correlate of growth, condition should then be related to the energy status of the larvae. Moreover, detectable changes in condition should occur only at levels of prey consumption intermediate between maintenance requirements and the positive and negative asymptotes at which the rate of change in condition (dC) relative to the rate of change in consumption (ingestion) (df) differs from zero ($dC/dF \neq 0$), as illustrated in Figure 1. The sensitivity of a given condition index will thus be related to both the minimum value of df required to elicit a measurable change in dC, and to the particular position of an individual larva on the consumption continuum. Moreover, it is expected that sensitivity will be maximal at consumption levels immediately above and below maintenance, and minimal or insensitive within the range of consumption values satisfying maintenance requirements, or near the upper and lower asymptotes, characterized by starvation or *ad libitum* regimes (Figure 1). The width of the maintenance range is likely to differ depending on the condition measure used (narrower for short-term and wider for long-term energy substrates) and the size of the animal (narrow for small and wider for large organisms).

3.1. Laboratory Studies

Until recently, the influence of starvation on larval survival was the primary focus of research in larval fish ecology, and most laboratory experiments were therefore designed to evaluate the effect of total food deprivation on growth and condition. Relatively few studies specifically examined the sensitivity of condition indices to sustained or intermittent differences in food levels. As a consequence, knowledge of the sensitivity of commonly used condition indices to food availability is weak, unsystematic, and frequently confused. We review below the most reliable of these, with the goal of (1) assessing the extent to which they support or reject the existence of the relationship depicted in Figure 1, and (2) attempting to quantify the relationship if it is supported.

Wyatt (1972) reported linear relationships between the logarithm of body height/length against time for plaice larvae fed for 7 d in the laboratory. The slope of these relationships increased in a non-linear fashion as food densities increased (50, 100, 200, 500 and 1000 *Artemia* nauplii l^{-1}) and approached an asymptote at the highest prey concentration. Approximately 78% of the variability in mean height/length could be predicted from food levels offered in these experiments. The observed non-linear increase in condition is consistent with that predicted for ration levels above maintenance (Figure 1). The point at which $C = C_{min}$ cannot be defined, because improvement of condition (positive growth) was recorded at all food concentrations. Therefore under the conditions experienced in these experiments, maintenance requirements were probably near or below 50 nauplii l^{-1}, the lowest food level provided. Using a mean weight of 1.85 μg for an *Artemia* nauplius (Table 1 of MacKenzie *et al.*, 1990 used for weight conversions), the maintenance food density under the conditions experienced was likely to be $\leq 92.5 \mu g\, l^{-1}$, and the asymptotic value $\geq 1850 \mu g\, l^{-1}$.

Laurence (1974) provided laboratory-reared haddock (*Melanogrammus aeglefinus*) larvae with natural plankton at densities of 10, 100, 500, 1000 and 3000 plankters (pl) l^{-1}. Mortality was 100% after 2 and 3 weeks at densities of 10 and 100 pl l^{-1}. Assuming rearing conditions were adequate in other regards, these prey levels may be assumed to result in consumption rates $C < C_{min}$. Unfortunately, the relationship between condition and maintenance feeding rates cannot be deduced from these experiments because condition was not reported for food levels in the range 10–100 pl l^{-1}. Standard length, the dry weight and the Fulton-K index (weight/length^3) increased at rations of 500, 1000, and 3000 pl l^{-1}. Growth rates, expressed as changes in both standard length and dry weight, exhibited a consistent increase with increasing food density. These results, too, are consistent with the predicted non-linear response of condition to changes in consumption (prey density). The relatively small differences in the slope of standard length and dry weight at prey densities of 1000 and 3000 pl l^{-1} suggest that the upper asymptote occurs within this range. C_{min} appears to have occurred at prey densities of 100–500 pl l^{-1}. The plankton offered to the larvae was dominated by nauplii of *Acartia clausi*, *Centropages hamatus* and *Balanus balanoides*. Based on the average weight of an *Acartia tonsa* nauplius (0.26 μg), the maintenance concentration under the conditions employed in these experiments is estimated to be in the range of 26–130 $\mu g\, l^{-1}$, and the upper asymptote value $>260 \mu g\, l^{-1}$.

Buckley (1979) reported that the RNA/DNA ratio of larvae of cod, 7–11 d posthatch, was positively correlated with the concentrations of natural plankton (0, 200, 1000 prey l^{-1}) offered in laboratory rearings. The fact that larvae maintained at prey densities of 200 prey l^{-1} exhibited a RNA/DNA ratio typical of starved larvae suggests that 200 prey l^{-1} may be near the maintenance level for larvae of this species and age under the rearing conditions employed. Buckley *et al.* (1984) examined the effect of three temperatures (5, 7 and 9°C) and three rotifer (*Brachionis plicatilis*) densities (200, 500 and 1000 pl l^{-1}) on the RNA/DNA ratio of laboratory-reared larvae of sand lance (*Ammodytes* spp.). A two-way analysis of variance indicated that RNA content and RNA/DNA ratios were significantly ($P < 0.01$) affected by rotifer concentrations and the age of the larvae. The multiple linear regression,

$$RNA/DNA = 1.20 \text{ rotifers ml}^{-1} + 0.03 \text{ Age} + 1.40 \ (r^2 = 0.76) \quad (5)$$

explained 76% of the variability in the RNA/DNA ratio. Instantaneous protein growth rates for all treatments ranged from -6.7 to 12.2% d^{-1}, and were found to be between -3.0 and 3.0% d^{-1} for the larvae fed at densities of 200 pl l^{-1}. Assuming that all energy above maintenance requirements was channelled into protein growth, it is reasonable to conclude that C_{min} occurred near the prey density of 200 prey l^{-1} which yielded positive and negative protein growth rates. Given a mean weight of 0.16 μg for a rotifer nauplius, this corresponds to 32 μg l^{-1}.

Pedersen *et al.* (1990) measured the effect of two different daily rations (high = 80 and low = 15 prey larva^{-1} d^{-1}) on the growth and digestive capacity of three groups of larval herring. *Acartia tonsa* nauplii were fed during the first week after first-feeding (4–11 d posthatch) and a mixture of nauplii and copepodites thereafter. The first group, exposed to a high ration for 31 d after yolk absorption, had higher trypsin and trypsinogen content than did a second group fed at low ration for 10 d and then switched to high ration, and a third group fed at low ration throughout the experiment. Between 4 and 14 d posthatch, all the three groups increased in length but decreased in protein content, indicating a decoupling of growth in length and weight. Between 14 and 35 d posthatch, the first group fed throughout at high ration grew at the highest rate and the second group (switched from low to high ration) showed compensatory growth in both standard length and protein content when compared with the high ration group. However, the trypsin activity of the second group failed to reach the high levels found in animals exposed continuously to high ration. The third group, which was fed a low ration throughout, exhibited reduced growth rate (3.81 μg protein d^{-1}, characteristic of slightly above maintenance level), as opposed to approximately 12 μg protein d^{-1} in the other two groups. These findings suggest that under the experimental conditions employed, trypsin and trypsinogen levels, and growth in length and protein content, were sensitive to ration differences in the range of 65 prey larva^{-1} d^{-1}. Given the larval stocking densities and the container size used, the high and low rations corresponded to food concentrations of 100 and 536 prey l^{-1}, or 26 and 140 μg l^{-1} (assuming a mean weight of 0.26 μg for

Acartia tonsa nauplius). However, the actual prey concentrations were likely to be much higher than these, because copepodites were provided in addition to nauplii.

The results of the above laboratory studies suggest that prey concentrations of 25–35 μg l^{-1} are sufficient to meet maintenance requirements and that the shape of the curve relating condition to prey levels at prey densities above maintenance values will be as predicted in Figure 1. The results also suggest that the upper asymptote will be reached at prey levels of 140–1850 μg l^{-1}. Values reported in the range of concentrations leading to below maintenance rations level (negative growth) were inadequate to assess the relationship between ration and condition. Therefore, the shape of the relationship between condition and prey concentration at below maintenance ration levels remains hypothetical. While the data are presently too limited to provide adequate resolution of its shape throughout the full prey concentration continuum, the curve depicted in Figure 1 can probably be approximated to a sigmoid curve, given the narrow range of prey concentrations yielding near maintenance rations. The uncertainty in the above estimates derives primarily from the small number of prey concentrations surveyed, and the large variance typically associated with individual estimates of food density in such experiments. Other possible sources of uncertainty include the inherent variability in ingestion rates of individual larvae at a given prey density (Kiørboe and Munk, 1986), and the variable calorific intake achieved from different prey types. It must be noted, too, that the application of these laboratory findings to field situations should be approached with caution (see MacKenzie *et al.*, 1990; MacKenzie and Leggett, 1991).

3.2. Field Studies

The problem of resolving small-scale patchiness renders the quantification of food levels to which larvae are exposed in the field far more difficult than is the case in laboratory experiments (Rothschild and Rooth, 1982; Houde, 1982). In addition, recent studies of the relationship between prey densities, microscale turbulence levels and prey ingestion rates (MacKenzie and Leggett, 1991) indicate that prey densities alone may be a poor predictor of feeding rates in larval fishes. As a consequence, only a very small number of studies provide insights into the shape and quantitative character of the relationship between larval condition and measures of food availability, and of the sensitivity of the various indices that have commonly been applied in field studies. Shelbourne (1957), O'Connell (1980), O'Connell and Paloma (1981), Koslow *et al.* (1985), Buckley and Lough (1987), Håkanson (1989b) and Canino *et al.* (1991)

were successful and Owen *et al.* (1989), Powell *et al.* (1990), Ferron (1991) and McGurk *et al.* (1992, 1993) unsuccessful in detecting a positive relationship between larval condition (assessed by different means) and various indices of *in situ* food concentrations. Below, we report the results from some of these studies which provide some insight into the sensitivity of condition indices to prey concentrations measured at sea.

O'Connell and Paloma (1981) reported that the liver glycogen content of field-collected northern anchovy larvae was more sensitive to changes in food levels than were histological scores. The authors concluded that glycogen represents the first line of energy reserves in the larvae, whereas histological scores are more representative of the extent of food deprivation and hence of the magnitude of the emaciation suffered. These conclusions are consistent with the expected higher sensitivity (and probably narrower maintenance range) of a short-term energy source such as glycogen, relative to longer-term energy stores such as protein and lipids which underlie changes in histological scores.

Powell *et al.* (1990) assessed morphological condition, as described in Powell and Chester (1985), on spot larvae sampled along a transect across the Mississippi river plume front, in order to determine their nutritional condition in relation to local food availability. Despite large intrastation variability in condition, these authors reported a weak negative relationship between the proportion of larvae clasified to be in starving condition and the proportion of larvae with food in their gut. This was believed to result from the fact that measures of gut fullness and condition respond on different time scales: gut content on a scale of hours, and morphometric condition on a scale of days. In this study then, condition was independent of gut content, as measured at the time of sampling, notwithstanding the statistical relationship between the two. This finding emphasizes the importance of considering the time response of condition indices (to be discussed in the next section) in addition to sensitivity, if the relationship between feeding success and condition is to be assessed reliably.

Suthers *et al.* (1992) used otolith indices (a measure of recent growth), triglyceride levels, dry weight, standard length and anal body depth of juvenile cod sampled on the Scotian Shelf to generate condition indices based on residuals of univariate regressions of these values on standard length. When these indices were compared with estimates of zooplankton biomass ($mg\,m^{-3}$) sampled coincident with the fish, triglyceride and morphometric condition indices were found to be insensitive to the food levels experienced, despite an order of magnitude variability in food abundance. The authors concluded that the animals sampled were capable of integrating small-scale variability in prey abundance. An alternative explanation could be that the maintenance range (Figure 1) was wider than the variability in the food levels encountered. This is plausible since larger and older fish have a higher rate of turnover of the main energy substrates (including triglycerides), and a greater capability of adjusting their metabolism to the quantity of energy available (Laurence, 1977), therefore minimizing any growth or condition changes resulting from differences in prey densities encountered. The greater specialization of storage organs such as the liver in older fish is also known to contribute to greater access to, and stability of, short-term energy sources such as glycogen (Margulies, 1993).

Canino et al. (1991) measured the RNA/DNA ratio of walleye pollock larvae in relation to prey concentrations and ingestion at four stations and on two different cruises in Shelikof Strait (Gulf of Alaska). *Pseudocalanus* spp. nauplii and eggs comprised the bulk of the diets, and occurred at concentrations of 12.5–23.1 nauplii l^{-1} (8.8–16.2 $\mu g\, l^{-1}$). Computing the critical RNA/DNA ratio Eq. (3) indicated that all larvae on one cruise, and 89–92% on the other cruise, exhibited negative protein growth rates indicative of starvation. Daily rations estimated for the same two cruises ranged from 7.2 to 27.9 prey $larva^{-1}\, d^{-1}$. While only average gut fullness were reported for each of the four sampling sites, average gut fullness was positively related to food availability (nauplii l^{-1}), and to the RNA/DNA ratio of the fish larvae collected at the same site. Based on laboratory bioenergetic calculations provided by Yamashita and Bailey (1989), Canino et al. (1991) also determined that a 5.5 mm pollock larvae would require 69 nauplii d^{-1} ($\approx 25.5\ \mu g\, d^{-1}$) to obtain the weight-specific ration of 21.5–29% of body dry weight d^{-1} required for maintenance and growth. Considering a search volume of 4.57 $l\, d^{-1}$, a daily feeding time of 16 h, and a capture efficiency of 15.2%, Canino et al. (1991) determined that pollock larvae would require prey concentrations of 190 prey l^{-1} or 70 $\mu g\, l^{-1}$ (0.37 $\mu g\, prey^{-1}$) to obtain this weight-specific ration. These values greatly exceeded most of the prey concentrations recorded. The authors concluded that starvation and food-limited growth were inevitable at all of the sampling sites visited.

Ferron (1991) determined the RNA/DNA ratios of mackerel (*Scomber scombrus*) larvae (6.5–7.5 mm) and related these to *in situ* measures of food concentrations. Larvae and their food (microzooplankton) were sampled hourly in the upper mixed layer, at a fixed station and at discrete depths (0–5, 5–10, and 10–15 m) for five consecutive days. The resulting time series revealed the advection of two distinct water masses and feeding environments past the station. The first, which persisted for 3 d, was characterized by food concentrations that increased gradually from 31 to 58 copepod nauplii l^{-1}, followed by a rapid decline (55% decrease over 4 h). The second water mass, which persisted for the remaining 2 d of sampling, was characterized by food levels ranging from 26 to 45 nauplii

l^{-1}. The RNA/DNA ratios of the larvae sampled from these discrete water masses showed food-limited positive protein growth rates (based on Buckley's (1984) model Eq. (1)), in 98% of the cases examined. No significant differences in the frequency histograms of RNA/DNA ratios were detected between the two feeding environments. However, vertical migrations of larvae between the three depth layers sampled became evident shortly after the decline in food concentrations which occurred at the transition between the first and second water masses. Since 7.0 mm mackerel larvae are fully competent foragers (Ware and Lambert, 1985; Buckley et al., 1987), this suggests an increased foraging activity in a food-reduced environment. About 70% of all nauplii sampled were *Oithona similis*, 25% were *Pseudocalanus minutus*, and 5% *Temora longicornis*. Assuming an average dry weight of $0.7 \mu g$ nauplius^{-1} (Buckley et al., 1987), the range of prey concentrations (26–58 nauplii l^{-1}) experienced by the larvae was $18.2–40.6 \mu g \, l^{-1}$.

McGurk et al. (1992) measured the RNA/DNA ratios of larvae of Pacific herring and of sand lance (*Ammodytes hexapterus*) collected in two embayments of the Gulf of Alaska. They observed that 11–23% of the first-feeding herring larvae and 45% of the first-feeding sand lance larvae had RNA/DNA ratios typical of starving animals, despite the fact that the two feeding environments were characterized by prey levels (16–84 prey l^{-1} or $11.2–58.8 \mu g \, l^{-1}$) known to support positive growth rates in the laboratory (Kiørboe and Munk, 1986; Buckley et al., 1987). Moreover, the residuals of regressions of ln(RNA/DNA) on larval length were uncorrelated with prey concentrations or water temperature, whereas positive correlations are expected between these variables based on laboratory results (Buckley, 1982, 1984; Buckley et al., 1984). McGurk et al. (1992) proposed that the nutritional condition of first-feeding larvae must be determined by the interaction of prey concentrations, temperature and the ability of the larvae to feed effectively. A further recent example of the lack of correlation between larval condition and food availability is provided by the study of McGurk et al. (1993), who collected larvae of Pacific herring from four distinct cohorts in Auke Bay (Alaska). The authors assessed larval condition using multivariate morphometric techniques developed by McGurk (1985a, b). Although the food densities measured (27–30 nauplii l^{-1}, $19–21 \mu g \, l^{-1}$) were at all times sufficient to support at least 4% larval growth in weight d^{-1}, and were an order of magnitude higher than food rations eliciting starvation in the laboratory (Kiørboe and Munk, 1986), it was estimated that between 8 and 35% of the herring larvae exhibited condition values characteristic of starving larvae.

There does, however, appear to be a good agreement between the condition of larvae as would be predicted from the maintenance range developed from the review of laboratory experiments (25–35 $\mu g \, l^{-1}$), and the *in situ* prey concentrations reported by Canino *et al.* (1991), Ferron (1991), and McGurk *et al.* (1992, 1993). Therefore, Canino *et al.* (1991) reported a high proportion of starving larvae at prey concentrations of 8.8–16.2 $\mu g\,l^{-1}$ which laboratory studies suggest are below maintenance. Ferron (1991) found little evidence of starvation at prey concentrations of 18.2–40.6 $\mu g\,l^{-1}$ which laboratory studies suggest are equal to or above maintenance levels. McGurk *et al.* (1992, 1993) reported 8–45% starving larvae at prey concentrations of 11.2–58.8 $\mu g\,l^{-1}$ and 19–21 $\mu g\,l^{-1}$, which range from below to above maintenance levels. This agreement is satisfactory given the numerous sources of error known to be associated with estimates of prey concentrations both in the laboratory and *in situ*. However, the frequent absence of significant relationships between condition and food availability (McGurk *et al.*, 1993) suggests other exogenous or endogenous factors influence feeding success and condition *in situ*. One such exogenous factor, which has not yet been examined in relation to larval condition, is the possible effect of microscale turbulence on feeding success (MacKenzie and Leggett, 1991).

Rosenthal and Alderdice (1976), who examined the effects of a broad variety of environmental stressors (abiotic and biotic) on marine fish eggs and larvae, concluded that sublethal effects were often limited to a very similar array of organismal responses, many of which were biochemical in origin. These biochemical changes were often translated, later in time, into histological, morphological, physiological, and ethological responses. Given this background, it is reasonable to expect that biochemical indicators of fish condition will be most sensitive to environmental stress (including food availability), and that sensitivity would then proceed in reduced order to cellular (histological) and organismal (morphometric) level indicators. Although the number of laboratory or field studies which allow evaluation of the sensitivity of condition indices to different food rations is limited, the trend of increased sensitivity as one proceeds from organismal, to tissue, and finally to the cellular levels indices is apparent. Within the array of biochemical measures of condition now available, a trend is also apparent. Short-term energy indicators such as digestive enzymes, glycogen and lipids appear to be more sensitive to food ration than are nucleic acids and proteins. Further research is required to confirm these trends and provide reliable measures of sensitivity for the several condition indices currently employed.

4. TIME RESPONSE (LATENCY AND DYNAMICS)

From a consideration of the direct effects of various food rations on condition, we now turn to the measurement of the time response of condition in relation to intermittent food levels. In contrast to predation, which is a near-instantaneous process, the response of larval condition and growth to variation in prey abundance is slower, and the relationship between cause and effect is likely to involve a time lag. Therefore, unless repeated measurements are taken on the same specimens (a logistic impossibility in field studies), the time response of condition to changes in food availability must be known in order to predict the relationship between present condition and feeding history, or to forecast future larval condition from a knowledge of the present feeding status. The time response of condition is also required in order to estimate the duration of different starvation categories, and to determine daily starvation mortality at sea (Theilacker, 1986; Theilacker and Watanabe, 1989).

We define latency as the time required for a given change in food availability to be reflected as a significant change in the particular index of condition used. This lag is a consequence of the time required for the energy balance of the animal to adjust to the new conditions, and for chemical reactions to occur which allow energy to be transformed, and then utilized or stored depending on the physiological needs of the moment. Latency can be viewed as equivalent to sensitivity over time. It is expected that sensitivity and latency will be negatively correlated (higher sensitivity—shorter latency). The dynamics of condition is defined as the rate of change in condition after the response is first detected. These concepts of latency and dynamics are depicted in Figure 2, in which the condition of fed and starved controls is illustrated over time, along with a delayed-fed treatment in which the larvae were starved for 6 d before being fed. In the example given, latency ($L = \Delta t$) is equivalent to the lag (3 d) between the time at which food was first given ($t = 6$), and the time a noticeable difference in condition (above the starvation control level) was detected ($t + \Delta t = 9$). Dynamics (D) refer to the rate of change dC/dt after the response has been elicited. Below we review laboratory studies which have systematically addressed the questions of latency and dynamics using delayed-fed and delayed-starved experiments. Other indirect measures, which are available from laboratory or field studies, are also used to generate time response predictions for different condition measures.

4.1. Morphometric Measurements

Theilacker (1978), who used a SWDA of 11 morphometric variables (see Section 2.1 and Table 1) to relate food level to condition, found that 87% of the fed and 94% of starved jack mackerel larvae could be differentiated after 3 d of starvation in the laboratory. However, when the starved

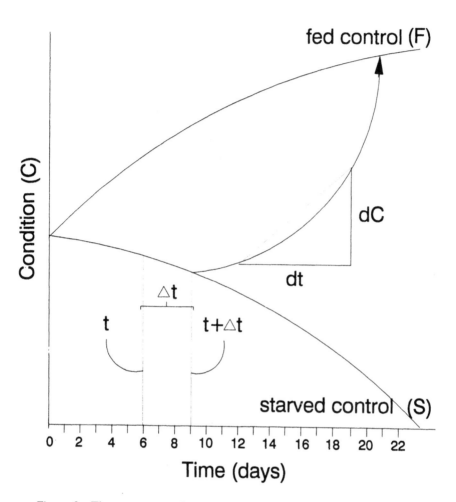

Figure 2 Time response of condition following feeding *ad libitum* (F) and starvation (S) controls. The arrow refers to the condition trajectory of specimens for which feeding was delayed for 6 d ($t = 6$) before being resumed. Latency ($L = \Delta t$) is equivalent to the time interval between food introduction ($t = 6$ d) and the time at which a significant increase in condition above the starvation control was recorded ($t + \Delta t = 9$ d). The dynamics ($D = \mathrm{d}C/\mathrm{d}t$) is the rate of increase in condition after the response has been elicited. In the example illustrated an exponential response of condition to food is used. Other shapes are, of course, possible.

group was further divided into moderately starved (starved for 1 or 2 d) and severely starved (starved for 3 d) larvae, the proportion correctly classified declined to 83% for the fed, 96% for the moderately starved, and only 78% for the severely starved larvae. Powell and Chester (1985) collected similar data for spot larvae, and found that 84% of the fed and 83% of the starved animals could be distinguished by a SWDA. However, when the moderately and severely starved specimens were separated on the basis of 50% mortality values, only 80% of the fed, 50% of the moderately starved, and 63% of the severely starved larvae were correctly classified. These results suggest that for the larvae studied, morphological changes have a short latency (<1 d) following starvation. However, after a significant decline over the first 2 d, the morphometric indices become less affected by prolonged starvation (3 d). Hence, after this interval, moderately and severely starved animals remain morphologically indistinct.

Wright and Martin (1985) measured standard length, eye diameter, head length, head depth, body height at pectorals and body height at the anus, on larvae of striped bass exposed to various delayed-fed and delayed-starved conditions in the laboratory. Five groups of 8 d posthatch larvae were exposed to 1000 *Artemia* nauplii l^{-1} for 2, 4, 6, 8 and 10 d, and then starved to determine their rate of deterioration. Eight other groups were starved for 1, 2, 3, 4, 5, 7, 9 and 11 d, before being offered food at 1000 nauplii l^{-1} to measure their rate of recovery. The experiment, which lasted 14 d, included two controls in which the animals were starved or fed for the total duration of the experiment. Wright and Martin (1985) reported the standard length of each group of larvae over time, and fitted parabolic curves to each treatment group. Most of the curves obtained were similar in shape to those illustrated in Figure 3. Dynamics showed a slow response following the application of the treatment (starvation or feeding) followed by a steeper slope when the values approached the control levels. In the early delayed-starved treatment F2 (fed for 2 d and starved) and F4 (fed for 4 d and starved), Wright and Martin (1985) found that standard length continued to increase for some days after food withdrawal before a shrinkage was recorded. If food was removed later, standard length showed no increase, and began to decrease with an average latency of 1–2 d. The dynamics of deterioration was approximately 0.3 mm d^{-1} for most of the delayed-starved treatments. Among the delayed-fed treatments, 5 d of starvation represented the maximum starvation time from which full recovery to fed-control levels was possible (compensatory growth). The dynamics of recovery was in the order of 0.3 mm d^{-1} for most delayed-fed treatments applied prior to 5 d, but was much lower (0.1 mm d^{-1}) for larvae starved for 7 d or more before being fed. The six morphological variables and all

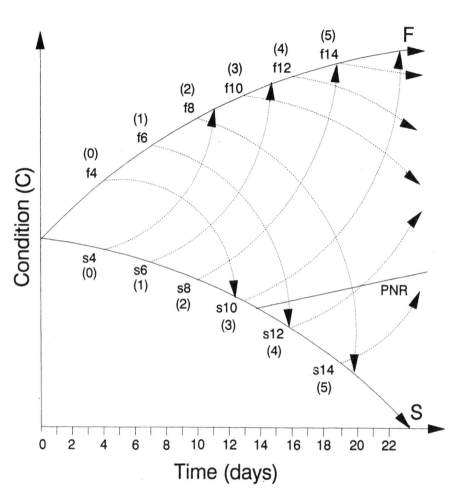

Figure 3 Proposed model for the time response of morphological indices and histological scores following different feeding regimes. The arrows refer to the condition trajectories of larvae exposed to different feeding conditions. F, feeding *ad libitum* control; S, starvation control, PNR = point-of-no-return and defines the condition level below which the larvae will die even if they are fed (also defined as the point of irreversible starvation). Delayed-fed treatments are labelled s*n* (s4–s14), where *n* is number of days feeding was delayed before being resumed. Delayed-starved treatments are labelled f*n* (f4–f14) where *n* is number of days starvation was delayed before being resumed. The numbers in brackets, above and below, refer to latency in days, and increase with the length of time feeding or starvation was initially delayed.

possible ratios were included in a SWDA following Theilacker (1978), and the values for the canonical variables reported in Martin and Wright (1987). The results showed that continuously fed and starved (control) specimens could be distinguished after only 2 d of feeding delay; overlapping centroids demonstrated, however, that recovering and starving specimens could not be discriminated from the different delayed-starved and delayed-fed treatments when passing through transition phases.

4.2. Histological Measurements

Theilacker (1978) showed that 83% of fed, 44% of moderately starved, and 83% of severely starved jack mackerel larvae could be correctly classified based on histological scores (Table 1). The fact that only 44% of the larvae in the intermediate class were correctly classified using histological scores, compared with 96% correct classification based on the analysis of morphometric data (see above), suggests an earlier response (shorter latency) of morphological changes to starvation. However, a greater resolution of histological scores within the class of moderately starved larvae was also demonstrated, since 19% were wrongly graded as healthy, 44% were correctly classified as moderately starved, and 37% were wrongly classified as severely starved (Table 1). It is expected that the 19% which remained in good condition after 1–2 d of starvation would be able to sustain a longer period of starvation than the 37% which were in poor condition. Theilacker (1978) also obtained good correspondence between morphological and histological assessments of condition (Table 1), the agreement being best (92.9%) for larvae past the point of irreversible starvation (3 d starvation), and lower for larvae starved for 1 and 2 d (40.7%). This suggests a different dynamic for the two indices.

Kashuba and Matthews (1984) recorded six different histological scores (midgut, pancreas, liver, muscle, cartilage, and notochord) on shad (*Dorosoma* spp.) larvae sampled weekly in a reservoir over a 2-month period. They noted a decrease in pancreas, midgut and liver condition following a drop in food availability at the time the larvae were sampled, and a decrease in muscle and notochord scores a week later. The authors concluded that pancreas, midgut and liver scores were indicators of early stages of starvation (latency < 1 week), while muscle and notochord scores were more indicative of later stages of starvation (latency > 1 week).

Martin *et al.* (1985) and Martin and Wright (1987), who analysed the dynamics of laboratory-reared larval striped bass histology, reported significant changes in Theilacker's (1978) histological scores after 2 d of starvation. By defining a boundary histological score to delimit animals in good and poor condition, they were able to determine that more time was required for larvae to recover from starvation than to deteriorate from fed to starvation levels. For both deteriorating or recovering fish, the total time required to go from control to boundary level condition also increased with the duration of the initial starvation or feeding conditions. These linear relationships (computed from data provided in Table 3 of Martin and Wright, 1987) are described by the following equations:

$$\text{Recovery time (days)} = 0.58 \text{ feeding delay (days)} + 2.07$$
$$(n = 6, \ r^2 = 0.918) \tag{6}$$

$$\text{Deterioration time (days)} = 0.59 \text{ starvation delay (days)} + 0.53$$
$$(n = 5, \ r^2 = 0.927) \tag{7}$$

Although very few data points were reported, the slopes of the two lines are very similar while the intercept is much higher for recovery time than for deterioration time. Based on the shape of the curves of delayed-fed treatments provided by Martin and Wright (1987), the dynamics of histological scores appear to follow the morphological model illustrated in Figure 3. If we picture the histological boundary level as a line running horizontally and equidistant between fed and starved controls, the increasing time required to recover or deteriorate must result from increased latency and divergence of the control lines with increasing feeding or starvation delays. Martin and Wright (1987) found that the first tissues to show signs of deterioration were the last to recover. This makes it difficult, even when assessing the status of several tissues, to determine with certainty whether the larvae were deteriorating or recovering at the time of capture.

Setzler-Hamilton et al. (1987) sampled larval striped bass weekly over several years, from two different rivers. They obtained a good correspondence between the per cent of larvae having starved morphometry (computed from SWDA using the same variables as Wright and Martin, 1985) and the average histological score of a subsample. The peaks in histological scores corresponded with troughs in the proportion of larvae classified as starving on the basis of their morphometry, with little lag between the two. The 1 week interval between sampling was, however, too long to allow a precise estimate of the degree of synchrony.

Oozeki et al. (1989) detected significant changes in the epithelium cell heights of the anterior, mid, and posterior regions of the gut, and in the liver and pancreatic cells of stone flounder larvae, 2–3 h after onset of starvation. These quantitative changes were rapid in the first 2–3 d

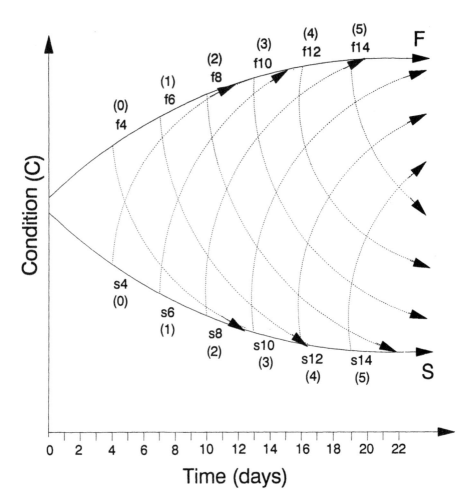

Figure 4 Proposed model for the time response of cell height measures and biochemical indices of condition following different feeding regimes. The arrows refer to the condition trajectories of larvae exposed to different feeding conditions. F, feeding *ad libitum* control, S, starvation control. Delayed-fed treatments are labelled s*n* (s4–s14), where *n* is number of days feeding was delayed before being resumed. Delayed-starved treatments are labelled f*n* (f4–f14) where *n* is number of days starvation was delayed before being resumed. The numbers in brackets, above and below, refer to latency in days, and increase with the length of time feeding or starvation was initially delayed. The PNR level was not plotted on this figure because the measures of condition following this type of response (mainly rate indicators) do not provide direct estimates of irreversible starvation.

following the onset of starvation, but stabilized and showed no further change until death. This dynamic was similar for all tissue cells examined, with the exception of the epithelium cell height of the anterior part of the gut, where the initial decrease was more gradual. The curves obtained had the same shape as those illustrated in Figure 4. The apparent difference between the dynamics of traditional histological scores (Figure 3) and cell height measurements (Figure 4) probably reflects the fact that the former integrate the values of a set of qualitative scores, while the latter represents a single quantitative measure. Therefore, average cell height responds more rapidly to changes in food level than traditional histological scores. Cell height, however, also attains its minimum or maximum value faster, and exhibits more resilience thereafter.

Theilacker and Watanabe (1989) delayed feeding for 0–4 d in post-yolk-sac larvae of northern anchovy kept in the laboratory, and measured their standard length and midgut cell height. Standard length at age and growth rates differed significantly between the five treatments such that the age of a 5 mm larvae could range from 9 to 19 d. The growth curves were more or less linear for the different treatments, and never converged, so differing from the model outlined in Figure 3. However, the decrease in midgut cell height reported for the delayed-fed groups showed a rapid decrease after 1 d of starvation and a lesser change after 2, 3 and 4 d. This conforms to the dynamics reported by Oozeki *et al.* (1989) and the model illustrated in Figure 4. Margulies (1993) reported significant changes in histological scores of black skipjack tuna (*Euthynnus lineatus*) less than 24 h after the onset of starvation in the laboratory, with liver hepatocytes responding in as little as 12–15 h. For experiments conducted over 6 d, the histological scores did not change in fed controls, while in starved treatments, a rapid drop was recorded within the first 24 h followed by a more gradual decline thereafter. Larvae fed after 48 h of starvation were able to recover to fed-control levels, but those provided with food after 72 h failed to recover (irreversible starvation). The shape of the recovery curve followed the model outlined in Figure 3, while the shape of the deterioration curve more closely approximated that illustrated in Figure 4.

4.3. Biochemical Measurements

Martin *et al.* (1984) measured total fatty acids in larval striped bass exposed to various starvation/feeding regimes in the laboratory. In addition to feeding and starvation controls, they also used four starvation treatments which consisted of delayed feeding for 4, 5, 6, and 7 d. Peaks in fatty acid levels were detected at 6 d after hatching in animals subjected to starvation. This peak was attributed to the catabolism of lipoproteins derived from yolk remains. Changes in growth rates were evident at 7 d but not at 5 d delayed feeding. There was no indication of the accumulation of fatty acids prior to 10 d posthatching, suggesting that the utilization of dietary sources was complete for at least 4 d after the peak seen at 6 d posthatch. Commenting on these results, Martin and Wright (1987) concluded that early feeding influenced fatty acid levels in ways which allowed separation of feeding and non-feeding larvae, but that fatty acid levels could neither be used as an indication of present nutritional status nor of the capacity of the larvae to withstand starvation because they were, in essence, indicators of the duration of the delay in first-feeding.

Martin et al. (1984) provided valuable information on total fatty acid dynamics and on fatty acid composition of striped bass larvae during starvation. It is important to note, however, that striped bass larvae differ from many marine fish larvae in that the egg contains large quantities of neutral fat stores. These are preferentially retained during the embryonic period and are utilized for up to 20 d after hatching, and well beyond the onset of first-feeding (Eldridge et al., 1982, 1983). This contrasts with herring and cod in which most neutral lipid stores are catabolized during early larval development (Fraser et al., 1988). The dynamics of neutral and polar lipids, first proposed as a possible indicator of the nutritional status of marine fish larvae by Fraser et al. (1987), remains poorly known. Given this reality, Håkanson (1989a) used data on neutral and polar lipids from his work on copepods (Håkanson, 1984) to predict neutral lipid changes following starvation in larvae of northern anchovy. In copepods, the total triglyceride content of starved specimens was consumed after 3 d, while their wax ester content showed significant change only 1 week after cessation of feeding. Unfortunately, the fact that fish larvae may differ in their lipid content, composition, diet and metabolism makes any comparison with copepod values questionable. Nevertheless, Håkanson (1989a) reported a predictable decrease in triglyceride (TAG) levels following 4 d of starvation in larval northern anchovy reared in the laboratory. However, TAG levels were $<1.0 \mu g$ larva^{-1} for animals up to 18 d posthatching, and changes in TAG levels after the onset of starvation were small, $<1.0 \mu g$. Large changes (1–6 μg) in TAG content were observed only in larvae that had been fed or starved beginning 18 d following hatching. Håkanson (1989a) made no attempt to calculate neutral lipid utilization rates.

The latency of changes in nucleic acid levels of larval fish relative to changes in food availability has been assumed to be on the order of a few days (Buckley, 1984; Buckley and Lough, 1987). Buckley (1979, 1980) detected changes in the RNA content (also expressed as changes in the RNA/DNA ratio) of reared larval cod and winter flounder 2 d and 4 d respectively following removal of food. However, in both studies, the frequency of sampling was 2 d, so that the minimum latency period remains undefined. Larval fathead minnows (*Pimephales pomelas*) exposed to sublethal levels of hydrogen cyanide exhibited a measurable reduction in their RNA, protein content and RNA/DNA ratios after only 24 h of exposure (Barron and Adelman, 1985). This short latency may result from the fact that toxicants act directly on oxidative processes affecting RNA synthesis, as opposed to changes in the food environment which indirectly influence metabolism through changes in energy substrate pools. Here too, a higher sampling resolution may have revealed an even shorter latency.

Wright and Martin (1985) conducted laboratory experiments involving larval striped bass to investigate the latency and dynamics of changes in the RNA/DNA ratio following various delayed-fed and delayed-starved conditions (see Section 4.1 on morphometrics). From the deterioration and recovery curves obtained (which were similar to those illustrated in Figure 4) they determined that the mean latency time for detectable changes in RNA/DNA ratios was 0.66 d (15.8 h) for larvae starved after being fed, and 0.81 d (19.4 h) for larvae fed after being starved. The recovery dynamic, obtained from their plots, was approximately $0.4 \, \text{RNA/DNA units d}^{-1}$, which is twice as fast as the deterioration dynamic ($\approx 0.2 \, \text{units d}^{-1}$). Wright and Martin (1985) concluded that the RNA/DNA ratio was sufficiently sensitive to allow striped bass larvae fed for 8 d and starved for 6 d to be distinguished from those which had been starved for 7 d and fed for 7 d. We disagree with their conclusion since it is very difficult, using their curves, to distinguish whether the larvae sampled at any given time were recovering or degenerating without a knowledge of their present and past condition and/or feeding environment. This restriction also applies to all other indices and to the models proposed in Figures 3 and 4.

Clemmesen (1987) obtained a linear relationship between the RNA/DNA ratio and the starvation interval (measured in degree-days) in groups of laboratory-reared herring and turbot larvae:

$$\text{RNA/DNA} = 2.353 - 0.009 \text{ starvation interval (degree-days)}$$
$$(r^2 = 0.39, \, n = 21) \qquad (8)$$

The addition of data for fed larvae and for dry weight measures to the relationship allowed Clemmesen (1987) to explain 62% of the observed variability in the length of the starvation interval (expressed in degree-days), thus allowing evaluation of the prior feeding history:

$$\text{Starvation interval} = 66.373 + 27.638 \log \text{dry weight/larvae}$$
$$- 53.387 \text{ RNA/DNA } (r^2 = 0.62, n = 163) \qquad (9)$$

However this model, when applied to sole larvae (Richard et al., 1991), explained only 28% of the variability in the starvation intervals. Moreover, while the relationship described in Eq. (9) could be used to determine the duration of starvation at a given temperature, it cannot discriminate whether the condition of an individual larva was improving or deteriorating at the time of sampling. The RNA/DNA ratio deterioration curves provided by Clemmesen (1987) for herring larvae are consistent with the model illustrated in Figure 4, with a sharp decline (0.2 units d^{-1}) during the first 4 d of starvation. The subsequent minimum starvation-control values were, however, less clear because very few data were reported in the 1.0–1.5 range.

Ueberschär and Clemmesen (1992) detected a significant ($P < 0.05$) decrease (43%) in the RNA/DNA ratio of starved herring larvae after 3 d of starvation. This was followed by a reduction in the rate of decrease and a stabilization in the ratio at values of approximately 1.0 at 10 and 13 d after the onset of starvation. The rate of decrease (dC/dt) in the RNA/DNA ratio ranged from -0.8 units d^{-1} between day 0 and day 3, to 0.2 units d^{-1} between day 3 and day 6 and 0.1 units d^{-1} thereafter. Although the sampling was conducted at 3-d intervals, the deterioration curve obtained followed the shape depicted in Figure 4.

In adult fish, the dynamics of the RNA/DNA ratio exhibit different rates of increase and decrease following starvation and refeeding. Mugiya and Oka (1991) detected significant RNA/DNA ratio differences in juvenile rainbow trout after 2 d of starvation, such that 50% RNA/DNA ratio levels were reached on the third day. However, RNA/DNA ratio recovery was more gradual and levels equal to those observed in control fed fish were attained only after 4 d.

Lied et al. (1983) measured ribosomal RNA and DNA content, and the rate of incorporation of ^{14}C-phenylalanine from ribosome extracts of epaxial muscle in 2-year-old cod. No significant changes in the rRNA, DNA and RNA/DNA ratio were detected after 3 d of starvation. However, the amino acid incorporating activity of the muscle was reduced to 55% of the control level observed in fed fish. No significant change in DNA content was recorded (constant number of cells) after 5 and 8 d of starvation, despite the fact that rRNA had declined to 87 and 72%, and the RNA/DNA ratio to 81 and 68% of their original values, respectively, indicating a decrease in the number of ribosomes per cell.

Amino acid incorporating activity declined to 39 and 13%, respectively, of the levels of the fed controls during the same time interval. Upon refeeding of fish starved for 8 d, the rRNA, rRNA/DNA ratio and the amino acid incorporating activity were completely restored within 12 h. These results indicate that, in juvenile cod, recovery of muscle protein synthetic activity after starvation and refeeding is much faster than is its decrease following starvation. More importantly, they illustrate the fact that protein synthesis is related to both the number of ribosomes present in the muscle (quantity of rRNA) and their specific activity, and that these two variables can vary independently (Lied et al., 1982). Roselund and Lied (1986) studied the effect of different protein energy levels on the muscle growth of saithe and rainbow trout. Their results indicate that the rate of amino acid incorporation $min^{-1} mg^{-1}$ RNA was relatively constant at all but the two lowest levels. These findings indicate that the reduced protein synthesis of starving fish was related both to a decrease in rRNA and to a reduced efficiency of amino acid incorporation. In fish fed at higher protein levels, the higher protein synthesis achieved was attributed to higher rRNA levels alone. Loughna and Goldspink (1984) also concluded that reduction in protein synthesis caused by long-term fasting was related to a reduction in the number of white muscle ribosomes and a lowering of their activity.

Miglavs and Jobling (1989) exposed four groups of juvenile Arctic charr (*Salvelinus alpinus*) maintained in the laboratory to periods of food satiation (S) and food restriction (R) for various lengths of time. The first group was satiated for 16 weeks, (S16), the second was fed a restricted diet for 8 weeks and then sampled, (R8–S0), the third was fed a restricted diet for 8 weeks and then satiated for 4, (R8–S4), and the fourth group was fed restricted for 8 weeks and satiated for 8, (R8–S8). The fish maintained on restricted diets and then provided with an excess of food showed compensatory growth and specific growth rates that were significantly higher than those predicted from the RNA/DNA ratio-growth rate relationship established for the satiated group (S16). Two possibilities could explain these results: (1) a differential ribosome protein synthetic activity level, or (2) a change in the relationship between anabolic and catabolic processes. Since high somatic growth rate is likely to be associated primarily with higher levels of protein synthesis and very low rates of protein degradation, the first explanation appears more plausible.

If the same decoupling of protein synthesis and RNA content (ribosome number) occurs in larval fish, it is likely that in severely starved animals that have reached a low rRNA level further decreases in protein synthetic capacity will be reflected only in a reduction in amino acid incorporation rates. In larval fish, the dynamics of changes in protein levels are likely to occur more rapidly since the rate of protein synthesis (and potentially degradation) is reported to be 20 times faster than in the adult fish (Hansen *et al.*, 1989). RNA levels measured in starved fish larvae typically decrease to values around 1.0 but never reach zero, suggesting a conservation of at least some RNA (Ueberschär and Clemmesen, 1992). This conservation is reflected in frequency histograms of RNA/DNA ratios which are typically skewed to the right and truncated at values below 1.0 (Buckley, 1984; Buckley and Lough, 1987; Ferron, 1991). Concurrent measurement of protein is needed with nucleic acids in order to determine RNA/DNA threshold values reflecting negative growth (starvation) and positive growth.

A latency between changes in RNA/DNA ratios and protein growth rate is expected, since RNA synthesis precedes any increase in protein growth rate. Haines (1980) reported a 7-d lag between changes in the RNA/DNA ratio and weight increase of adult black crappies (*Pomoxis nigromaculatus*) collected at weekly intervals in the field. Buckley (1984) and Buckley and Lough (1987) inferred from their studies that the RNA/DNA ratio reflected growth 2–4 d prior to capture. This statement is misleading since RNA/DNA ratio changes should precede and not follow changes in protein growth. Hovenkamp (1990) observed a relationship between growth rates over the last 5 d (as determined from the width of otolith daily increments) and protein growth rates as predicted from Buckley's model (Eq. (1)) for groups of plaice larvae collected at two different sites in the North Sea.

Pedersen and Hjelmeland (1988) provided the first data on the latency and dynamics of proteolytic enzyme activity following food intake. Their sampling of starved herring larvae within 75 min following food ingestion revealed significant increases in trypsin levels during that short time interval. Further, the measurement of trypsin activity 3 h after food intake, and after a second meal was offered, showed no further changes suggesting a persistence of high trypsin levels after a single meal and an extensive reuse of the enzyme to facilitate the digestion of several meals. Larvae that were starved following the ingestion of a single meal exhibited declines in trypsin activity to pre-ingestion levels after 24–48 h, consistent with a 24–48 h protection against autodigestion of the enzyme in the gut. Following more than 6 d of starvation, the trypsin secretory response to a food pulse decreased substantially in spite of the presence of significant quantities of pancreatic trypsinogen, suggesting a defective release of the enzyme proform and a loss of digestive capabilities under extreme food deprivation. Ueberschär and Clemmesen (1992) detected a 90.2% decrease in trypsin activity following 3 d of starvation in herring larvae, but no further changes for the remaining 10 d when the larvae were deprived of food. Proteolytic enzyme dynamics followed the model proposed in Figure 4, but the initial rate of change following onset of feeding or starvation was much higher than is typical for nucleic acids. Latency was also much shorter, with significant responses being recorded on the order of hours.

The above review of latency and dynamics of various condition measures demonstrates that most indices conform to one of the two models proposed (Figures 3 and 4). Figure 3, characterized by a slow initial response, appears to describe the dynamics of morphometric indices and histological methods that employ qualitative scores. The model illustrated in Figure 4, on the other hand, more appropriately describes the dynamics of histological cell heights, nucleic acids and proteolytic enzymes following intermittent food energy inputs. The indices conforming to the model illustrated in Figure 3 have the disadvantage of responding over long time scales, but have the advantage of accurately assessing long-term effects such as irreversible starvation. Indices that follow the model illustrated in Figure 4 have the advantage of responding over shorter time scales, but their disadvantage lies in their inability to discriminate between individuals situated at the extremes (fed and starved control levels). This dichotomy in the time response of condition indices must be considered when selecting indices to be used in the field testing of hypotheses. The generalized responses illustrated in Figures 3 and 4 provide little more than a framework and a first step toward the dynamic modelling of condition and the forecasting of condition trajectories from various feeding and starvation regimes. Figure 5 simulates two condition trajectories, using the framework provided in Figure 4. Similar simulation results were obtained by Sclafani (1992) and Sclafani *et al.* (1993) for individual changes in the vertical distribution of larval cod as a function of their condition. While the models proposed here are heuristically useful, their direct application should be approached with caution, since dynamics and latency are expected to be size- and species-dependent. Laboratory calibration is therefore called for before such models are applied to different field situations. We are currently conducting such calibrations.

5. LABORATORY VERSUS FIELD ESTIMATES

There are several examples in which larvae reared in the laboratory were found to vary morphometrically, histologically or biochemically from specimens collected at sea. Since most calibration studies on condition are carried out in laboratory tanks or enclosures, this becomes a major concern when laboratory results are extrapolated to field studies. For

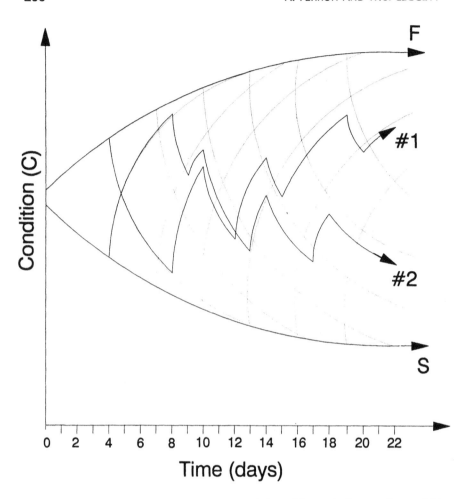

Figure 5   Condition trajectories simulated from Figure 4, for two larvae (#1 and #2) exposed for 22 d to different intermittent feeding conditions. The scenarios are for larvae #1: fed 4 d, starved 4 d, fed 2 d, starved 2 d, fed 2d, starved 1 d, fed 4 d, starved 1 d, and fed 2 d. For larvae #2: starved 4 d, fed 4 d, starved 1 d, fed 1 d, starved 3 d, fed 1 d, starved 3 d, fed 1 d, and starved 4 d. No latency was accounted for between each change in food level.

morphometric measurements, Hempel and Blaxter (1963) and Blaxter (1971) compared the Fulton-K condition factor (weight/length^3) of herring larvae sampled at sea to those reared in the laboratory. The shrinkage associated with death before fixation was found to be greater in sea-caught larvae (20%) as compared to laboratory-reared larvae (12%). The resulting inflated condition of wild larvae gave the incorrect perception that larvae sampled at sea were in better condition than those maintained in the laboratory. However, at lengths above 12 mm, Blaxter (1971) noted that laboratory-reared herring larvae tended to have a greater body depth for a given length and therefore a higher Fulton-K index than did sea-caught larvae. Many older larvae sampled at sea had a condition factor that was consistently below the experimentally determined starvation level. Three possible explanations were given by Blaxter (1971) for these differences: (1) different growth characteristics between the two groups, (2) the condition index was sensitive to physical damage suffered by the animal during capture, (3) a selective capture at sea of weak or moribund larvae.

Balbontin *et al.* (1973) reinforced the view that laboratory and field specimens differed morphometrically after finding that laboratory-reared herring larvae, ranging in size between 35 and 85 mm, had significantly heavier and deeper bodies and larger heads than did wild larvae of the same size. This assessment was based on measurements of total, standard, and pre-anal lengths, along with head and suborbital head heights. Ehrlich (1975) reported that wild plaice juveniles had a dry weight more than twice that of laboratory animals of the same length. Arthur (1976) detected significant differences in relative body depth at the pectorals (body depth/standard length) between ocean-caught and laboratory-reared northern anchovy larvae. He showed a clear divergence between wild and laboratory larvae at lengths over 7.0 mm. While laboratory larvae increased allometrically in relative body depth and weight over their entire larval life, the ocean-caught larvae showed a decrease in the same measure at a mid-larval stage (15–20 mm) followed by an increase at later stages. These differences were attributed to differential food availability in the two environments.

Laurence (1979) suggested that the Fulton-K index, while useful for adult fish, might not be applicable for larval fish because of allometric growth causing the index to become a function of length. He based his assertion on length–weight relationships obtained from seven species of marine fish larvae reared in the laboratory which showed an average length exponent of 4.15 (range 3.76–4.77), far from the exponent of 3.0 expected from an isometric index. Frank and McRuer (1989) obtained a length exponent of 3.03 for field-collected haddock larvae and suggested that the high length exponent (4.48) obtained by Laurence (1979) for the same species could be explained by the fact that these measurements were obtained on laboratory-reared animals which are known to differ morphologically from field-collected specimens. In support of their findings they noted the exponent values of length obtained by Economou (1987) for five species of field-collected gadoid larvae, which ranged from 2.92 to 3.22. However, Koslow *et al.* (1985) obtained a length exponent of 3.55 for field-collected cod, another gadoid. For herring larvae, Marshall et al. (1937) obtained a length exponent of 4.52 and Sameoto (1972) a length exponent of 4.49 for field-collected specimens, which is comparable to the Ehrlich et al. (1976) estimate of 4.57 for laboratory-reared animals of the same species. Moreover Balbontin et al. (1973) calculated length–weight relationships for both laboratory-reared and field-collected herring larvae, and found that although the intercepts differed significantly, the slopes (3.61 and 3.54) did not. Therefore, while isometry could be demonstrated for some field-caught gadoids and not for their laboratory counterparts, herring larvae showed consistent allometry for both wild and laboratory-reared specimens.

The reliability of morphometric indices has often been questioned because of the failure to correct adequately for differential shrinkage occurring in preserved specimens (Theilacker, 1986; Theilacker and Watanabe, 1989). Although shrinkage of laboratory collected larvae is usually constant after preservation, those collected in the field shrink more and show greater variability due to abrasion and damage suffered in the net (Theilacker, 1980a). The average shrinkage measured by Theilacker (1980a) for different body parameters in larval northern anchovy was 3% resulting from net capture alone, and a further 8% resulting from preservation in formalin. This shrinkage has also been measured in terms of length (Hay, 1981, 1982) and dry weight (Hay, 1984) for larval Pacific herring larvae. Theilacker (1980a), Hay (1982) and McGurk (1985a) all found that both the duration of the tow and the time elapsed between capture and preservation were important in determining the shrinkage suffered by collected larvae. Theilacker (1986) found, for instance, that body height could vary by 0–23% in jack mackerel larvae after net treatment ranging from 5 to 20 min. Delayed fixation as short as 3 min after capture could lead to a total length shrinkage of more than 30% in herring larvae (Hay, 1981). Larvae killed prior to fixation also shrank more than larvae killed by fixation (Hay, 1981). Litvak and Leggett (1992) noticed that capelin (Mallotus villosus) larvae which were dead prior to preservation curled substantially more than those killed by preservation. They developed a "curl factor" to detect and correct for the presence of animals that were dead at the time of capture. Salinity and the concentration of formalin are also known to have an impact on shrinkage at preservation. In general, high formalin concentration and high salinity tend to maximize length reduction and minimize weight loss, while low formalin concentrations and low salinity tend to minimize length reduction but maximize weight loss (Hay, 1984). Shrinkage due to fixation is an important consideration in the calculation of length–weight relationships. For example, Hay (1984), working with Pacific herring larvae, obtained a length exponent of 2.92 for fixed and 4.12 for unfixed specimens. Caution must also be exercised when using average mean values to correct for shrinkage due to preservation because shrinkage can vary greatly between individuals.

McGurk (1985a), who studied the effect of net collection on differential shrinkage in various morphometric measurements applied on Pacific herring larvae, found that the body dimension most affected by capture was the head, which shrank more in depth and length than other parts. This resulted in an inflation of several morphometric measures and an overestimation of condition. Multivariate analyses neither eliminated the effects of differences in shape between field-collected and laboratory-reared larvae nor corrected for the distortion caused by net capture. The problem of shrinkage due to fixation can be avoided by videotaping live anaesthetized specimens and collecting the required measurements from the digitized images (McClatchie et al., pers. commun.). This technique does not, however, overcome the difficulties associated with damage and distortion of important body parts, or of shrinkage of body dimensions, resulting from mixing and abrasion in the net (McGurk, 1985b).

Turning to histological measurements, O'Connell (1980) observed that some ocean-caught northern anchovy larvae exhibited necrosis of the midgut mucosa, although these characteristics were never observed in laboratory-reared specimens. He explained this difference by the fact that larvae starved in laboratory experiments are never fed and hence do not experience gut function, while wild larvae, in contrast, may experience brief periods of starvation and refeeding. Theilacker (1986) found that ocean-caught jack mackerel larvae exhibited four tissue conditions not seen in the laboratory. These were related to the presence of lesions in the brain, luminar vacuoles in the midgut, total degeneration of the midgut mucosal cells, and a wavy configuration of the muscle fibres. However, their influence on condition was considered negligible because these characteristics were seen only in a small proportion of the fish examined, and they constituted only a small fraction of the characteristics used to categorize condition. A dark staining of the pancreas, on the other hand, which was very common in wild larvae, was never encountered in laboratory-reared larvae. This prevented the use of that sensitive organ for grading condition in field-collected specimens (Theilacker, 1986). Oozeki et al. (1989) noticed that wild-caught stone flounder larvae generally had larger midgut cells than did their fed laboratory counterparts. Liver cells, however, were intermediate in size between fed and starved laboratory specimens.

A further important consideration when applying histological assessments to fish larvae sampled at sea is the autolytic tissue decomposition that can occur within 2 to 3 min after death (Theilacker and Watanabe, 1989; Owen et al., 1989). These changes demand that the time between death and preservation be kept to a minimum if reliable estimates of histological condition are to be obtained. Based on observations of specimens collected at sea, Owen *et al.* (1989) proposed the use of the midgut cell height (Theilacker and Watanabe, 1989) which appears to be unaffected by up to 5–6 min delays between capture and preservation.

A small number of authors have reported differences between the biochemical status of laboratory-reared and field-caught larvae. Balbontin *et al.* (1973) found that the proportions of neutral lipids, polar lipids and protein were significantly lower in field-collected herring larvae than in laboratory-reared larvae of similar size (55–85 mm). This caused the triglyceride/total lipid ratio to be much higher in laboratory animals than in wild specimens. Ehrlich (1975) observed the same phenomenon in wild plaice juveniles which were heavier at a given length, and contained smaller percentages of water, triglycerides and carbon, but higher ash and carbohydrates content than did laboratory-reared animals. Fraser *et al.* (1987) obtained comparable lipid composition and lipid class proportions between herring larvae reared in large enclosures and those collected at sea. However, results from enclosure experiments, when compared with those from experiments conducted in 18-l containers (Tocher *et al.*, 1985a, b), showed that starved larvae from the 18-l tanks had lower triacylglycerol catabolism than did those fed at low levels in enclosures. This suggests that container size can affect larval composition independent of the food provided. Theilacker (1980b) reported similar tank effects on morphological measurements in northern anchovy larvae reared in containers of different size. Håkanson (1989a) found that cholesterol levels were slightly higher, and triglyceride contents much higher, in laboratory-reared northern anchovy larvae as compared with specimens collected at sea. The fact that the larvae in the laboratory were fed only rotifers, which are known to be nutritionally inferior to natural plankton, is one possible reason for these differences.

Wild winter flounder larvae did not appear to differ in their RNA, DNA, and protein content from specimens reared in the laboratory (Buckley, 1980). But when comparing length–chemical constituent relationships, Buckley (1981) found that wild specimens had higher protein, DNA, RNA and total dry weight than did their laboratory counterparts. Sand lance (*Ammodytes* spp.) specimens sampled at-sea also showed higher RNA/DNA ratios than did cultured larvae (Buckley *et al.*, 1984). Buckley (1984) offered three possible explanations: (1) higher food availability at sea, (2) poor food quality in laboratory rearings, or (3) differential survival favouring faster growing individuals at-sea. Raae *et al.* (1988) observed similar differences in RNA/DNA ratios of cod larvae, with the individuals reared in laboratory tanks having, on average, lower ratios than those reared in outdoor ponds. There is also evidence that the types of food typically used in laboratory rearings may have an influence on larval fish condition. For example, *Artemia* nauplii, which are known to be less nutritional than natural plankton, were suspected of being responsible for the suboptimal growth conditions in studies conducted by Clemmesen (1987), Richard *et al.* (1991) and Davis and Olla (1992).

Heath (1992) argued that one advantage of biochemical over histological and morphometric condition indices is that these estimates are independent of net damage. However, it has been suggested that rapid enzyme action (lipases, RNases or proteases) shortly after death can cause deterioration of the sample (Bulow, 1974) and bias its compositional determination. Ferguson and Drahushchak (1989) demonstrated experimentally that this was not the case for nucleic acid content of rainbow trout muscle tissue. Routine methods usually preclude preservation of the specimens in liquid nitrogen immediately after capture and preferably before death. Ferron (1991) found that sorting of live from dead larvae prior to freezing, in order to avoid enzyme action, can lead to an overestimation of the condition of the specimens collected. For these reasons great care must be taken when sorting the samples and interpreting the results from field studies employing biochemical indices.

6. SIZE AND AGE SPECIFICITY

The confounding effect of age and size on the determination of larval fish condition is a major concern for all categories of indices. Hempel and Blaxter (1963), Blaxter (1971) and Vilela and Zijlstra (1971) measured the Fulton-K index (weight/length^3) of herring larvae sampled in different areas of the North Sea and in different years. The index decreased from hatching to resorption of the yolk, followed by a plateau at lengths between 8 and 15 mm and then a gradual increase which resulted from allometric growth, ossification and deposition of energy reserves in larvae >15 mm. This pattern was consistent for groups of larvae sampled at different sites and in different years, and persistent even when calculated separately for 1- and 2-mm length intervals. Sameoto (1972) also reported similar changes in the Fulton-K index of field collected herring larvae.

To overcome the allometry problems associated with the use of the Fulton-K index on herring larvae, Chenoweth (1970) computed a relative condition index based on the ratio of the observed weight to the expected weight which was empirically derived from the length–weight relationship of the population from which the samples were collected. Seasonal differences in relative condition could be detected based on this index, but these differences were small and could not be compared with those

from other studies for which a different length exponent was demonstrated. Ehrlich *et al.* (1976) provided calculations of both relative and
Fulton-K condition indices for herring larvae. However, while the
Fulton-K index was found to increase with size, the relative index showed
a slow decrease. These authors suggested the use of the Fulton-K index
only within very limited size ranges. Body height also increased with
size and the Fulton-K index, but changes in body height during starvation
could not be used as a starvation index because of the large differences in
this measure between individuals of the same size. Von Westernhagen
and Rosenthal (1981) demonstrated a decrease in the Fulton-K index of
field-collected Pacific herring, from hatching to the complete resorption
of the yolk-sac. This decrease was attributed to an increase in total length
concurrent with a decrease in yolk-sac volume which lead to a decrease in
total weight. The index further increased with size after the onset of
first-feeding. This was attributed to allometric growth (Von Westernhagen and Rosenthal, 1981). Laurence (1974) reported a similar increasing trend in the Fulton-K index of haddock larvae reared in the
laboratory. In contrast, Neilson *et al.* (1986) reported a decline in this
index during the 21 d following initiation of feeding in laboratory-reared
larvae of cod. This decline was not correlated with the different prey
concentrations offered.

  Cone (1989) reviewed the use of three morphological condition indices
in adult fish (Fulton-K, relative condition factor, and relative weight). He
concluded that these indices all suffered from faulty assumptions that
could lead to erroneous conclusions. He recommended, in their place, a
contrast between parameters derived from the length–weight regression
characterizing the different groups to be compared. This suggestion
generated significant controversy over the properties and the wise use of
morphological condition indices (see Springer *et al.*, 1990). Cone's reply
to these comments (which also appear in Springer *et al.* (1990)) included
an important distinction between condition and form, condition being a
difference in weight at a given length, and form being the rate of increase
in weight as a function of length. Cone also argued that the relative
weight should only be used for comparing individuals within the same
size-class. Although Cone's comments were applied to adult fishes, his
message can be adapted in the context of fish larvae.

  Several investigators, who have employed stepwise discriminant analyses of morphological variables (Theilacker, 1978; Powell and Chester,
1985; Martin and Wright, 1987), have suggested that such analysis should
be restricted to small size ranges, because most of the ratios used as
inputs to the analyses are highly correlated and share size information.
Koslow *et al.* (1985), McGurk (1985b) and Powell and Chester (1985)
used multivariate approaches to accommodate comparisons of condition

between larvae of different sizes. However, these authors, too, cautioned that it was preferable to restrict morphometric analysis to small size ranges in order to avoid allometric problems. It seems that a stepwise discriminant analysis using principal components of shape after extracting principal components of size (McGurk, 1985b) could more effectively overcome problems associated with comparison of animals of different size.

Theilacker (1980a) showed that shrinkage of northern anchovy larvae following net capture is size dependent. Smaller larvae (4.0 mm) tend to be more sensitive to net damage and, consequently, exhibit greater shrinkage (19%) than larger (18.0 mm) and more robust larvae (8%). Hay (1981, 1984) reported similar length-dependent shrinkage in net-sampled herring larvae.

O'Connell (1976) expressed confidence that the histological grading of larval northern anchovy was independent of size or age effects, the one exception being larvae which retained remnants of yolk or pancreatic zymogen. These specimens were treated as a separate group. O'Connell and Paloma (1981), who compared glycogen levels in the liver with the overall histological grading of small anchovy larvae (<6.4 mm), found high glycogen in robust specimens and low glycogen in emaciated specimens. This relationship did not, however, persist in larger specimens (>6.4 mm). These differences were rationalized by a possible greater tolerance of larger larvae to fluctuations in plankton regimes, as a consequence of the wider range of prey size they can eat. This diversity could make larger larvae less susceptible to a scarcity of some planktonic prey. Both histological and morphometric analyses performed by Theilacker (1986) indicated that starvation was a major source of mortality during the interval between first feeding and 6 d posthatching in wild jack mackerel larvae, but not for older specimens. Stone flounder larvae, fed *ad libitum* in the laboratory, exhibited a gradual increase in pancreatic, liver, and posterior gut cell height, 30 d after hatching (Oozeki *et al.*, 1989). These changes were explained on the basis that these structures were the sites of active storage following food intake. Midgut cell height, in contrast, followed the same increasing trend with age but only following an abrupt decline 2 d after hatching. A similar decline was noticed in the anterior gut cell height at 20 d after hatching. The midgut cell height, employed by Theilacker and Watanabe (1989) as an index of starvation, could not be used for anchovy larvae >6.0 mm because of the folding of the midgut at that stage. Margulies (1993) used histological grading of the liver and gut to infer that starvation mortality was very important in field-collected scombridids, but that vulnerability to starvation was highly dependent on developmental stage. Rice *et al.* (1987) and Sieg *et al.* (1989) postulated that the ability to detect starvation in field-caught fish larvae depended on the development stage of the animal studied. They further suggested that older stages were much less susceptible to starvation than were first-feeding larvae. Recently, Sieg (1992a, b) has suggested that the level of organ development is more important in determining resistance to starvation than is size or age, and that only the condition of specimens belonging to similar developmental stages (or physiological ages) should be compared.

Age and size specificity has been a matter of concern in the application of most biochemical condition indices used to date (Buckley, 1984; Bulow, 1987; Clemmesen, 1989; Richard et al., 1991; Suthers et al., 1992). Although the order of utilization of major body constituents may remain constant following starvation in individuals of different sizes, the increased stored energy reserves accumulated from the post-yolk-sac stage to the juvenile stage inevitably provides a greater "buffer" against changes in macromolecular composition (Love, 1970; Balbontin et al., 1973; Ehrlich, 1974a, b, 1975; Sieg, 1992b). These energy reserves in older fish result from an increase in the surplus energy available for growth as metabolism and activity levels decrease and storage organs become more fully developed with age and size (Blaxter and Ehrlich, 1974; Sieg, 1992a, b). Increased digestive and assimilation efficiencies of the intestine and increased efficiency of energy deposition in the liver are also characteristic of older fish larvae (Sieg, 1992a, b). One consequence of this is the increase in the liver-somatic index (LSI) with size in fish. This LSI becomes a useful indicator of food intake and energy storage in juvenile and adult fish (Heidinger and Crawford, 1977; Bulow et al., 1978; Black and Love, 1986). The concept of physiological age has been recently introduced by Sieg (1992a) to illustrate this development-dependent starvation resistance.

Balbontin et al. (1973) demonstrated that neutral fats (triglyceride), carbohydrates, and total lipids did not vary significantly with size in wild herring larvae ranging from 55 to 85 mm in length. Protein and ash content did, however, change with size in laboratory-reared larvae of similar age, and the triglyceride/total lipid ratio was higher in smaller larvae, suggesting that faster growing individuals had used up a greater proportion of their neutral fat stores. This age-dependent starvation was documented by Blaxter and Ehrlich (1974), who reported an increasing time to reach irreversible starvation in plaice and herring larvae as they developed. The sinking rate of herring larvae was also shown to increase as they aged as a consequence of the build-up of muscle tissue, decreases in water content, and ossification of the skeleton. However, when starved, their buoyancy increased regardless of their size.

Gatten et al. (1983) reported that Atlantic herring larvae reared in enclosures exhibited, with increasing age, distinct phases in the proportion of triglycerides (TAG) comprising total lipids. A first phase, which occurred between hatching and 20 d, and was characterized by a very low percentage (1–2%) of TAG; a second phase, between 20 and 60 d characterized by TAG comprising about 15–20% of the total lipids, and finally a third phase, which occurred in larvae older than 60 d, and was characterized by the proportion of TAG increasing to more than 60%. During the same developmental period, polar lipids progressively declined as a proportion of total lipids, and decreased from about 30 to 10% between 40 and 90 d posthatching. In addition, the exponential increase in total lipid levels observed in the following 60 d was mirrored in an exponential increase in growth rate and was mainly attributed to increasing TAG levels. Based on the composition of their fatty acids, Gatten et al. (1983) determined that these TAG were primarily obtained from zooplankton rich in wax-esters. Fukuda et al. (1986) observed that the high rate of triacylglycerol (TAG) accumulation which occurred throughout Pacific herring larval development decreased at the onset of metamorphosis. Commenting on these findings, Fraser (1989) attributed the slower rate of TAG accumulation at metamorphosis to the fact that greater quantities of exogenous lipids were required for the high levels of cell multiplication and differentiation which accompany metamorphosis. This developmental stage was also characterized by lower protein/DNA ratios. The quantities of cholesterol, polar lipids and TAG were reported to increase non-linearity with length in northern anchovy larvae (Håkanson, 1989a, b). Moreover, polar lipids and cholesterol, which are constituents of cell membranes, were found to be linearly related to larval dry weight, confirming their role as useful denominators to account for size differences when using TAG/sterol and TAG/cholesterol ratios.

   Although the effects of age and size on the chemical composition of larval fish are typically confounded, the quantity of any major biochemical constituent is commonly more strongly correlated to size than to age (Ehrlich, 1974a,b; Buckley, 1981; Clemmesen, 1987; Richard et al., 1991). For nucleic acids and protein, relationships between dry weight (DW) and RNA, DW and DNA, and DW and protein, are often highly significant and are found to be within the same range for different species (Buckley and Lough, 1987; Clemmesen, 1987). However, when the quantity of RNA is divided by the quantity of DNA present in a fish, it is expected that the RNA/DNA ratio resulting will become independent of size, since DNA is typically more tightly coupled to size than is RNA. Bulow (1987), in his review of RNA/DNA ratios as indicators of growth in fish, recommended that the use of this ratio should be restricted to comparison of discrete size and life history stages. However, Buckley (1980) showed that the RNA/DNA ratio of winter flounder larvae was not significantly affected by size up to the point of metamorphosis, and it is only at that stage that a significant increase in the RNA/DNA ratio was observed. Further, the relationship between protein growth rate, RNA/DNA ratio, and temperature depicted in Eq. (1), and established by Buckley (1984) using laboratory-reared larvae of eight different species, was not affected by size, except for individuals near or past metamorphosis. Buckley and Lough (1987) showed that in haddock and cod larvae sampled at different sites on Georges Bank, size had no significant effect on their RNA/DNA ratios. Hovenkamp (1990) found no relationship between RNA/DNA ratios and size or developmental stage in plaice larvae sampled in the North Sea.

In contrast, Clemmesen (1992) found that the RNA/DNA ratio of individual Atlantic herring larvae older than 10 d posthatch increased linearly with size in both starved and fed specimens, and at a rate which depended on the length of the starvation period. She therefore established RNA/DNA ratio–length relationships for different starvation periods (starved 2–3 d, 3–4 d, 4–6 d, 6–9 d), and recommended the use of a starvation range, defined as two standard deviations from each of these regression lines, as a means of determining the extent of starvation for individual larvae sampled at sea. A primary reason for the significant influence of size on RNA/DNA ratios reported by Clemmesen (1992), as contrasted with the absence of such a relationship in earlier studies, was the application of the ratio to individual fish larvae. Chambers (1993) and Pepin and Miller (1993) have recently cautioned against the use of aggregated versus individual based analyses of early life history traits. This caution appears to apply equally to indices of condition.

Richard et al. (1991) reported that the RNA/DNA ratio of sole larvae was consistently higher in fed than in starved specimens, but that the differences between the two groups diverged more slowly with increasing age. The rate of decrease (deterioration dynamics) in RNA levels following starvation was also found to be dependent on developmental stage. This caused Richard et al. (1991) to suggest the restriction of the RNA/DNA ratio comparisons to narrow size and age classes.

No significant differences were detected in the RNA/DNA ratios of fed and starved yolk-sac larvae analysed by Buckley (1979, 1980, 1981), Clemmesen (1987), Robinson and Ware (1988) and Richard et al. (1991). However, RNA/DNA ratios were observed to decline progressively in both groups as the larvae aged. Although we do not expect the presence of food to have a significant impact on the growth and rate of yolk utilization of larvae before they feed exogenously, the decreasing RNA/DNA ratio in yolk-sac larvae is contrary to the expectation of positive growth during a stage of intensive tissue build-up. In this connection, McGurk and Kusser (1992) found that RNA/DNA ratios were sensitive to the method of analysis and, in particular, ethidium bromide determinations conducted without appropriate nucleic acid purification yielded lower values than methods using purification steps, especially in yolk-sac larvae. The decreasing RNA/DNA ratio trends observed during yolk-sac resorption could therefore result from procedural artefacts caused by interfering substances present in the yolk. This is supported by studies using purified material. Clemmesen (1989, 1992) found that starved and fed Atlantic herring larvae had constant RNA/DNA ratios (no trends present) both before yolk exhaustion and up to an age of 9 d posthatching. The RNA/DNA ratios were also similar between the two groups of fed and starved specimens and could not be used to detect starving animals earlier than 10 d posthatching. Steinhart and Eckman (1992), too, found no significant differences in RNA/DNA ratios of fed and starved yolk-sac whitefish (*Coregonus* spp.) larvae, even though some of the specimens analysed showed clear signs of feeding. Mathers *et al.* (1993) found that RNA levels were insensitive to various feeding rations in rainbow trout fry before they reached 30 d posthatch. These authors associated unchanged RNA levels with the fact that mixed feeding (endogenous and exogenous) supplied the energy during the period preceding total yolk absorption at 40 d posthatch. The existence of stable RNA/DNA ratios and active protein catabolism (negative growth) during the interval between hatching and yolk absorption is consistent with the absence of cyclic ring deposition on the otolith prior to yolk-sac exhaustion in most marine fish larvae (Campana and Neilson, 1985).

In many marine fish species, the larval mouth opens during or shortly after yolk absorption, and the digestive tract differentiates rapidly to enable the larvae to feed on planktonic prey. At this stage, the larvae possess a functional intestine with three morphologically distinct parts, the fore-, mid-, and hindgut. However, the larvae typically remain stomachless for some time, often until metamorphosis (Govoni, 1980; Vu, 1983; Lauff and Hofer, 1984; Govoni *et al.*, 1986). For these reasons, the main proteolytic enzymes found in the gut are of the trypsin type, and their activity is concentrated in the mid and posterior part of the intestine (Vu, 1983) where proteins are absorbed via the pinocytotic capacity of the epithelial cells (Govoni *et al.*, 1986; Deplano *et al.*, 1991; Walford and Lam, 1993). The quantity of these proteolytic enzymes increases with age as a consequence of increases in food intake and the size of the secreting organs (pancreas and liver). Pepsin-like enzyme activity only becomes important when the stomach differentiates (Walford and Lam, 1993). In first-feeding larvae, exogenous proteolytic enzymes obtained from the diet appear to be essential, since the larvae are unable to produce their own (Lauff and Hofer, 1984). The absence of dietary sources of these enzymes has been identified as one inadequacy of some artificial diets.

Hjelmeland *et al.* (1984) reported a large increase in trypsin activity 4 and 5 d posthatching, followed by a decline shortly after in cod larvae. This peak was independent of feeding status, and appeared to be associated with a period of physiological conditioning prior to first-feeding. This preconditioning precluded the use of tryptic enzyme activity levels as an index of nutritional condition in cod larvae younger than 5 d posthatch. Pedersen *et al.* (1987) identified three phases of tryptic enzyme activity during the early development of herring larvae. The first yolk-sac phase was characterized by increasing enzyme activity. The second declining phase was observed 7–12 d after hatching. A third phase (13–24 d after hatching) was characterized by increasing trypsin activity and was related to the amount of food offered to the larvae. These results were subsequently confirmed by Pedersen *et al.* (1990). The first and second phases of trypsin activity observed in herring are similar to those described by Hjelmeland *et al.* (1984) for cod larvae. In herring (Pedersen *et al.*, 1987, 1990), the largest differences in trypsin activity between fed and starved treatments were observed during the second phase. Cousin *et al.* (1987) found that proteolytic enzymes activity was equivalent in starved and fed turbot larvae 5 d posthatching, suggesting that 5-day-old larvae were incapable of controlling the release of these enzymes in quantities proportional to the quantity of food ingested. Clemmesen and Ueberschär (1993) observed a linear increase in tryptic enzymes activity with size in both starved and fed herring larvae. In this case, the enzyme activity–length relationship for starved larvae had a lower slope than for fed larvae, and this slope was related to the length of starvation.

The possibility that the timing of major physiological events may occur at "target sizes" rather than at "target ages" as suggested by Chambers and Leggett (1987) may also mean that major changes in fish body composition are associated with developmental thresholds. This is exemplified in the changes in chemical composition that accompany metamorphosis in bonefish (*Albula* spp.) leptocephalus larvae (Pfeila and Luna, 1984), and in the DNA-dry weight ratio changes seen before and after metamorphosis in sole (Bergeron *et al.*, 1991). Biochemical changes associated with developmental thresholds are also consistent with the theory of saltatory ontogeny (Balon, 1984), which predicts that development in fishes does not proceed as a continuous series of small histological changes, but rather via short periods of major change in form and function of the organs, separated by longer steady-state intervals. During these intervals, different tissues forming an organ are believed to align their rate of development in order to become functional simultaneously.

7. SPECIES-SPECIFICITY

Biochemical pathways and processes are much more universal than are anatomical characters. As a consequence, a general declining trend in species-specific variability in condition-related responses is seen as one progresses from morphometric, to histological, to biochemical indices. Hence, great care must be exercised in drawing conclusions from between-species comparisons based on common measures, especially when employing morphometric indices. For example, Ehrlich *et al.* (1976) noticed that the top of the head of Atlantic herring larvae changed from convex to concave during starvation. As a consequence, changes in the eye height/head height ratio provided a reliable indication of starvation. However, in plaice no such change occurred. Each species must be considered independently when morphological assessment is made, because shape differences and differential growth of various body parts is characteristic for each species. Laboratory calibration is therefore essential for each species considered.

Histological indices of condition have now been developed for larvae of a considerable number of species: yellowtail flounder (Umeda and Ochiai, 1975), northern anchovy (O'Connell, 1976, 1980; O'Connell and Paloma, 1981; Owen *et al.*, 1989; Theilacker and Watanabe, 1989), jack mackerel (Theilacker, 1978, 1986), stone flounder (Oozeki *et al.*, 1989), *Vinciguerria* spp. (Sieg, 1992a,b), black skipjack tuna (*Euthynnus lineatus*), frigate tuna (*Auxis* spp.), and sierra (*Scomberomorus sierra*) (Margulies, 1993). While some species-specific differences have been noted, the great majority of histological changes in gut epithelium, digestive organs and muscle tissue during starvation were common to all species examined (Kashuba and Matthews, 1984; Theilacker, 1986; Oozeki *et al.*, 1989). It must be emphasized, however, that when subjective histological classifications are used, interspecies differences may represent little more than interlaboratory differences in scoring. Interspecies and interlaboratory calibrations are, therefore, needed. Quantitative and objective measures such as the midgut cell height (Theilacker and Watanabe, 1989) might provide a solution to the problem of interspecies comparisons, especially if the index is found to show low species-specificity. To date, the number of studies reporting such quantitative measures is too limited to allow a definitive assessment of species-specificity.

The lipid composition of fish eggs exhibits a high degree of species-specificity. Tocher and Sargent (1984) reported that the lipid-rich eggs of capelin and sand lance had a higher proportion of neutral lipids (mainly triglycerides) relative to total lipids, than did eggs of cod, herring, and saithe, which were characterized by lower overall lipid levels but a higher proportion of polar lipids (mainly phospholipids). The utilization of these lipid classes during the embryonic and larval stages also varies between species. For example, in cod the only lipid class to decline significantly during embryonic development is the phospholipids (Fraser et al., 1988). Neutral lipids reserves are utilized only during the larval period. The rate of their utilization increases shortly after the yolk is depleted. Similarly, in striped bass, phospholipids were completely transferred from the yolk to the embryo within 5 d of hatching. Neutral lipids (triglycerides and wax esters), present mainly in the oil globule, were catabolized for up to 20 d posthatching (Eldridge et al., 1982, 1983). In contrast, in species that have relatively large quantities of neutral lipids (in the form of wax esters or triglycerides) in their eggs (red drum, capelin, sand lance or salmonids), these neutral lipid sources are typically catabolized early and throughout their embryonic stage (Vetter et al., 1983; Tocher and Sargent, 1984; Cowey et al., 1985). Vetter et al. (1983) calculated, in red drum, that neutral lipids accounted for approximately 98% of the energy requirements of the embryonic stage. If triglycerides are used preferentially for short-term energy needs and polar lipids, which are structural, are not affected by starvation, the species-specific lipid composition of marine fish eggs could seriously bias the reliability of TAG/cholesterol ratios as indicators of condition. Therefore, triglycerides levels in newly hatched larvae are determined by their levels in the egg content, and will influence the dynamics of their change during starvation and the length of time these changes will be detectable prior to complete exhaustion of the lipid stores.

Because of the central role of nucleic acids in protein synthesis, the RNA/DNA ratio is potentially a reliable index for interspecific comparisons of growth and condition as evidenced by the wide range of taxa to which it has been applied (Dagg and Littlepage, 1972; Dortch et al., 1983; Ota and Landry. 1984; Wright and Hetzel, 1985; DeFlaun et al., 1976; Wang and Stickle, 1986; DeBevoise and Taghon, 1988; Clarke et al., 1989; Berdalet and Dortch, 1991; Juinio et al., 1992; see also the review by Frantzis et al., 1993). Clemmesen (1987, 1988) claimed that the rate of decrease (deterioration dynamics) in the RNA/DNA ratio of fish larvae during starvation was species-independent. However, the generality of the relationship between protein growth rate, RNA/DNA ratio and temperature (Buckley, 1984; Eq. (1)) determined for eight marine species was recently questioned by Bergeron and Bouhlic (1994) who found it to be unreliable when applied to laboratory-reared sole larvae. The rate of protein synthesis, as measured by quantifying the amino acid incorporation in the muscle, was also found to differ between and among juveniles of saithe, cod, rainbow trout, and Atlantic herring, and to be positively correlated with their respective growth rate (Roselund *et al.*, 1983).

With respect to proteolytic enzymes, species having a functional stomach at, or immediately after, hatching (e.g. salmonids, Timeyko and Novikov, 1987) routinely produce pepsin-like enzymes, while other species which lack a stomach during all or part of their larval stage produce trypsin-like enzymes (Lauff and Hofer, 1984). The capacity to produce endogenous proteolytic enzymes also seems to differ considerably between species even at comparable stages of digestive tract differentiation (Lauff and Hofer, 1984; Cousin *et al.*, 1987; Munilla-Moran *et al.*, 1990).

8. PROCESSING TIME, COSTS, AND REQUIREMENTS

The large sample sizes that typically confront biologists who sample larval fish at sea, coupled with the time and cost demands associated with their processing, require that the procedures adopted provide the required data resolution at the minimum cost possible. Setzler-Hamilton *et al.* (1987) reported that the time required to process samples in order to obtain indices of condition for larval striped bass increased as follows: (1) morphometric measurements, (2) RNA/DNA ratio analysis, (3) fatty acid content, and (4) histological scoring. It must be remembered, however, that processing requirements prior to preservation can add considerably to the total cost of developing condition indices. While special care must be taken when preserving samples for morphometric and histological analyses, larvae used for biochemical analyses need only to be sorted from the bulk sample and frozen individually. When prepreservation costs are included, Setzler-Hamilton *et al.* (1987) determined that the most expensive indices are generally those based on fatty acids determinations, followed by indices based on nucleic acids. Indices based on histological and morphometric measures are generally the least expensive to develop. They therefore recommended the use of morphological and histological indices on the basis of their good correlation with condition, and their low average costs.

Many researchers who have employed morphological indices of condition admit to the potential of biochemical and histological assessments of condition, but note that these methods are too time-consuming and costly to be used routinely in ichthyoplankton surveys (Ehrlich *et al.*, 1976; Theilacker, 1978, 1986; Koslow *et al.*, 1985; Powell and Chester, 1985). One solution is to use less sensitive but less costly indices on all samples and to validate the data obtained from these results via histological or biochemical assessments based on a subsample (Theilacker, 1986). For the future, given the real limits on improvements in the precision of morphological indices, the most promising developments in this area are likely to be related to methods for reducing the processing time now associated with histological and biochemical indices. For example, the midgut cell height index developed by Theilacker and Watanabe (1989) and similar quantitative histological assessments such as those proposed by Oozeki *et al.* (1989) have proved to be equally sensitive, but less costly, than more traditional histological scores. However, given the rapid development of new molecular approaches, it is likely that the greatest gains in cost reduction will occur in the analysis of biochemically based indices of condition and growth. Automated continuous-flow systems (Caldarone and Buckley, 1991) have already substantially reduced the time required to process samples, such that a 100-sample carousel can be processed overnight (Sterzel *et al.*, 1985). This is, however, not achieved without concurrent increase in unit costs. The potential for development of fluorescent probes specific to all major biomolecules (Haugland, 1992) could dramatically reduce the costs associated with the quantification of lipids, proteins, carbohydrates, enzymes and nucleic acids, and make possible the development of related indices from very small amounts of tissue and the use of relatively inexpensive equipment. Fluorescence techniques, in addition to being sensitive and specific, can also be automated and may eventually be performed at sea.

Given the recent emphasis on the individual as the unit of study in fish early life history and recruitment studies (Chambers and Leggett, 1987; Chambers *et al.*, 1989; Chambers, 1993; Van Winkle *et al.*, 1993) and the high level of between-individual variation inherent in measures of condition, it is important that new methods be applicable to individual larvae. This has not been a problem with morphological and histological methods. However, the application of new biochemical techniques to individual fish larvae has been hampered by the fact that the quantity of material required for the analyses frequently required the pooling of samples (Buckley, 1984; Hjelmeland *et al.*, 1984; Fraser *et al.*, 1987). Recent advances in methodology have now made it possible to develop indices based on lipids (Håkanson, 1989a), nucleic acids (Clemmesen, 1988, 1989, 1993), and proteolytic enzymes (Ueberschär, 1988; Ueberschär and Clemmesen, 1993; Ueberschär *et al.*, 1992) for individual fish larvae. Several of these measurements can be obtained concurrently on the same specimens, and multivariate biochemical condition indices generated. These individual-based measurements have also highlighted the erroneous conclusions which resulted from the earlier pooling of samples (Clemmesen, 1988; Richard *et al.*, 1991; Ueberschär *et al.*, 1992; Bergeron and Bouhlic, 1994).

9. SUMMARY AND RECOMMENDATIONS

Table 2 provides a summary of key attributes of each category of condition index reviewed. From the table and the discussion above, it is clear that the sources of variability that influence both the nature and the interpretation of index values remain poorly known for all categories. The sensitivity, the latency and the dynamics of condition have also been very poorly described and the majority of the attributes given in Table 2 are still not well defined. This gap in knowledge is a major limitation to the application of these methods to studies of larval survival and recruitment. All indices exhibited some effect of size, and species-specificity, that must be considered when measuring condition over a range of larval sizes and across species. Processing time, costs and requirements are important considerations in improvement of current condition indices and for the development of new techniques.

Measures of condition integrate feeding success over time, and give an indication of the probability of starvation and survival. During the last two decades, the emphasis of research into the condition of larval fish has changed from a search for correlations with year-class strength (Chenoweth, 1970; Blaxter, 1971; Vilela and Zijlstra, 1971) to much more focused analyses of the impact of food quantity and quality on condition. However, as Neilson *et al.* (1986) have observed, while the number of studies in which one or more measures of condition have been employed continues to grow rapidly, very few have attempted, and even fewer have succeeded, to link directly measures of condition to probability of survival, which is often the stated goal in such studies. Rather the thrust of much of the work recently published has been the development of new and more accurate measures of condition. In such a context, it is perhaps important to restate the reality that the initial, and still central purpose of measuring condition is to predict reliably survival probabilities under given food regimes in order that observed larval abundances can be used as early predictors of recruitment (Frank and McRuer, 1989).

The ideal condition index should be capable of providing a direct estimate of starvation mortality for larvae that have passed the point-of-no-return, and a reliable estimate of susceptibility to starvation or starvation-induced predation for the specimens determined to be at intermediate or high levels of condition. Most of the condition indices reviewed here exhibit a decline shortly after onset of starvation. However, only histological scores and protein measurements continued to change beyond the point of irreversible starvation, a requirement for quantification of the proportion of larvae in a sample clearly destined to die regardless of future feeding conditions. The few estimates of the proportion of starving larvae in sampled populations which do exist

Table 2 Summary of key attributes for the seven categories of condition indices reviewed.

	Morphological indices	Histological (scores)	Histological (cell heights)	Nucleic acid and protein	Lipids	Digestive enzymes	Metabolic enzymes
Reliability to detect condition changes	Yes, through multivariate analysis	Yes, through multivariate analysis	Yes	Yes	No, because are likely to be influenced by gut content	No, because are likely to be influenced by gut content	Yes, but with limited data
Sources of variability other than nutrition	Some identified, none studied	Some identified, none studied	Some identified, none studied	Several identified, some studied	Some identified, few studied	Some identified, few studied	Some identified, none studied
Sensitivity	Believed to be low	Believed to be low	Believed to be moderate	Believed to be moderate	Believed to be moderate	Believed to be high	Unknown
Latency	Days	Days	Hours–day	Hours–day	Hours–day	Hours	Unknown
Dynamics	Slow initial changes	Slow initial changes	Moderate initial changes	Moderate initial changes	Moderate initial changes	Rapid initial changes	Unknown
Laboratory versus field differences	Large	Moderate	Small	Small	Large	Unknown	Unknown
Size and age-specificity	Large	Large	Moderate	Low	Large	Large	Unknown
Species-specificity	Large	Large	Small	Small	Moderate	Large	Unknown
Relative processing time	Short	Long	Long	Long	Long	Long	Long
Relative processing costs	Low	Moderate	Moderate	High	High	High	High
Processing requirements prior to analysis	High due to differential shrinkage	High due to autolysis	High due to autolysis	Low	Low	Low	Low

suggest that instantaneous starvation mortality may be relatively low (5–35%, O'Connell, 1980; Theilacker, 1986) although it may occasionally account for a high proportion of the mortality experienced by larvae (Theilacker, 1986). It is reasonable to assume, however, from the data available that most larvae will have been exposed to food-limited growth conditions, and will therefore be characterized as being in moderate or good condition. If this proves true, morphological indices are likely to be only marginally useful when applied to field studies, because they do not differentiate between animals that differ slightly in this degree of starvation. Methods based on analyses of triglycerides and proteolytic enzymes, in contrast, respond quickly to changes in feeding, and their short response time reduces to a very narrow window in time the period during which moderately starved animals can be identified. Their effective use therefore requires a high resolution sampling which is frequently difficult to achieve at sea. Indices based on histological cell heights and nucleic acid and protein measurements exhibited more gradual changes to both starvation and recovery, therefore providing a longer window of opportunity for detecting moderately starved animals. Selecting the "ideal index" thus becomes a question of matching the sensitivity, the latency, and dynamics of the index to the needs of the hypothesis to be tested, and to the *a priori* knowledge of the likely magnitude and duration of food deprivation. In the absence of such insight, indices based on histological, nucleic acid and protein assessments would appear to offer the greatest overall return when applied to larvae that are likely to be in intermediate condition.

Providing that a sufficiently fine-scale sampling of the food environment is achieved, and the advection of the water masses in which larvae occur is well described, the past, present, and future prey field of larval fish can be quantified. Further, through selection of the appropriate index, it should be possible to quantify the proportion of larvae in a given water mass that has died during a defined interval in time, provided the sensitivity and the dynamics of the index are known. These estimates of larval mortality can be validated by monitoring, coincidently, larval abundances corrected for dispersal losses, which can best be obtained from Lagrangian sampling (Heath, 1992). When dealing with larvae in intermediate condition, relating present condition to previous prey fields, or future condition to present prey fields, will also require a knowledge of the latency, the dynamics, and the sensitivity of the index. Factors other than food concentration known to influence feeding success (e.g. turbulence, larval fish behaviour) should also be evaluated in order to improve the predictions. Finally it is worth noting that the time and space scales of larval sampling programmes must be developed with reference to the characteristics of the condition index to be employed. This has rarely been done, and the result is likely to be a serious bias of the data (Taggart and Frank, 1990).

It will be almost impossible to investigate systematically the relationship between condition and survival in larval fish, unless and until the determination of condition becomes a routine component of larval surveys. Moreover, future research on condition indices is likely to be more productive if it is devoted to the refinement of existing methods, rather than to search for the ideal condition index. From the review presented here, it should be obvious that there is no "best index" for all circumstances. Rather, a suite of indices chosen with reference to their particular characteristics, and incorporated into a properly scaled programme is more likely to yield the answers that are sought.

ACKNOWLEDGEMENTS

Financial support for this study was provided by grants from the Natural Sciences and Engineering Research Council of Canada (NSERC), from Fonds pour la formation de Chercheurs et l'Aide à la Recherche (FCAR) to W.C. Leggett, and from OPEN (the Ocean Production Enhancement Network), one of 15 Networks of Centres of Excellence supported by the Government of Canada through NSERC. A. Ferron was supported by doctoral fellowships from NSERC and GIROQ (Groupe Interuniversitaire de Recherches Océanographiques du Québec).

REFERENCES

Arthur, D.K. (1976). Food and feeding of larvae of three fishes occurring in the California current, *Sardinops sagax*, *Engraulis mordax*, and *Trachurus symmetricus*. *Fishery Bulletin U.S.* **74(3)**, 517–530.

Bailey, K. M. and Houde, E.D. (1989). Predation on eggs and larvae of marine fishes and the recruitment problem. *Advances in Marine Biology* **25**, 1–83.

Bailey, K.M. and Spring, S.M. (1992). Comparison of larval, age-0 juvenile and age-2 recruit abundance indices of walleye pollock, *Theragra chalcogramma*, in the western Gulf of Alaska. *ICES Journal of Marine Science* **49**, 297–304.

Balbontin, F., De Silva, S.S. and Ehrlich, K.F. (1973). A comparative study of anatomical and chemical characteristics of reared and wild herring. *Aquaculture* **2**, 217–240.

Balon, E.K. (1984). Reflections on some decisive events in the early life of fishes. *Transactions of the American Fisheries Society* **113**, 178–185.

Barron, M.G. and Adelman, I.R. (1984). Nucleic acid, protein content, and growth of larval fish sublethally exposed to various toxicants. *Canadian Journal of Fisheries and Aquatic Sciences* **41**, 141–150.

Barron, M.G. and Adelman, I.R. (1985). Temporal characterization of growth of fathead minnow, *Pimephales promelas*, larvae during sublethal hydrogen cyanide exposure. *Comparative Biochemistry and Physiology* 81C(2), 341–344.

Beacham, T.D. and Murray, C.B. (1985). Effect of female size, egg size, and water temperature on developmental biology of chum salmon *Oncorhynchus keta* from the Nitinat river, British Columbia. *Canadian Journal of Fisheries and Aquatic Sciences* 42, 1755–1765.

Bengtson, D.A., Barkman, R.C. and Berry, W.J. (1987). Relationship between maternal size, egg diameter, time of spawning season, temperature and length at hatch of Atlantic silverside *Menidia menidia*. *Journal of Fish Biology* 31, 697–704.

Berdalet, E. and Dortch, Q. (1991). New double-staining technique for RNA and DNA measurement in marine phytoplankton. *Marine Ecology Progress Series* 73, 295–305.

Bergeron, J.-P. and Boulhic, M. (1994). Rapport ARN/ADN et évaluation de l'état nutritionnel et de la croissance des larves de poissons marins: Un essai de mise au point expérimentale chez la sole *Solea solea*. *ICES Journal of Marine Science* (in press).

Bergeron, J.-P., Boulhic, M. and Galois, R. (1991). Effet de la privation de nourriture sur la teneur en ADN de la larve de sole *Solea solea*. *ICES Journal of Marine Science* 48, 127–134.

Black, D. and Love, R.M. (1986). The sequential mobilization and restoration of energy reserves in tissues of Atlantic cod during starvation and refeeding. *Journal of Comparative Physiology B* 156, 469–479.

Blaxter, J.H.S. (1971). Feeding and condition of Clyde herring larvae. *Rapports et Procès-verbaux des Réunions. Conseil international pour l'Exploration de la Mer* 160, 128–136.

Blaxter, J.H.S. and Ehrlich, K.F. (1974). Changes in behaviour during starvation of herring and plaice larvae. *In* "The Early Life History of Fish" (J.H.S. Blaxter, ed.), pp. 575–588. Springer-Verlag, Berlin.

Blaxter, J.H.S. and Hempel, G. (1963). The influence of egg size on herring larvae *Clupea harengus*. *Rapports et Procès-verbaux des Réunions. Conseil international pour l'Exploration de la Mer* 28, 211–240.

Boujard, T. and Leatherland, J.F. (1992). Circadian rhythms and feeding time in fishes. *Environmental Biology of Fishes* 35, 109–131.

Bradford, M.J. (1992). Precision of recruitment predictions from early life stages of marine fishes. *Fishery Bulletin U.S.* 90, 439–453.

Buckley, L.J. (1979). Relationship between RNA–DNA ratio, prey density, and growth rate in Atlantic cod *Gadus morhua*. *Journal of the Fisheries Research Board of Canada* 36, 1497–1502.

Buckley, L.J. (1980). Changes in ribonucleic acid, desoxyribonucleic acid, and protein content during ontogenesis in winter flounder *Pseudopleuronectes americanus*, and effect of starvation. *Fishery Bulletin U.S.* 77(3), 703–708.

Buckley, L.J. (1981). Biochemical changes during ontogenesis of cod *Gadus morhua* and winter flounder *Pseudopleuronectes americanus* larvae. *Rapports et Procès-verbaux des Réunions. Conseil international pour l'Exploration de la Mer* 178, 547–552.

Buckley, L.J. (1982). Effects of temperature on growth and biochemical composition of larval winter flounder *Pseudopleuronectes americanus*. *Marine Ecology Progress Series* 8, 181–186.

Buckley, L.J. (1984). RNA–DNA ratio: an index of larval fish growth in the sea. *Marine Biology* 80, 291–298.

Buckley, L.J. and Lough, R.G. (1987). Recent growth, biochemical composition, and prey field of larval haddock *Melanogrammus aeglefinus* and Atlantic cod *Gadus morhua*, on Georges Bank. *Canadian Journal of Fisheries and Aquatic Sciences* **44**, 14–25.

Buckley, L.J. and McNamara, P. (1993). Estimation of zooplankton and ichthyoplankton growth and condition using nucleic acid probe techniques. *U.S. Globec News* **4**, 12.

Buckley, L.J., Turner, S.I., Halavik, T.A., Smigielski, A.S., Drew, S.M. and Laurence, G.C. (1984). Effects of temperature and food availability on growth, survival, and RNA–DNA ratio of larval sand lance *Ammodytes americanus*. *Marine Ecology Progress Series* **15**, 91–97.

Buckley, L.J., Halavik, T.A., Smigielski, A.S. and Laurence, G.C. (1987). Growth and survival of the larvae of three species of temperate marine fishes reared at discrete prey densities. *American Fisheries Society Symposium* **2**, 82–92.

Buckley, L.J., Smigielski, A.S., Halavik, T.A. and Laurence, G.C. (1990). Effects of water temperature on size and biochemical composition of winter flounder *Pseudopleuronectes americanus* at hatching and feeding initiation. *Fishery Bulletin U.S.* **88**, 419–428.

Buckley, L.J., Smigielski, A.S., Halavik, T.A., Caldarone, E.M., Burns, B.R. and Laurence, G.C. (1991a). Winter flounder *Pseudopleuronectes americanus* reproductive success. I. Among-location variability in size and survival of larvae reared in the laboratory. *Marine Ecology Progress Series* **74**, 117–124.

Buckley, L.J., Smigielski, A.S., Halavik, T.A., Caldarone, E.M., Burns, B.R. and Laurence, G.C. (1991b). Winter flounder *Pseudopleuronectes americanus* reproductive success. II. Effects of spawning time and female size on size, composition and viability of eggs and larvae. *Marine Ecology Progress Series* **74**, 125–135.

Bulow, F.J. (1970). RNA–DNA ratios as indicators of recent growth rates of a fish. *Journal of the Fisheries Research Board of Canada* **27**, 2343–2349.

Bulow, F.J. (1974). A review of the literature concerning the relationship between nucleic acids and the growth rates of fish. *The Tennessee Technical Journal* **9**, 17–23.

Bulow, F.J. (1987). RNA–DNA ratios as indicators of growth in fish: A review. *In* "The Age and Growth of Fish" (R.C. Summerfelt and G.E. Hall, eds), pp. 45–64. Iowa State University Press, Ames, Iowa.

Bulow, F.J., Coburn, C.B. and Cobb, C.S. (1978). Comparisons of two bluegill populations by means of the RNA–DNA ratio and liver-somatic index. *Transactions of the American Fisheries Society* **107(6)**, 799–803.

Caldarone, E. and Buckley, L.J. (1991). Quantitation of DNA and RNA in crude tissue extracts by flow injection analysis. *Analytical Biochemistry* **199**, 137–141.

Campana, S.E. and Neilson, J. (1985). Microstructure of fish otoliths. *Canadian Journal of Fisheries and Aquatic Sciences* **42**, 1014–1032.

Canino, M.F., Bailey, K.E. and Incze, L.S. (1991). Temporal and geographic differences in feeding and nutritional condition of walleye pollock larvae *Theragra chalcogramma*, in Shelikof strait, Gulf of Alaska. *Marine Ecology Progress Series* **79**, 27–35.

Chambers, R.C. (1993). Phenotypic variability in fish populations and its representation in individual-based models. *Transactions of the American Fisheries Society* **122(3)**, 404–414.

Chambers, R.C. and Leggett, W.C. (1987). Size and age at metamorphosis in marine fishes: an analysis of laboratory-reared winter flounder *Pseudopleuronectes americanus* with a review of variation in other species. *Canadian Journal of Fisheries and Aquatic Sciences* 44, 1936–1947.

Chambers, R.C., Leggett, W.C. and Brown, J.A. (1989). Egg size, female effects, and the correlations between early life history traits of capelin, *Mallotus villosus*: an appraisal at the individual level. *Fishery Bulletin U.S.* 87, 515–523.

Checkley, D.M. Jr (1984). Relation of growth to ingestion for larvae of Atlantic herring *Clupea harengus* and other fish. *Marine Ecology Progress Series* 18, 215–224.

Chenoweth, S.B. (1970). Seasonal variations in condition of larval herring in Boothbay area of the Maine coast. *Journal of the Fisheries Research Board of Canada* 27, 1875–1879.

Clarke, A., Rodhouse, P.G., Holmes, L.J. and Pascoe, P.L. (1989). Growth rate and nucleic acid ratio in cultured cuttlefish *Sepia officinalis* (Mollusca: Cephalopoda). *Journal of Experimental Marine Biology and Ecology* 133, 229–240.

Clarke, M.E., Calvi, C., Domeier, M., Edmonds, M. and Walsh, P.J. (1992). Effect of nutrition and temperature on metabolic enzymes activities in larval and juvenile red drum, *Sciaenops ocellatus*, and lane snapper, *Lutjanus synagris*. *Marine Biology* 112, 31–36.

Clemmesen, C. (1987). Laboratory studies on RNA–DNA ratios of starved and fed herring *Clupea harengus* and turbot *Scophthalmus maximus* larvae. *Journal du Conseil international pour l'Exploration de la Mer* 43, 122–128.

Clemmesen, C. (1988). A RNA and DNA fluorescence technique to evaluate the nutritional condition of individual marine fish larvae. *Meeresforsch* 32, 134–143.

Clemmesen, C. (1989). RNA–DNA ratios of laboratory-reared and wild herring larvae determined with a highly sensitive fluorescence method. *Journal of Fish Biology* 35(Suppl. A), 331–333.

Clemmesen, C. (1992). The effect of food availability, age or size on the RNA/DNA ratio of laboratory reared individually measured herring larvae. *International Council for the Exploration of the Sea, Council Meeting, 1992/ L:33*.

Clemmesen, C. (1993). Improvements in the fluorometric determination of the RNA and DNA content of individual marine fish larvae. *Marine Ecology Progress Series* 100, 177–183.

Clemmesen, C. and Ueberschär, B. (1993). Application of RNA/DNA ratio and tryptic enzyme activity on laboratory-reared and wild-caught herring larvae-short communication. *In* "Fish Ecotoxicology and Ecophysiology" (T. Braunbeck, W. Hanke and H. Segner, eds), pp. 227–232. VCH Verlagsgesellschaft mbH, D-6940 Weinheim.

Cone, R.S. (1989). The need to reconsider the use of condition indices in fishery science. *Transactions of the American Fisheries Society* 118, 510–514.

Cousin, J.C.B., Balouet, G. and Baudin-Laurencin, F. (1986). Altérations histologiques observées chez les larves de turbot *Scophthalmus maximus* en élevage intensif. *Aquaculture* 52, 173–189.

Cousin, J.C.B., Baudin-Laurencin, F. and Gabaudan, J. (1987). Ontogeny of enzymatic activities in fed and fasting turbot, *Scophthalmus maximus*. *Journal of Fish Biology* 30, 15–33.

Cowey, C.B., Bell, J.G., Knox, D., Fraser, A. and Youngson, A. (1985). Lipids and lipid antioxidant systems in developing eggs of salmon *Salmo salar*. *Lipids* 20, 567–572.

Cushing, D.H. (1972). The production cycle and the numbers of marine fish. *Symposium of the Zoological Society of London* **29**, 213–232.

Dabrowski, K. (1982). Proteolytic enzyme activity decline in starving fish alevins and larvae. *Environmental Biology of Fishes* **7**(1), 73–76.

Dabrowski, K. and Glogowski, J. (1977). Studies on the role of exogenous proteolytic enzymes in digestion processes in fish. *Hydrobiologia* **54**, 129–134.

Dagg, M.J. and Littlepage, J.L. (1972). Relationships between growth rate and RNA, DNA, protein and dry weight in *Artemia salina* and *Euchaeta elongata*. *Marine Biology* **17**, 162–170.

Davis, M.W. and Olla, B.L. (1992). Comparison of growth, behavior and lipid concentrations of walleye pollock *Theragra chalcogramma* larvae fed lipid-enriched, lipid-deficient and field-collected prey. *Marine Ecology Progress Series* **90**, 23–30.

DeBevoise, A.E. and Taghon, G.L. (1988). RNA/DNA ratios of the hydrothermal-vent vestimentiferans *Ridgeia piscesae* and *R. phaeophiale* indicate variations in growth rates over small spatial scales. *Marine Biology* **97**, 421–426.

DeFlaun, M.F., Paul, J.H. and Davis, D. (1976). Simplified method for dissolved DNA determinations in aquatic environments. *Applied Environmental Microbiology* **52**(4), 654–659.

Delauney, F., Marty, Y., Moal, J. and Samain, J.-F. (1992). Growth and lipid class composition of *Pecten maximus* larvae grown under hatchery conditions. *Journal of Experimental Marine Biology and Ecology* **163**, 209–219.

De March, B.G.E. (1991). Genetic, maternal, and tank determinants of growth in hatchery-reared juvenile Arctic charr *Salvelinus alpinus*. *Canadian Journal of Zoology* **69**, 655–660.

Denman, K.L. and Powell, T.M. (1984). Effects of physical processes on planktonic ecosystems in the coastal ocean. *Oceanography and Marine Biology Annual Review* **22**, 125–168.

Deplano, M., Connes, R., Diaz, J.P. and Barnabé, G. (1991). Variation in the absorption of macromolecular proteins in larvae of the sea bass *Dicentrarchus labrax*, during transition to the exotrophic phase. *Marine Biology* **110**, 29–36.

DeVlaming, V.L., Sage, M. and Tiegs, R. (1975). A diurnal rhythm of pituitary prolactin activity with diurnal effects of mammalian and teleostean prolactin on total body lipid deposition and liver lipid metabolism in teleost fishes. *Journal of Fish Biology* **7**, 717–726.

Dortch, Q., Roberts, T.L., Clayton, J.R. and Ahmed, S.I. (1983). RNA/DNA ratios and DNA concentrations as indicators of growth rate and biomass in planktonic marine organisms. *Marine Ecology Progress Series* **13**, 61–71.

Economou, A. (1987). Ecology of survival in some gadoid larvae of the northern North Sea. *Environmental Biology of Fishes* **19**(4), 241–260.

Ehrlich, K.F. (1974a). Chemical changes during growth and starvation of larval *Pleuronectes platessa*. *Marine Biology* **24**, 39–48.

Ehrlich, K.F. (1974b). Chemical changes during growth and starvation of herring larvae. *In* "The Early Life History of Fish" (J.H.S. Blaxter, ed.), pp. 301–323. Springer-Verlag, Berlin.

Ehrlich, K.F. (1975). A preliminary study of the chemical composition of sea-caught larval herring and plaice. *Comparative Biochemistry and Physiology* **51B**, 25–28.

Ehrlich, K.F., Blaxter, J.H.S. and Pemberton, R. (1976). Morphological and histological changes during the growth and starvation of herring and plaice larvae. *Marine Biology* **35**, 105–118.

Eldridge, M.B., Whipple, J.A. and Bowers, M.J. (1982). Bioenergetics and growth of striped bass, *Morone saxatilis*, embryos and larvae. *Fishery Bulletin U.S.* **80(3)**, 461–474.

Eldridge, M.B., Joseph, J.D., Taberski, K.M. and Seaborn, G.T. (1983). Lipid and fatty acid composition of the endogenous energy sources of striped bass *Morone saxatilis* eggs. *Lipids* **18(8)**, 510–513.

Farber-Lorda, J. (1991). Multivariate approach to the morphological and biochemical differentiation of Antarctic krill *Euphausia superba* and *Thysanoessa macrura*. *Deep Sea Research* **38**, 771–779.

Farbridge, K.J. and Leatherland, J.F. (1987). Lunar cycles of coho salmon *Oncorhynchus kisutch*. II. Scale amino acid uptake, nucleic acids, metabolic reserves and plasma thyroid hormones. *Journal of Experimental Biology* **129**, 179–189.

Ferguson, M.M. and Danzmann, R.G. (1990). RNA/DNA ratios in white muscle as estimates of growth in rainbow trout held at different temperatures. *Canadian Journal of Zoology* **68**, 1494–1498.

Ferguson, M.M. and Drahushchak, L.R. (1989). Effects of tissue collection and storage methods on nucleic acid determinations in white muscle of fishes. *Transactions of the American Fisheries Society* **118**, 709–713.

Ferron, A. (1991). État nutritionnel des larves de maquereau *Scomber scombrus* en rapport avec la disponibilité de nourriture et la structure physique côtière dans le sud-ouest du Golfe Saint-Laurent. MSc thesis, Département d'Océanographie, Université du Québec à Rimouski, Rimouski, Québec.

Fogarty, M.J. (1993). Recruitment in randomly varying environments. *ICES Journal of Marine Science* **50**, 247–260.

Fogarty, M.J., Sissenwine, M.P. and Cohen, E.B. (1991). Recruitment variability and the dynamics of exploited marine populations. *Trends in Ecology and Evolution* **6(8)**, 241–246.

Foster, A.R., Houlihan, D.F., Hall, S.J. and Burren, L.J. (1992). The effects of temperature acclimation on protein synthesis rates and nucleic acid content of juvenile cod *Gadus morhua*. *Canadian Journal of Zoology* **70**, 2095–2102.

Foster, A.R., Houlihan, D.F. and Hall, S.J. (1993). Effects of nutritional regime on correlates of growth rate in juvenile Atlantic cod, *Gadus morhua*: Comparison of morphological and biochemical measurements. *Canadian Journal of Fisheries and Aquatic Sciences* **50**, 505–512.

Frank, K.T. and McRuer, J.K. (1989). Nutritional status of field-collected haddock *Melanogrammus aeglefinus* larvae from southwestern Nova Scotia: an assessment based on morphometric and vertical distribution data. *Canadian Journal of Fisheries and Aquatic Sciences* **46**(Suppl.1), 125–133.

Frantzis, A., Grémare, A. and Vétion, G. (1993). Taux de croissance et rapports ARN/ADN chez le bivalve dépositivore *Abra ovata* nourri à partir de différents détritus. *Oceanologica acta*, **16(3)**, 303–313.

Fraser, A.J. (1989). Triacylglycerol content as a condition index for fish, bivalve, and crustacean larvae. *Canadian Journal of Fisheries and Aquatic Sciences* **46**, 1868–1873.

Fraser, A.J., Tocher, D.R. and Sargent, J.R. (1985). Thin-layer chromatography-flame ionization detection and the quantitation of marine neutral lipids and phospholipids. *Journal of Experimental Marine Biology and Ecology* **88**, 91–99.

Fraser, A.J., Sargent, J.R., Gamble, J.C. and MacLachlan, P. (1987). Lipid class and fatty acid composition as indicators of the nutritional condition of larval Atlantic herring. *American Fisheries Society Symposium* **2**, 129–143.

Fraser, A.J., Gamble, J.C. and Sargent, J.R. (1988). Changes in lipid content, lipid class composition and fatty acid composition of developing eggs and unfed larvae of cod *Gadus morhua*. *Marine Biology* **99**, 307–313.

Frolov, A.V. and Pankov, S.L. (1992). The effect of starvation on the biochemical composition of the rotifer *Brachionus plicatilis*. *Journal of the Marine Biological Association of the United Kingdom* **72**, 343–356.

Fukuda, M., Nakano, H. and Yamamoto, K. (1986). Biochemical changes in Pacific herring during early developmental stages. *Hokkaido Daigaku Faculty of Fisheries Bulletin* **37**, 30–37.

Fyhn, H.J. (1989). First feeding of marine fish larvae: are free amino acids the source of energy? *Aquaculture* **80**, 111–120.

Fyhn, H.J. and Serigstad, B. (1987). Free amino acids as energy substrate in developing eggs and larvae of the cod *Gadus morhua*. *Marine Biology* **96**, 335–341.

Gallager, S.M., Mann, R. and Sakaki, G.C. (1986). Lipid as an index of growth and viability in three species of bivalve larvae. *Aquaculture* **56**, 81–103.

Gatten, R.R., Sargent, J.R. and Gamble, J.C. (1983). Diet-induced changes in fatty acid composition of herring larvae reared in enclosed ecosystems. *Journal of the Marine Biological Association of the United Kingdom* **63**, 575–584.

Gleeson, M. and Maughan, R.J. (1986). A simple enzymatic fluorometric method for the determination of triglycerides in 10 μl of serum. *Clinica Chimica Acta* **156**, 97–103.

Goolish, E.M. and Adelman, I.R. (1987). Tissue-specific cytochrome oxidase activity in largemouth bass: the metabolic costs of feeding and growth. *Physiological Zoology* **69**, 454–464.

Goolish, E.M. and Adelman, I.R. (1988). Tissue-specific allometry of an aerobic respiratory enzyme in large and small species of cyprinid (Teleostei). *Canadian Journal of Zoology* **66**, 2199–2208.

Goolish, E.M., Barron, M.G. and Adelman, I.R. (1984). Thermoacclimatory response of nucleic acid and protein content of carp muscle tissue: influence of growth rate and relationship to glycine uptake by scales. *Canadian Journal of Zoology* **62**, 2164–2170.

Govoni, J.J. (1980). Morphological, histological, and functional aspects of alimentary canal and associated organ development in larval *Leiostomus xanthurus*. *Revue Canadienne de Biologie* **39**(2), 69–80.

Govoni, J.J., Boehlert, G.W. and Watanabe, Y. (1986). The physiology of digestion in fish larvae. *Environmental Biology of Fishes* **16**(1–3), 59–77.

Haines, T.A. (1973). An evaluation of RNA–DNA ratio as a measure of long-term growth in fish populations. *Journal of the Fisheries Research Board of Canada* **30**, 195–199.

Haines, T.A. (1980). Seasonal patterns of muscle RNA–DNA ratio and growth in black crappie, *Pomoxis nigromaculatus*. *Environmental Biology of Fishes* **5**(1), 67–70.

Håkanson, J.L. (1984). The long and short term feeding condition in field-caught *Calanus Pacificus*, as determined from the lipid content. *Limnology and Oceanography* **29**(4), 794–804.

Håkanson, J.L. (1989a). Analysis of lipid components for determining the condition of anchovy larvae, *Engraulis mordax*. *Marine Biology* **102**, 143–151.

Håkanson, J.L. (1989b). Condition of larval anchovy *Engraulis mordax* in the southern California Bight, as measured through lipid analysis. *Marine Biology* **102**, 153–159.

Hansen, P.E., Lied, E. and Børresen, T. (1989). Estimation of protein synthesis in fish larvae using an *in vitro* polyribosome assay. *Aquaculture* **79**, 85–89.

Harris, G.P. (1980). Temporal and spatial scales in phytoplankton ecology. Mechanisms, methods, models and management. *Canadian Journal of Fisheries and Aquatic Sciences* **37**, 877–900.

Harris, R.K., Nishiyama, T. and Paul, A.J. (1986). Carbon, nitrogen and caloric content of eggs, larvae, and juveniles of the walleye pollock, *Theragra chalcogramma*. *Journal of Fish Biology* **29**, 87–98.

Haugland, R.P. (1992). "Handbook of Fluorescent Probes and Research Chemicals", 5th edn. Molecular Probes Inc. Eugene, OR 97402-0414, USA.

Hay, D.E. (1981). Effects of capture and fixation on gut contents and body size of Pacific herring larvae. *Rapports et Procès-verbaux des Réunions. Conseil international pour l'Exploration de la Mer* **178**, 395–400.

Hay, D.E. (1982). Fixation shrinkage of herring larvae: effects of salinity, formalin concentration, and other factors. *Canadian Journal of Fisheries and Aquatic Sciences* **39**, 1138–1143.

Hay, D.E. (1984). Weight loss and change of condition factor during fixation of Pacific herring, *Clupea harengus pallasi*, eggs and larvae. *Journal of Fish Biology* **25**, 421–433.

Heath, M.R. (1992). Field investigations of the early life stages of marine fish. *Advances in Marine Biology* **28**, 1–153.

Heidinger, R.C. and Crawford, S.D. (1977). Effect of temperature and feeding rate on the liver-somatic index of the largemouth bass, *Micropterus salmoides*. *Journal of the Fisheries Research Board of Canada* **34**, 633–638.

Hempel, G. and Blaxter, J.H.S. (1963). On the condition of herring larvae. *Rapports et Procès-verbaux des Réunions. Conseil international pour l'Exploration de la Mer* **154**, 35–40.

Hewitt, R.P., Theilacker, G.H. and Lo, N.C.H. (1985). Causes of mortality in young jack mackerel. *Marine Ecology Progress Series* **26**, 1–10.

Hislop, J.R.G. (1988). The influence of maternal length and age on the size and weight of the eggs and the relative fecundity of the haddock, *Melanogrammus aeglefinus*, in British waters. *Journal of Fish Biology* **32**, 923–930.

Hjelmeland, K. and Jørgensen, T. (1985). Evaluation of radioimmunoassay as a method to quantify trypsin and trypsinogen in fish. *Transactions of the American Fisheries Society* **114**, 619–621.

Hjelmeland, K., Huse, I., Jørgensen, T., Moløvik, G. and Raae, J. (1984). Trypsin and trypsinogen as indices of growth and survival potential of cod *Gadus morhua*, larvae. *In* "The Propagation of Cod *Gadus morhua* Part 1" (E. Dahl, D.S. Danielssen, E. Moksness and P. Solemdal, eds), pp. 189–202. Flødevigen Biological Station, Arendal, Norway.

Hjelmeland, K., Pedersen, B.H. and Nilssen, E.M. (1988). Trypsin content in intestines of herring larvae, *Clupea harengus*, ingesting inert polystyrene spheres or live crustacea prey. *Marine Biology* **98**, 331–335.

Hjort, J. (1914). Fluctuations in the great fisheries of northern Europe viewed in the light of the biological research. *Rapports et Procès-verbaux des Réunions. Conseil permanent international pour l'Exploration de la Mer* **20**, 1–228.

Horne, E.P.W. and Platt, T. (1984). The dominant space and time scales of variability in the physical and biological fields on continental shelves. *Rapports et Procès-verbaux des Réunions. Conseil international pour l'Exploration de la Mer* **183**, 8–19.

Houde, E.D. (1982). Micro and fine-scale biology. *In* "Fish Ecology III—A Foundation for REX a Recruitment Experiment" (B.J. Rothschild and C. Rooth, eds), pp. 96–122. University of Miami, Miami, FL.

Houde, E.D. (1987). Fish early life dynamics and recruitment variability. *American Fisheries Society Symposium* **2**, 17–29.

Houde, E.D. (1989). Comparative growth, mortality, and energetics of marine fish larvae: temperature and implied latitudinal effects. *Fishery Bulletin U.S.* **87**, 471–495.

Houde, E.D. and Schekter, R.C. (1983). Oxygen uptake and comparative energetics among eggs and larvae of three subtropical marine fishes. *Marine Biology* **72**, 283–293.

Hovenkamp, F. (1990). Growth differences in larval plaice *Pleuronectes platessa* in the southern Bight of the North Sea as indicated by otolith increments and RNA–DNA ratios. *Marine Ecology Progress Series* **58**, 205–215.

Hovenkamp, F. and Witte, J. IJ. (1991). Growth, otolith growth and RNA–DNA ratios of larval plaice *Pleuronectes platessa* in the North Sea 1987 to 1989. *Marine Ecology Progress Series* **70**, 105–116.

Hunter, J.R. (1976). Report of a colloquium on larval fish mortality studies and their relation to fishery research, January 1975. *NOAA Technical Report, National Marine Fisheries Service, CIRC-395, 5 p.*

Hunter, J.R. (1981). Feeding ecology and predation of marine fish larvae. *In* "Marine Fish Larvae" (R. Lasker, ed.), pp. 34–77. University of Washington Press, Seattle, WA.

Hunter, J.R. (1984). Inferences regarding predation on the early life stages of cod and other fishes. *In* "The Propagation of Cod *Gadus morhua*, Part 2" (E. Dahl, D.S. Danielssen, E. Moksness and P. Solemdal, eds), pp. 533–552. Flødevigen Biological Station, Arendal, Norway.

Juinio, M.A.R., Cobb, J.S., Bengtson, D. and Johnson, M. (1992). Changes in nucleic acids over the molt cycle in relation to food availability and temperature in *Homarus americanus* postlarvae. *Marine Biology* **114**, 1–10.

Jürss, K., Bittorf, Th., and Vökler, Th. (1986). Influence of salinity and food deprivation on growth, RNA/DNA ratio and certain enzyme activities in rainbow trout *Salmo gairdneri*. *Comparative Biochemistry and Physiology* **83B**(2), 425–433.

Jürss, K., Bittorf, Th., Vökler, Th. and Wacke, R. (1987). Effects of temperature, food deprivation and certain enzyme activities in rainbow trout *Salmo gairdneri*. *Comparative Biochemistry and Physiology* **87B**, 241–253.

Kamler, R. (1992). "Early Life History of Fish. An Energetics Approach." Fish and Fisheries Series 4, Chapman & Hall, New York.

Kashuba, S.A. and Matthews, W.J. (1984). Physical condition of larval shad during spring-summer in a southwestern reservoir. *Transactions of the American Fisheries Society* **113**, 199–204.

Kayes, T. (1978). Effects of hypophysectomy and beef growth hormone replacement therapy on morphometric and biochemical indicators of growth in the fed versus starved black bullhead *Ictalurus melas*. *General Comparative Endocrinology* **35**, 419–431.

Keast, A. and Eadie, J.M. (1985). Growth depensation in year-0 largemouth bass: the influence of diet. *Transactions of the American Fisheries Society* **114**, 204–213.

Kiørboe, T. (1989). Growth in fish larvae. Are they particularly efficient? *Rapports et Procès-verbaux des Réunions. Conseil international pour l'Exploration de la Mer* **191**, 383–389.

Kiørboe, T. and Munk, P. (1986). Feeding and growth of larval herring, *Clupea harengus*, in relation to density of copepod nauplii. *Environmental Biology of Fishes* **17**(2), 133–139.

Kiørboe, T., Munk, P. and Richardson, K. (1987). Respiration and growth of larval herring *Clupea harengus*: relation between specific dynamic action and growth efficiency. *Marine Ecology Progress Series* **40**, 1–10.

Klungsøyr, J., Tilseth, S., Wilhelmsen, S., Falk-Petersen, S. and Sargent, J.R. (1989). Fatty acid composition as an indicator of food intake in cod larvae *Gadus morhua* from Lofoten, northern Norway. *Marine Biology* **102**, 183–188.

Knutsen, G.M. and Tilseth, S. (1985). Growth, development, and feeding success of Atlantic cod larvae *Gadus morhua* related to egg size. *Transactions of the American Fisheries Society* **114**, 507–511.

Koslow, J.A. (1992). Fecundity and the stock-recruitment relationship. *Canadian Journal of Fisheries and Aquatic Sciences* **49**, 210–217.

Koslow, J.A., Brault, S., Dugas, J., Fournier, R.O. and Hughes, P. (1985). Condition of larval cod *Gadus morhua* off southwest Nova Scotia in 1983 in relation to plankton abundance and temperature. *Marine Biology* **86**, 113–121.

Lasker, R. (1962). Efficiency and rate of yolk utilization by developing embryos and larvae of the Pacific sardine, *Sardinops caerulea*. *Journal of the Fisheries Research Board of Canada*, **19(5)**, 867–875.

Lasker, R. (1981). The role of a stable ocean in larval fish survival and subsequent recruitment. *In* "Marine Fish Larvae" (R. Lasker, ed.), pp. 80–87. University of Washington Press, Seattle, WA.

Lasker, R. and Zweifel, J.R. (1978). Growth and survival of first-feeding northern anchovy larvae *Engraulis mordax* in patches containing different proportions of large and small prey. *In* "Spatial Patterns in Plankton Communities" (J.H. Steele, ed.), pp. 329–354. Plenum, New York.

Lauff, M. and Hofer, R. (1984). Proteolytic enzymes in fish development and the importance of dietary enzymes. *Aquaculture* **37**, 335–346.

Laurence, G.C. (1974). Growth and survival of haddock *Melanogrammus aeglefinus* larvae in relation to planktonic prey concentration. *Journal of the Fisheries Research Board of Canada* **31**, 1415–1419.

Laurence, G.C. (1977). A bioenergetic model for the analysis of feeding and survival potential of winter flounder *Pleuronectes americanus* larvae during the period from hatching to metamorphosis. *Fishery Bulletin U.S.* **75(3)**, 529–546.

Laurence, G.C. (1979). Larval length–weight relations for seven species of northwest Atlantic fishes reared in the laboratory. *Fishery Bulletin U.S.* **76(4)**, 890–895.

LeCren, E.D. (1951). The length–weight relationship and seasonal cycle in gonad weight and condition in the perch *Perca flavescens*. *Journal of Animal Ecology* **20**, 201–219.

Leggett, W.C. (1986). The dependence of fish larval survival on food and predator densities. *In* "The Role of Freshwater Outflow in Coastal Marine Ecosystems" (S. Skreslet, ed.), pp. 117–137. *NATO ASI Series*, Vol. **G7**.

Lied, E. and Rosenlund, G. (1984). The influence of the ratio of protein energy to total energy in the feed on the activity of protein synthesis *in vitro*, the level of ribosomal RNA and the RNA–DNA ratio in white trunk muscle of Atlantic cod *Gadus morhua*. *Comparative Biochemistry and Physiology* **77A**, 489–494.

Lied, E., Lund, B. and Von Der Decken, A. (1982). Protein synthesis *in vitro* by epaxial muscle polyribosomes from cod, *Gadus morhua*. *Comparative Biochemistry and Physiology* **72B**, 187–193.

Lied, R., Rosenlund, G., Lund, B. and Von Der Decken, A. (1983). Effect of starvation and refeeding on *in vitro* protein synthesis in white trunk muscle of Atlantic cod *Gadus morhua*. *Comparative Biochemistry and Physiology* **76B**, 777–781.

Litvak, M.K. and Leggett, W.C. (1992). Age and size-selective predation on larval fishes: the bigger-is-better hypothesis revisited. *Marine Ecology Progress Series* **81**, 13–24.

Lone, K.P. and Ince, B.W. (1983). Cellular growth responses of rainbow trout *Salmo gairdneri* fed different levels of dietary protein, and an anabolic steroid ethylestrenol. *General Comparative Endocrinology* **49**, 32–49.

Loughna, P.T. and Goldspink, G. (1984). The effects of starvation upon protein turnover in red and white myotomal muscle of rainbow trout, *Salmo gairdneri*. *Journal of Fish Biology* **25**, 223–230.

Love, R.M. (1970). "The Chemical Biology of Fishes", Vol. 1. Academic Press, New York.

Love, R.M. (1980). "The Chemical Biology of Fishes", Vol. 2: *Advances 1968–1977*. Academic Press, New York.

Lowery, M.S. and Somero, G.N. (1990). Starvation effects on protein synthesis in red and white muscle of barred sand bass *Paralabrax nebulifer*. *Physiological Zoology* **63**, 630–648.

Lowery, M.S., Roberts, S.J. and Somero, G.N. (1987). Effects of starvation on the activities and localization of glycolytic enzymes in the white muscle of the barred sand bass *Paralabrax nebulifer*. *Physiological Zoology* **60**, 538–549.

MacKas, D.L., Denman, K.L. and Abbott, M.R. (1985). Plankton patchiness: biology in the physical vernacular. *Bulletin of Marine Sciences* **37**(2), 652–674.

MacKenzie, B.R. and Leggett, W.C. (1991). Quantifying the contribution of small-scale turbulence to the encounter rates between larval fish and their zooplankton prey: effects of wind and tide. *Marine Ecology Progress Series* **73**, 149–160.

MacKenzie, B.R., Leggett, W.C. and Peters, R.H. (1990). Estimating larval fish ingestion rates: can laboratory derived values be reliably extrapolated to the wild? *Marine Ecology Progress Series* **67**, 209–225.

Margulies, D. (1993). Assessment of the nutritional condition of larval and early juvenile tuna and spanish mackerel (Pisces: Scombridae) in the Panama Bight. *Marine Biology* **115**, 317–330.

Marsh, E. (1986). Effects of egg size on offspring fitness and maternal fecundity in the orangethroat darter *Etheostoma spectabile* (Pices: Percidae). *Copeia* **1986**(1), 18–30.

Marshall, S.M., Nicholls, A.G. and Orr, A.P. (1937). On the growth and feeding of the larval and post-larval stages of the Clyde herring. *Journal of the Marine Biological Association of the United Kingdom* **22**, 245–267.

Martin, F.D. and Wright, D.A. (1987). Nutritional state analysis and its use in predicting striped bass recruitment: laboratory calibration. *American Fisheries Society Symposium* **2**, 109–114.

Martin, F.D., Wright, D.A. and Means, J.C. (1984). Fatty acids and starvation in larval striped bass *Morone saxatilis*. *Comparative Biochemistry and Physiology* **77B**(4), 785–790.

Martin, F.D., Wright, D.A., Means, J.C. and Setzler-Hamilton, E.M. (1985). Importance of food supply to nutritional state of larval striped bass in the Potomac river estuary. *Transactions of the American Fisheries Society* **114**, 137–145.

Mathers, E.M., Houlihan, D.F. and Cunningham, M.J. (1992). Nucleic acid concentrations and enzyme activities as correlates of growth rate of the saithe *Pollachius virens*: growth-rate estimates of open-sea fish. *Marine Biology* **112**, 363–369.

Mathers, E.M., Houlihan, D.F., McCarthy, I.D. and Burren, L.J. (1993). Rates of growth and protein synthesis correlated with nucleic acid content in fry of rainbow trout, *Oncorhynchus mykiss*: effects of age and temperature. *Journal of Fish Biology* **43**, 245–263.

May, R.C. (1971). Effects of delayed feeding on larvae of the grunion, *Leuresthes tenuis*. *Fishery Bulletin U.S.* **69**(2), 411–425.

May, R.C. (1974). Larval mortality in marine fishes and the critical period concept. *In* "The Early Life History of Fish" (J.H.S. Blaxter, ed.), pp. 3–19. Springer-Verlag, Berlin.

McEvoy, L.A. and McEvoy, J. (1991). Size fluctuation in the eggs and newly hatched larvae of captive turbot *Scophthalmus maximus*. *Journal of the Marine Biological Association of the United Kingdom* **71**, 679–690.

McGurk, M.D. (1985a). Effect of net capture on the postpreservation morphometry, dry weight, and condition factor of Pacific herring larvae. *Transactions of the American Fisheries Society* **114**, 348–355.

McGurk, M.D. (1985b). Multivariate analysis of morphometry and dry weight of Pacific herring larvae. *Marine Biology* **86**, 1–11.

McGurk, M.D. and Kusser, W.C. (1992). Comparison of three methods of measuring RNA and DNA concentrations of individual Pacific herring, *Clupea harengus pallasi*, larvae. *Canadian Journal of Fisheries and Aquatic Sciences* **49**, 967–974.

McGurk, M.D., Warburton, H.D., Galbraith, M. and Kusser, W.C. (1992). RNA–DNA ratio of herring and sand lance larvae from Port Moller, Alaska: comparison with prey concentration and temperature. *Fisheries Oceanography* **1**(3), 193–207.

McGurk, M.D., Paul, A.J., Coyle, K.O., Ziemann, D.A. and Haldorson, L.J. (1993). Relationships between prey concentration and growth, condition and mortality of Pacific herring, *Clupea pallasi*, larvae in an Alaska subarctic embayment. *Canadian Journal of Fisheries and Aquatic Sciences* **50**, 163–180.

Meffe, G.K. (1987). Embryo size variation in mosquitofish: optimality vs. plasticity in propagule size. *Copeia* **1987**(3), 762–768.

Meffe, G.K. (1990). Offspring size variation in eastern mosquitofish (*Gambusia holbrooki*: Poeciliidae) from contrasting thermal environments. *Copeia* **1990**(1), 10–18.

Miglavs, I. and Jobling, M. (1989). Effects of feeding regime on food consumption, growth rates and tissue nucleic acids in juvenile Arctic charr, *Salvelinus alpinus*, with particular respect to compensatory growth. *Journal of Fish Biology* **34**, 947–957.

Moon, T.W. (1983). Metabolic reserves and enzyme activities with food deprivation in immature American eels, *Anguilla rostrata*. *Canadian Journal of Zoology* **61**, 802–811.

Moon, T.W. and Johnston, I.A. (1980). Starvation and the activities of glycolytic and gluconeogenic enzymes in skeletal muscles and liver of the plaice, *Pleuronectes platessa*. *Journal of Comparative Physiology* **136**, 31–38.

Mugiya, Y. and Oka, H. (1991). Biochemical relationship between otolith and somatic growth in the rainbow trout *Oncorhynchus mykiss*: consequence of starvation, resumed feeding, and diel variations. *Fishery Bulletin U.S.* **89**, 239–245.

Munilla-Moran, R., Stark, J.R. and Barbour, A. (1990). The role of exogenous enzymes in digestion in cultured turbot larvae *Scophthalmus maximus*. *Aquaculture* **88**, 337–350.

Navarro, J.C. and Sargent, J.R. (1992). Behavioural differences in starving herring *Clupea harengus* larvae correlate with body levels of essential fatty acids. *Journal of Fish Biology* **41**, 509–513.

Neilson, J.D., Perry, R.I., Valerio, P. and Waiwood, K.G. (1986). Condition of Atlantic cod *Gadus morhua* larvae after the transition to exogenous feeding: morphometrics, buoyancy and predator avoidance. *Marine Ecology Progress Series* **32**, 229–235.

Nemeth, P.M., Hitchins, O.E., Solanki, L. and Cole, T.G. (1986). Fluorometric procedures for measuring triglyceride concentrations in small amounts of tissue and plasma. *Journal of Lipid Research* **27**, 447–452.

O'Connell, C.P. (1976). Histological criteria for diagnosing the starving condition in early post yolk sac larvae of the northern anchovy *Englaulis mordax*. *Journal of Experimental Marine Biology and Ecology* **25**, 285–312.

O'Connell, C.P. (1980). Percentage of starving northern anchovy *Engraulis mordax*, larvae in the sea as estimated by histological methods. *Fishery Bulletin U.S.* **78(2)**, 475–489.

O'Connell, C.P. (1981). Estimation by histological methods of the percent of starving larvae of the northern anchovy *Engraulis mordax* in the sea. *Rapports et Procès-verbaux des Réunions. Conseil international pour l'Exploration de la Mer* **178**, 357–360.

O'Connell, C.P. and Paloma, P.A. (1981). Histochemical indications of liver glycogen in samples of emaciated and robust larvae of the northern anchovy *Engraulis mordax*. *Fishery Bulletin U.S.* **79(4)**, 806–812.

Olivier, J.D., Holeton, G.F. and Chua, K.E. (1979). Overwinter mortality of fingerling smallmouth bass in relation to size, relative energy stores, and environmental temperature. *Transactions of the American Fisheries Society* **108**, 130–136.

Oozeki, Y., Ishii, T. and Hirano, R. (1989). Histological study of the effects of starvation on reared and wild-caught larval stone flounder, *Kareius bicoloratus*. *Marine Biology* **100**, 269–275.

Ota, A.Y. and Landry, M.R. (1984). Nucleic acids as growth rate indicators for early developmental stages of *Calanus pacificus*. *Journal of Experimental Marine Biology and Ecology* **80**, 147–160.

Owen, R.W., Lo, N.C.H., Butler, J.L., Theilacker, G.H., Alvarino, A., Hunter, J.R. and Watanabe, Y. (1989). Spawning and survival patterns of larval northern anchovy, *Engraulis mordax*, in contrasting environments—a site intensive study. *Fishery Bulletin U.S.* **87**, 673–688.

Panagiotaki, P. and Geffen, A.J. (1992). Parental effects on size variation in fish larvae. *Journal of Fish Biology* **41** (Suppl. B), 37–42.

Pedersen, B.H. and Hjelmeland, K. (1988). Fate of trypsin and assimilation efficiency in larval herring *Clupea harengus* following digestion of copepods. *Marine Biology* **97**, 467–476.

Pedersen, B.H., Nilssen, E.M. and Hjelmeland, K. (1987). Variations in the content of trypsin and trypsinogen in larval herring *Clupea harengus* digesting copepod nauplii. *Marine Biology* **94**, 171–181.

Pedersen, B.H., Ugelstad, I. and Hjelmeland, K. (1990). Effects of a transitory, low food supply in the early life of larval herring *Clupea harengus* on mortality, growth and digestive capacity. *Marine Biology* **107**, 61–66.

Pepin, P. (1991). Effects of temperature and size on development, mortality, and survival rates of the pelagic early life history of marine fish. *Canadian Journal of Fisheries and Aquatic Sciences* **48**, 503–518.

Pepin, P. and Miller, T.J. (1993). Potential use and abuse of general empirical models of early life history processes in fish. *Canadian Journal of Fisheries and Aquatic Sciences* **50**, 1343–1345.

Pepin, P., Shears, T.H. and DeLafontaine, Y. (1992). Significance of body size to the interaction between a larval fish *Mallotus villosus* and a vertebrate predator *Gasterosteus aculeatus*. *Marine Ecology Progress Series* **81**, 1–12.

Peterman, R.M., Bradford, M.J., Lo, N.C.H. and Methot, R.D. (1988). Contribution of early life stages to interannual variability in recruitment of northern anchovy *Engraulis mordax*. *Canadian Journal of Fisheries and Aquatic Sciences* **45**, 8–16.

Peterson, M.S. and Brown-Peterson, N. (1992). Growth under stressed conditions in juvenile channel catfish *Ictalurus punctatus* as measured by nucleic acids. *Comparative Biochemistry and Physiology* **103A(2)**, 323–327.

Pfeila, E. and Luna, A. (1984). Changes in biochemical composition and energy utilizzation during metamorphosis of leptocephalus larvae of the bonefish *Albula*. *Environmental Biology of Fishes*, **10(4)**, 243–251.

Platt, T. and Denman, K.L. (1975). Spectral analysis in ecology. *Annual Review of Ecology and Systematics* **6**, 189–210.

Powell, A.B. and Chester, A.J. (1985). Morphometric indices of nutritional condition and sensitivity to starvation of spot larvae. *Transactions of the American Fisheries Society* **114**, 338–347.

Powell, A.B., Chester, A.J., Govoni, J.J. and Warlen, S.M. (1990). Nutritional condition of spot larvae associated with the Mississippi river plume. *Transactions of the American Fisheries Society* **119**, 957–965.

Purcell, J.E. and Grover, J.J. (1990). Predation and food limitation as causes of mortality in larval herring at a spawning ground in British Columbia. *Marine Ecology Progress Series* **59**, 55–61.

Raae, A.J., Opstad, I., Kvenseth, P. and Th. Walther, B. (1988). RNA, DNA and protein during early development in feeding and starved cod *Gadus morhua*. *Aquaculture* **73**, 247–259.

Rice, J.A., Crowder, L.B. and Binkowski, F.P. (1987). Evaluating potential sources of mortality for larval bloater *Coregonus hoyi*: starvation and vulnerability to predation. *Canadian Journal of Fisheries and Aquatic Sciences* **44**, 467–472.

Rice, J.A., Miller, T.J., Rose, K.A., Crowder, L.B., Marschall, E.A., Trebitz, A.S. and DeAngelis, D.L. (1993). Growth rate variation and larval survival: inferences from an individual-based size-dependent predation model. *Canadian Journal of Fisheries and Aquatic Sciences* **50**, 133–142.

Richard, P., Bergeron, J.-P., Boulhic, M., Galois, R. and Person-Le Ruyet, J. (1991). Effect of starvation on RNA, DNA and protein content of laboratory-reared larvae and juveniles of *Solea solea*. *Marine Ecology Progress Series* **72**, 69–77.

Robinson, S.M.C. and Ware, D.M. (1988). Ontogenetic development of growth rates in larval Pacific herring, *Clupea harengus pallasi*, measured with RNA–DNA ratios in the strait of Georgia, British Columbia. *Canadian Journal of Fisheries and Aquatic Sciences* **45**, 1422–1429.

Rønnestad, I., Fyhn, H.J. and Gravningen, K. (1992a). The importance of free amino acids to the energy metabolism of eggs and larvae of turbot *Scophthalmus maximus*. *Marine Biology* **114**, 517–525.

Rønnestad, I., Finn, R.N., Groot, E.P. and Fyhn, H.J. (1992b). Utilization of free amino acids related to energy metabolism of developing eggs and larvae of lemon sole *Microstomus kitt* reared in the laboratory. *Marine Ecology Progress Series* **88**, 195–205.

Roselund, G. and Lied, E. (1986). Growth and muscle protein synthesis *in vitro* of saithe *Pollachius virens* and rainbow trout *Salmo gairdneri* in response to protein-energy intake. *Acta Agriculturae Scandinavica* **36**, 195–204.

Roselund, G., Lund, B., Lied, E. and Von Der Decken, A. (1983). Properties of white trunk muscle from saithe *Pollachius virens*, rainbow trout *Salmo gairdneri*, and herring *Clupea harengus*: protein synthesis *in vitro*, electrophoretic study of proteins. *Comparative Biochemistry and Physiology* **74B(3)**, 389–397.

Rosenthal, H. and Alderdice, D.F. (1976). Sublethal effects of environmental stressors, natural and pollutional, on marine fish eggs and larvae. *Journal of the Fisheries Research Board of Canada* **33**, 2047–2065.

Rothschild, B.J. (1986). "Dynamics of Marine Fish Populations." Harvard University Press, Cambridge, MA.

Rothschild, B.J. and Rooth, C. (1982). "Fish Ecology 111, a Foundation for REX a Recruitment Experiment." University of Miami, Miami, FL.

Sale, P.F. (1990). Recruitment of marine species: is the bandwagon rolling in the right direction? *Trends in Ecology and Evolution* **5(1)**, 25–27.

Sameoto, D.D. (1972). Distribution of herring *Clupea harengus* larvae along the southern coast of Nova Scotia with observations on their growth and condition factor. *Journal of the Fisheries Research Board of Canada* **29**, 507–515.

Sclafani, M. (1992). Vertical migration of marine larval fish: patterns, models and application to recruitment research. MSc thesis, Department of Oceanography, Dalhousie University, Halifax, NS.

Sclafani, M., Taggart, C.T. and Thompson, K.R. (1993). Condition, buoyancy and the distribution of larval fish: implications for vertical migration and retention. *Journal of Plankton Research* **15(4)**, 413–435.

Segner, H. and Möller, H. (1984). Electron microscopical investigations on starvation-induced liver pathology in flounders *Platichthys flesus*. *Marine Ecology Progress Series* **19**, 193–196.

Setzler-Hamilton, E.M., Wright, D.A., Martin, F.D., Millsaps, C.V. and Whitlow, S.I. (1987). Analysis of nutritional condition and its use in predicting striped bass recruitment: field studies. *American Fisheries Society Symposium* **2**, 115–128.

Shelbourne, J.E. (1957). The feeding and condition of plaice larvae in good and bad plankton patches. *Journal of the Marine Biological Association of the United Kingdom* **36**, 539–552.

Sieg, A. (1992a). A histological study on the nutritional condition of larval and metamorphosing fishes of the genus *Vinciguerria* (Photichthyidae, Pisces) sampled in two contrasting environments. *Journal of Applied Ichthyology* **8**, 154–163.

Sieg, A. (1992b). Histological study of organogenesis in the young stages of the mesopelagic fish *Vinciguerria* (Photichthyidae, Pisces). *Bulletin of Marine Sciences* **50(1)**, 97–107.

Sieg, A. (1993). Histological study on larval nutritional condition of the southwest Atlantic anchovy *Engraulis anchoita*, caught in three hydrographically differing frontal systems of the southwest-Atlantic. *International Council for the Exploration of the Sea, Council Meeting, 1993/L:57*. (mimeo.).

Sieg, A., Clemmesen, C. and Ueberschär, B. (1989). Comparison of biochemical and histological methods for the evaluation of the *in situ* nutritional condition of marine fish larvae. *International Council for the Exploration of the Sea, Council Meeting, 1989/L:4.* (mimeo.)

Sinclair, M. (1988). "Marine Populations: An Essay on Population Regulation and Speciation." University of Washington Press, Seattle.

Sissenwine, M.P. (1984). Why do fish populations vary? *In* "Exploitation of Marine Communities" (R.M. May, ed.), pp. 59–94. Springer-Verlag, New York.

Smith, P.E. (1985). Year-class strength and survival of 0-group clupeoids. *Canadian Journal of Fisheries and Aquatic Sciences* **42** (Suppl. 1), 69–82.

Smith, S. (1957). Early development and hatching. *In* "The Physiology of Fishes" (M. Brown, ed.), pp. 323–359. Academic Press, New York.

Soivio, A., Niemistö, M. and Bäckström, M. (1989). Fatty acid composition of *Coregonus muskun* Pallas: changes during incubation, hatching, feeding and starvation. *Aquaculture* **79**, 163–168.

Springer, T.A., Murphy, B.R., Gutreuter, S., Anderson, R.O., Miranda, L.E. and Jackson, D.C. and Cone, R.S. (1990). Properties of relative weight and other condition indices. *Transactions of the American Fisheries Society* **119**, 1048–1058.

Steele, J.H. (1978). "Spatial Pattern in Plankton Communities." Plenum Press, New York.

Steinhart, M. and Eckmann, R. (1992). Evaluating the nutritional condition of individual whitefish *Coregonus* spp. larvae by the RNA/DNA ratio. *Journal of Fish Biology* **40**, 791–799.

Sterzel, W., Bedford, P. and Eisenbrand, G. (1985). Automated determination of DNA using the fluorochrome Hoechst 33258. *Analytical Biochemistry* **147**, 462–467.

Stryer, L. (1981). "Biochemistry", 2nd edn. W.H. Freeman, San Francisco.

Sullivan, K.M. and Somero, G.N. (1980). Enzyme activities of fish skeletal muscle and brain as influenced by depth of occurrence and habits of feeding and locomotion. *Marine Biology* **60**, 91–99.

Sundby, S., Bjørke, H., Soldal, A.V. and Olsen, S. (1989). Mortality rates during the early life stages and year-class strength of northeast Arctic cod *Gadus morhua*. *Rapports et Procès-verbaux des Réunions. Conseil international pour l'Exploration de la Mer* **191**, 351–358.

Suthers, I.M., Fraser, A. and Frank, K.T. (1992). Comparison of lipid, otolith and morphometric condition indices of pelagic juvenile cod *Gadus morhua* from the Canadian Atlantic. *Marine Ecology Progress Series* **84**, 31–40.

Taggart, C.T. and Frank, K.T. (1990). Perspectives on larval fish ecology and recruitment processes. probing the scales of relationships. *In* "Large Marine Ecosystems, Patterns, Processes and Yields" (K. Sherman, L.M. Alexander and B.J. Gold, eds), pp. 151–164. American Association for the Advancement of Science, Washington, DC.

Theilacker, G.H. (1978). Effect of starvation on the histological and morphological characteristics of jack mackerel *Trachurus symmetricus* larvae. *Fishery Bulletin U.S.* **76(2)**, 403–414.

Theilacker, G.H. (1980a). Changes in body measurements of larval northern anchovy, *Engraulis mordax*, and other fishes due to handling and preservation. *Fishery Bulletin U.S.* **78(3)**, 685–692.

Theilacker, G.H. (1980b). Rearing container size affects morphology and nutritional condition of larval jack mackerel, *Trachurus symmetricus*. *Fishery Bulletin U.S.* **78(3)**, 789–791.

Theilacker, G.H. (1986). Starvation-induced mortality of young sea-caught jack mackerel *Trachurus symmetricus*, determined with histological and morphological methods. *Fishery Bulletin U.S.* **84(1)**, 1–17.

Theilacker, G.H. and Watanabe, Y. (1989). Midgut cell height defines nutritional status of laboratory raised larval northern anchovy *Engraulis mordax. Fishery Bulletin U.S.* **87**, 457–469.

Timeyko, V.N. and Novikov, G.G. (1987). Proteolytic activity in the digestive tract of Atlantic salmon, *Salmo salar*, during larval development. *Journal of Ichthyology* **27(4)**, 27–33.

Tocher, D.R. and Sargent, J.R. (1984). Analyses of lipids and fatty acids in ripe roes of some northwest European marine fish. *Lipids* **19(7)**, 492–499.

Tocher, D.R., Fraser, A.J., Sargent, J.R. and Gamble, J.C. (1985a). Fatty acid composition of phospholipids and neutral lipids during embryonic and early larval development in Atlantic herring, *Clupea harengus. Lipids* **20(2)**, 69–74.

Tocher, D.R., Fraser, A.J., Sargent, J.R. and Gamble, J.C. (1985b). Lipid class composition during embryonic and early larval development in Atlantic herring, *Clupea harengus. Lipids* **20(2)**, 84–89.

Ueberschär, B. (1988). Determination of the nutritional condition of individual marine fish larvae by analyzing their proteolytic enzyme activities with a highly sensitive fluorescence technique. *Meeresforsch* **32**, 144–154.

Ueberschär, B. and Clemmesen, C. (1992). A comparison of the nutritional condition of herring larvae as determined by two biochemical methods—tryptic enzyme activity and RNA/DNA ratio measurements. *ICES Journal of Marine Science* **49**, 245–249.

Ueberschär, B., Pedersen, B.H. and Hjelmeland, K. (1992). Quantification of trypsin with a radioimmunoassay in herring larvae *Clupea harengus*, compared with a highly sensitive fluorescence technique to determine tryptic enzyme activity. *Marine Biology* **113**, 469–473.

Umeda, S. and Ochiai, A. (1975). On the histological structure and function of digestive organs of the fed and starved larvae of the yellowtail, *Seriola quinqueradiata. Japanese Journal of Ichthyology* **21(4)**, 213–219.

Underwood, A.J. and Fairweather, P.G. (1989). Supply-side ecology and benthic marine assemblages. *Trends in Ecology and Evolution* **4(1)**, 16–19.

Van Winkle, W., Rose, K.A. and Chambers, R.C. (1993). Individual-based approach to fish population dynamics: an overview. *Transactions of the American Fisheries Society* **122(3)**, 397–403.

Vetter, R.D., Hodson, R.E. and Arnold, C. (1983). Energy metabolism in a rapidly developing marine fish egg, the red drum *Sciaenops ocellata. Canadian Journal of Fisheries and Aquatic Sciences* **40**, 627–634.

Vilela, M.H. and Zijlstra, J.J. (1971). On the condition of herring larvae in the central and southern North Sea. *Rapports et Procès-verbaux des Réunions. Conseil international pour l'Exploration de la Mer* **160**, 137–141.

Vlymen, W.J. (1977). A mathematical model of the relationship between larval anchovy *Engraulis mordax*, growth, prey microdistribution and larval behaviour. *Environmental Biology of Fishes* **2**, 211–233.

Von Westernhagen, H. and Rosenthal, H. (1981). On condition factor measurements in Pacific herring larvae. *Helgoländer Meeressunters* **34**, 257–262.

Vu, T.T. (1983). Étude histoenzymologique des activités protéasiques dans le tube digestif des larves et des adultes de bar, *Dicentrarchus labrax. Aquaculture* **32**, 57–69.

Walford, J. and Lam, T.J. (1993). Development of digestive tract and proteolytic enzyme activity in seabass *Lates calcarifer* larvae and juveniles. *Aquaculture* **109**, 187–205.

Wang, S.Y. and Stickle, W.B. (1986). Changes in nucleic acid concentration with starvation in the blue crab *Callinectes sapidus*. *Journal of Crustacean Biology* **6**, 49–56.

Ware, D.M. and Lambert, T.C. (1985). Early life history of Atlantic mackerel *Scomber scombrus* in the southern Gulf of St. Lawrence. *Canadian Journal of Fisheries and Aquatic Sciences* **42**, 577–592.

Westerman, M.E. and Holt, G.J. (1988). The RNA–DNA ratio: measurement of nucleic acids in larval *Sciaenops ocellatus*. *Contributions to Marine Sciences* **30** (Suppl.), 117–124.

Wicker, A.M. and Johnson, W.E. (1987). Relationship among fat content, condition factor, and first-year survival of Florida largemouth bass. *Transactions of the American Fisheries Society* **116**, 264–271.

Wilson, D.C. and Millemann, R.E. (1969). Relationships of female age and size to embryo number and size in the shiner perch, *Cymatogaster aggregata*. *Journal of the Fisheries Research Board of Canada* **26**, 2339–2344.

Wright, D.A. and Hetzel, E.W. (1985). Use of RNA:DNA ratios as an indicator of nutritional stress in the American oyster *Crassostrea virginica*. *Marine Ecology Progress Series* **25**, 199–206.

Wright, D.A. and Martin, F.D. (1985). The effect of starvation on RNA–DNA ratios and growth of larval striped bass, *Morone saxatilis*. *Journal of Fish Biology* **27**, 479–485.

Wyatt, T. (1972). Some effects of food density on the growth and behaviour of plaice *Pleuronectes platessa* larvae. *Marine Biology* **14**, 210–216.

Yamashita, Y. and Bailey, K.M. (1989). A laboratory study of the bioenergetics of larval walleye pollock, *Theragra chalcogramma*. *Fishery Bulletin U.S.* **87**, 525–536.

Yin, M.C. and Blaxter, J.H.S. (1986). Morphological changes during growth and starvation of larval cod *Gadus morhua* and flounder *Platichthys flesus*. *Journal of Experimental Marine Biology and Ecology* **104**, 215–228.

Yin, M.C. and Blaxter, J.H.S. (1987). Temperature, salinity tolerance, and buoyancy during early development and starvation of Clyde and North Sea herring, cod and flounder larvae. *Journal of Experimental Marine Biology and Ecology* **107**, 279–290.

Zastrow, C.E., House, E.D. and Saunders, E.H. (1989). Quality of striped bass *Morone saxatilis* eggs in relation to river source and female weight. *Rapports et Procès-verbaux des Réunions. Conseil international pour l'Exploration de la Mer* **191**, 34–42.

Zeitoun, I.H., Ullrey, D.E., Bergen, W.G. and Magee, W.T. (1977). DNA, RNA, protein and free amino acids during ontogenesis of rainbow trout *Salmo gairdneri*. *Journal of the Fisheries Research Board of Canada* **34**, 83–88.

The Biology of Seamounts

A.D. Rogers

Marine Biological Association of the United Kingdom,
The Laboratory, Citadel Hill, Plymouth, PL1 2PB, UK

1. INTRODUCTION

The presence of numerous seamounts in the world's oceans, especially in the Pacific, has only become known to the scientific community in the last

ADVANCES IN MARINE BIOLOGY VOL 30
ISBN 0–12–026130–8

50 years (Hess, 1946; Menard and Dietz, 1951; Menard and Ladd, 1963). C.L. Hubbs was one of the first biologists to work on the biology of seamounts and as early as 1959 he posed a number of fundamental problems regarding seamount faunas. These included: What species inhabit seamounts and with what regularity and abundance? How do species disperse to and become established on seamounts? Do seamounts represent stepping stones for transoceanic dispersal of species? Are demersal or pelagic fishes sufficiently abundant on seamounts to provide profitable fisheries? What factors are responsible for the abundance of life over seamounts?

These questions remain relevant today, and despite a large quantity of work since Hubbs (1959), most of them remain incompletely answered. Biological investigations of seamounts have been very scattered in terms of the geographical areas covered and aspects of biology studied. The quality of data has also varied considerably often according to particular taxonomic expertise of the investigators and on methods of data collection. Dissemination of data and attendant conclusions on the biology of seamounts has also been hampered by language barriers. For example, quantities of useful Russian data on seamounts have been inaccessible to English-speaking biologists due to the difficulties in obtaining complete translations of scientific papers.

The following review draws together the scattered literature on the biology of seamounts. It summarizes current knowledge on the effects of seamounts on pelagic ecosystems, factors that influence the structure of seamount communities, the establishment, maintenance and genetic isolation of populations on seamounts and on the effects of commercial exploitation on seamounts organisms. In summarizing current knowledge in these areas, many of which are related directly to the problems posed by Hubbs more than 30 years ago, this review also points out the areas in which data are lacking or contradictory and where further research is required. A map showing the approximate position of the seamounts discussed in this review is given in Figure 1.

Figure 1   Map showing approximate position of the seamounts discussed in this review. 1, Norfolk Ridge; 2, Cobb seamount; 3, Axial seamount; 4, Fieberling seamount; Fieberling II seamount; Hoke seamount; Jasper seamount; Stoddard seamount; Nidever Bank; Banco San Isidro; 5, Red Volcano; 6, Volcano 7; 7, San Salvador seamount; 8, Atlantis II seamount; 9, Corner Rise seamounts; 10, Great Meteor seamount; 11, Dacia seamount; 12, Josephine seamount; Gettysburg seamount; 13, Marsili seamount; 14, Vema seamount; 15, Equator seamount; 16, Minami-Kasuga seamount; 17, Conical seamount; 18, Southeast Hancock seamount; 19, Mid-Pacific Mountains; 20, Horizon Guyot; 21, Magellan Rise; 22, Emperor seamount chain; 23, Peepa seamount; 24, North Hawaiian Ridge; 25, Cross seamount; 26, Loihi seamount; 27, Challenger Plateau; 28, Ritchie Bank; 29, North Chatham Rise.

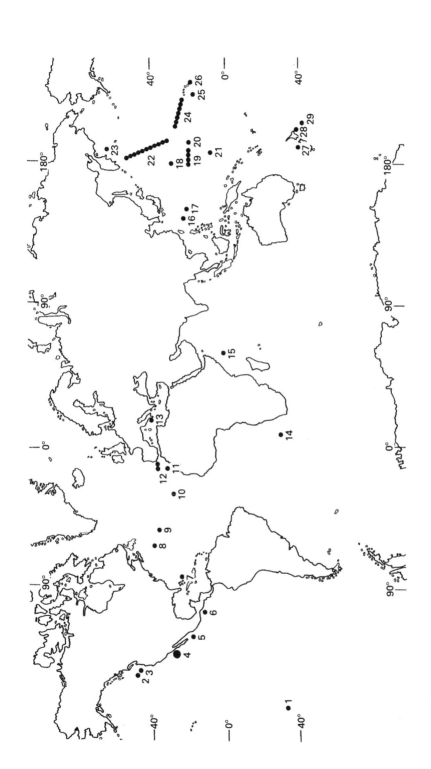

Recent reviews have covered aspects of the biology of seamounts especially those in Keating *et al.* (1987) but this literature has had limited circulation amongst biologists. Furthermore in recent years much important work has been published on seamounts especially in respect to current topography interactions, the biology of the soft benthos on seamounts and on the biology of commercially valuable species of fish associated with seamounts. The combination of this new data with that obtained in previous studies provides a complete picture of the current knowledge of the biology of seamounts which will be of use to marine biologists, fisheries biologists and oceanographers.

2. GEOLOGY AND OCEANOGRAPHY

Seamounts are undersea mountains which rise steeply from the sea bottom to below sea level. They have been defined as having an elevation of more than 1000 m with a limited extent across the summit (Menard, 1964; US Board of Geographic Names, 1981). Features that have elevations between 500 and 1000 m have been defined as knolls and those that have a relief of less than 500 m as hills (US Board of Geographic Names, 1981). These definitions have not been adhered to in literature on the biology, geology and oceanography of such features and most are referred to as seamounts regardless of size (i.e. Epp and Smoot, 1989).

Seamounts are a variety of shapes but are generally conical with a circular, elliptical or more elongate base. Examples of typical seamount shapes are "Conical seamount" (Fryer and Fryer, 1987; circular), Great Meteor seamount (Pratt, 1963; elliptical) and Horizon Guyot (Karig *et al.*, 1970; elongate) (see Figure 2 (a), (b) and (c)). They are usually of volcanic origin (Epp and Smoot, 1989) though some are formed by vertical tectonic movement along converging plate margins (see Fryer and Fryer, 1987). Seamounts often occur in chains or clusters known as provinces (Menard and Dietz, 1951; Menard, 1964) which may be associated with seafloor hotspots (Epp and Smoot, 1989). It has been estimated that there are over 30 000 seamounts with a height of over 1000 m in the Pacific ocean (Smith and Jordan, 1988), approximately 810 (over 100 m in height) in the Atlantic (Epp and Smoot, 1989) and an indeterminate number in the Indian ocean.

Seamounts provide a striking contrast to the surrounding flat sediment-covered abyssal plain. Their profiles can show declivities of up to 60° (Sagalevitch *et al.*, 1992), much greater than anywhere else in the deep sea. Hard substrata atypical of the deep-sea environment are common on seamounts and may take the form of calderas (Levin and Nittrouer,

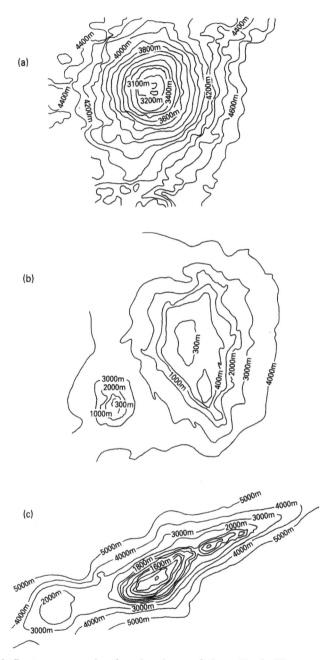

Figure 2 Contour maps showing the shape of three "typical" seamounts. (a) Conical seamount (after Fryer and Fryer, 1987). (b) Great Meteor seamount (after Hinz, 1969). (c) Horizon Guyot (after Smith *et al.*, 1989).

1987), terraces (Pratt, 1963; Hinz, 1969; Rad, 1974), pit craters (Levin and Nittrouer, 1987), canyons (Raymore, 1982), caves (Heydorn, 1969), pinnacles (Hughes, 1981; Raymore, 1982), knobs (Boehlert and Genin, 1987), crevices (Heydorn, 1969), rocks (Raymore, 1982), cobbles (Raymore, 1982) and marine organisms (Levin *et al.*, 1986). Hydrothermal precipitates may form crusts, mounds and chimneys on seamounts (Levin and Nittrouer, 1987). Some seamounts, known as guyots (Hess, 1946), have flat summits, formed by wave erosion when they were above sea level (Hinz, 1969). The tops of such seamounts are frequently covered in biogenic sediments such as foraminiferan sands (Pratt, 1963; Karig *et al.*, 1970; Hinz, 1969; Levin and Nittrouer, 1987). These may be supplemented by sediments of a volcanic origin which may be composed of a variety of materials such as fragments of basaltic glass and tephra (Natland, 1976). Lithogenic sediments, transported from the continental margin, may also be present and authigenic sedimentation, principally from the precipitation of ferromanganese oxides, may also form significant components of seamount sediments (Levin and Nittrouer, 1987). Hydrothermal sediments may be found on some young and active seamounts such as Red Volcano located near the East Pacific Rise (Lonsdale *et al.*, 1982).

Seamounts have complex effects on ocean circulation, which are poorly understood (Roden, 1987; Eriksen, 1991). This is because of the great diversity in seamount size, shape and distribution (in relation to neighbouring seamounts), the complexity of the currents impinging upon them and the importance of Coriolis forces and stratification on current–topography interactions (Roden, 1987; see also Brink, 1989; Zhang and Boyer, 1991).

Observations of the effects of seamounts on ocean circulation have been at a range of scales from the macroscale to effects in the immediate vicinity of a seamount. At a large scale, seamounts in the Emperor seamount chain have been shown to deflect both the Kuroshio extension and subarctic currents (Roden *et al.*, 1982, Roden and Taft, 1985; Vastano *et al.*, 1985; also see Roden (1991) for Fieberling seamount). At a smaller scale, effects of seamounts on ocean currents include the formation of trapped waves (Eriksen, 1982a, 1991; Brink, 1989; Genin *et al.*, 1989) and the reflection, amplification and distortion of internal waves (Bell, 1975; Wunsch and Webb, 1979; Eriksen, 1982b, 1985, 1991; Kaneko *et al.*, 1986). Diurnal and semidiurnal tides may be amplified over seamounts leading to fast tidal currents ($>40 \, \text{cm} \, \text{s}^{-1}$) around some seamounts (Chapman, 1989; Genin *et al.*, 1989; Noble and Mullineaux, 1989).

The production of jets and eddies may also be a feature of the interaction of seamounts with ocean currents (Vastano and Warren, 1976;

Roden, 1991). Eddies may be trapped over seamounts to form closed circulations known as Taylor columns (after G.I. Taylor who first studied the effects of obstacles on rotating flows (Taylor, 1917)). These are thought to occur when a steady current impinging on a seamount causes an uplifting of isotherms (upwelling). This compresses vortex lines and induces anticyclonic vorticity generating a closed eddy over the seamount (Huppert and Bryan, 1976). Observations of Taylor columns or Taylor column-like structures have been observed over several seamounts including the Great Meteor seamount (Meincke, 1971), Atlantis II seamount (Vastano and Warren, 1976), seamounts of the Emperor seamount chain (Cheney et al., 1980), the Corner Rise seamounts (Richardson, 1980), Fieberling seamount (Genin et al., 1989; Roden, 1991), Fieberling II, Hoke and Stoddard seamounts (Roden, 1991) and Cobb seamount (Dower et al., 1992). Taylor column-like effects have also been noted near smaller abyssal hills and "bumps" (Owens and Hogg, 1980; Gould et al., 1981). Taylor columns can last for considerable periods of time over seamounts. Richardson (1980) tracked the path of an anticyclonic eddy over the Corner Rise seamounts, using freely drifting buoys, for 6 weeks before it left the seamounts and drifted away.

3. THE EFFECTS OF SEAMOUNTS ON PELAGIC ECOSYSTEMS

3.1. Primary Production and Seamount Fisheries

The concentration of commercially valuable fish species around seamounts is well documented (Hubbs, 1959; Hughes, 1981; Uchida and Tagami, 1984; Parin and Prut'ko, 1985; Alton, 1986; Boehlert, 1986; Sasaki, 1986; Seki and Tagami, 1986; Yasui, 1986; Genin et al., 1988; Parin et al., 1990; Fonteneau, 1991; Gerber, 1993). It has been suggested that this is due to increased densities of prey organisms (i.e. macroplankton) over seamounts which in turn are caused by enhanced primary productivity due to topographic effects on local hydrographic conditions (see above). In oligotrophic waters it has been suggested that the uplifting of isotherms into the euphotic zone by seamounts, as a result of Taylor column formation, can introduce biogenes into nutrient-poor water and cause an increase in primary production (Genin and Boehlert, 1985; Tseytlin, 1985; Voronina and Timonin, 1986; Boehlert and Genin, 1987; Dower et al., 1992).

Evidence for enhanced primary productivity leading to concentrations of fish and zooplankton over seamounts is, however, sparse. There is evidence that waters over seamounts do show increased primary productivity. Lophukin (1986), for example, studied ATP concentrations across 12 Atlantic seamounts. In almost all cases a dome-like layer of upwelling water, poor in microplankton, was detected directly over the seamount whilst high concentrations of chlorophyll A, ATP and microplankton were detected over the seamount flanks. High concentrations of chlorophyll A above seamounts have also been detected by Bezrukov and Natarov (1976), Genin and Boehlert (1985) and Dower *et al.* (1992).

Genin and Boehlert (1985) measured temperature and chlorophyll A concentrations across the Minami-kasuga seamount on the Mariana ridge on three separate dates with the second and third samples being taken 2 and 17 days after the first. The first sample showed a cold temperature dome (upwelling) over the seamount with high concentrations of chlorophyll between 80 and 100 m. This was not due to decreased grazing by zooplankton, since concentrations of zooplankton were also elevated over the seamount. In the subsequent samples this cold dome, along with high concentrations of chlorophyll and zooplankton, was no longer present. It was suggested that, since conditions of enhanced primary productivity only lasted a few days, production could not be transferred to higher trophic levels, a process which may take several months (Tseytlin, 1985; Boehlert and Genin, 1987), and could not explain high numbers of fish and zooplankton around seamounts.

Dower *et al.* (1992) recorded current flow and measured light transmission (negatively correlated to chlorophyll A concentration) at different depths above Cobb seamount in the North East Pacific. Current meter readings indicated the presence of a Taylor column over the seamount which did not penetrate to the surface (a Taylor cone). Light transmission readings indicated elevated chlorophyll A levels above Cobb seamount for the 2.5 weeks of the study (122.4 mg m^{-2} chlorophyll A compared to <35 mg m^{-2} background levels). It is likely that such enhanced chlorophyll A concentrations represented an increase in algal biomass over Cobb seamount and therefore an increase in primary production. There was no evidence of upwelling of deep nutrient-rich water in this investigation, possibly due to the inadequacy of sampling methods.

Dower *et al.* (1992) attributed the enhanced productivity over Cobb seamount to the presence of a Taylor column with its attendant upwelling of nutrient-rich water located in the euphotic zone. This was possible because the summit of Cobb seamount, unlike many others which have been studied, penetrates well into the euphotic zone (24 m from the surface). Whether enhanced productivity on Cobb seamount could be sustained for a period long enough to be transferred to higher trophic levels is uncertain.

Other studies have not been able to demonstrate persistent high chlorophyll patches over seamounts and banks (i.e. Pelaez and McGowan, 1986). It is conceivable that, if upwelling occurred frequently, it may enhance productivity over a seamount, but it is more likely that such episodic events would enhance production dowstream of seamounts rather than in their immediate vicinity (Genin and Boehlert, 1985; see also Zaika and Kovalev, 1985). The effects of seasonality on upwelling and primary productivity over seamounts has not been investigated and it may be that seasonality is partially responsible for the large variation in the results and conclusions of such studies.

Some workers have denied the importance of the enhancement of primary productivity by current–topography interactions as a cause of accumulation of fish and macroplankton over seamounts (i.e. Zaika and Kovalev, 1985). Isaacs and Schwartzlose (1965) first hypothesized that dense populations of fish on seamounts fed upon zooplankton that were undergoing a normal diurnal vertical migration and were being swept onto seamounts or banks by prevailing currents (see Figure 3). They suggested that downwardly migrating zooplankton trapped over seamounts were more vulnerable to predation by predators that were themselves protected to some degree by seamounts from predation while feeding. Diurnally migrating plankton trapped over a seamount in conditions of increasing ambient light levels (and reflected light from the seamount surface) may be highly visible to visual predators because their body surfaces are often adapted for camouflage in low light conditions (see Marshall, 1979).

Acoustic data collected from the Banco San Isidro off the South West coast of the USA showed fish that were intercepting layers of vertically migrating plankton around the bank (Isaacs and Schwartzlose, 1965). Genin *et al* (1988) investigated feeding in epibenthic fish (*Sebastes* spp.) on a shallow bank in the Southern California Bight. The results of their investigation also indicated that in the morning *Sebastes* ascended from the seamount summit towards downwardly migrating plankters (mainly *Euphausia pacifica*). Studies of the stomach contents of these fish showed that they were feeding to a large extent on these diurnal migrators, especially when the plankton was advected by currents over the edge of the seamount on to the summit.

These studies would appear to support the hypothesis that seamounts support large pelagic and benthopelagic fish communities by trapping diurnally migrating plankton rather than by any increase of primary productivity resulting from current–topography interactions, which might increase levels of nutrients in oligotrophic waters found over and around seamounts. Though it is thought unlikely by some workers that this production could be transferred to higher trophic levels in pelagic or benthopelagic communities on seamounts, the evidence is contradictory.

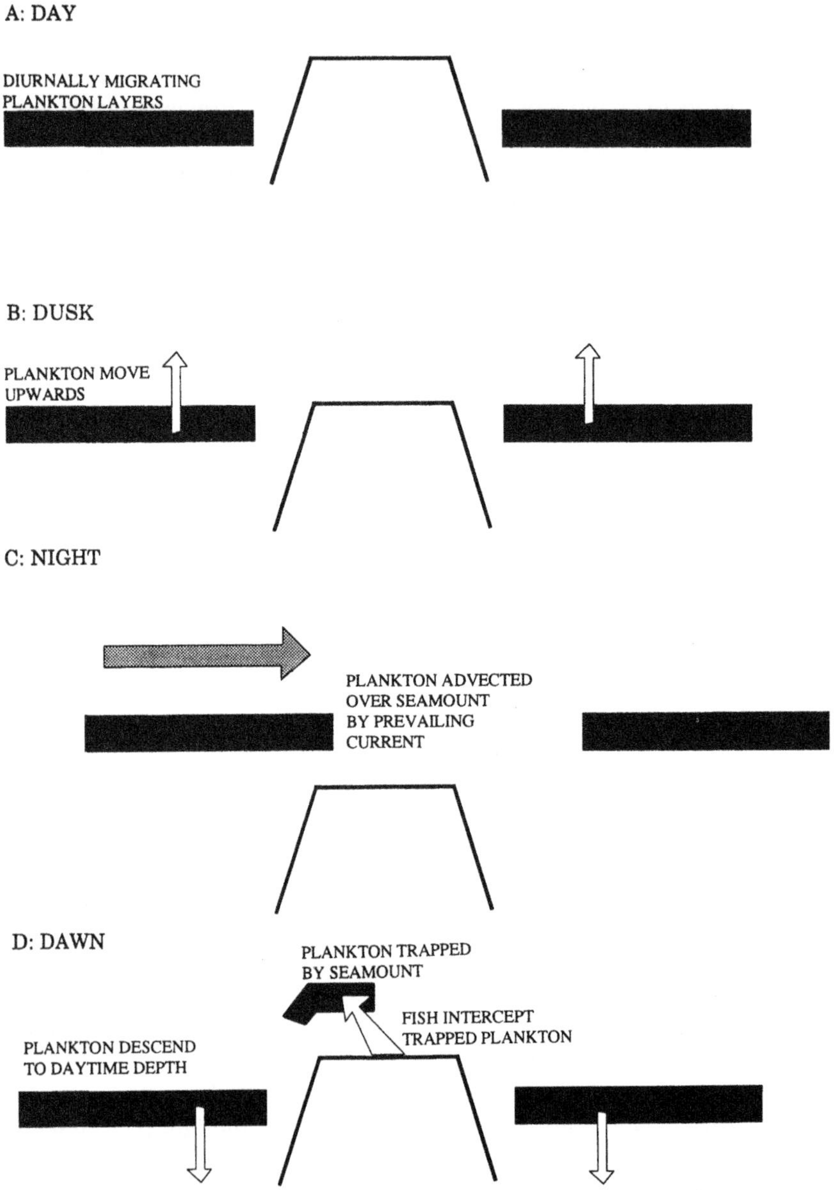

A: DAY

DIURNALLY MIGRATING
PLANKTON LAYERS

B: DUSK

PLANKTON MOVE
UPWARDS

C: NIGHT

PLANKTON ADVECTED
OVER SEAMOUNT
BY PREVAILING
CURRENT

D: DAWN

PLANKTON TRAPPED
BY SEAMOUNT

FISH INTERCEPT
TRAPPED PLANKTON

PLANKTON DESCEND
TO DAYTIME DEPTH

Figure 3 Diagram representing the theory that seamount fish populations feed on diurnally migrating layers of plankton (Isaacs and Schwartzlose, 1965; Genin *et al.*, 1988).

It is also likely that the concentration of fish species around seamounts occurs for reasons additional to the trapping of migrating plankton. For example, in most cases the role of seamounts in the reproduction and life histories of many fish species, as well as their behavioural ecology and their interactions with the seamount benthos, is unknown. It is clear that further studies are required in order to draw firmer conclusions about the role of seamounts in enhancing fish populations and increasing pelagic productivity.

3.2. The Structure of Pelagic Communities Over Seamounts

Some observations on the structure of elements of pelagic communities above seamounts show little difference between samples taken over seamounts and those taken in the surrounding oceanic waters; that is, protozoa (Moiseyev, 1986), zooplankton (Voronina and Timonin, 1986) and ichthyoplankton (Belyanina, 1985). However, the majority of studies indicate qualitative and quantitative differences between pelagic communities over seamounts and those in the surrounding oceans.

The biomass of planktonic organisms is often found to be increased over seamounts. Studies on the Minami-kasuga seamount and seamounts in the Atlantic Ocean have described elevated levels of chlorophyll and zooplankton, probably due to upwelling (see above; Genin and Boehlert, 1985; Lophukin, 1986). Observations of numbers of copepods over the Hawaiian Ridge have also shown increased density over several seamounts that was attributed to upwelling (Fedosova, 1974). Nellen (1973) noted a concentration of the larvae of neritic fish species (and a decrease in the abundance of the larvae of midwater fish) over the Great Meteor seamount in the Atlantic. Zaika and Kovalev (1985) quote a number of Russian papers which appear to show a two- to eight-fold increase in plankton biomass over seamounts.

Some studies have shown a decrease in the abundance of plankton over seamounts. Nellen (1973) reported a lower abundance of migrating mesopelagic fish larvae and plankton over the Great Meteor seamount compared with surrounding waters. Boehlert and Seki (1984) also noted a decrease in the abundance of oceanic plankton over the Southeast Hancock seamount and Genin et al. (1988) noted an almost complete lack of migrating euphausiids over the Nidever Bank in the North Pacific. This decrease in the abundance of migrating plankton over seamounts could be caused by the grazing-out of migrators by demersal predators located on the seamount or by the displacement of migrators around the seamount during the day while they were at depth (Nellen, 1973; Genin et al., 1988; Boehlert and Mundy, 1993). Genin et al. (1988) compared abundance data for strong, intermediate and weakly migrating plankton, both on the Nidever bank and over deep water. They found a large difference in the variance and patchiness index of strong migrators between seamount and deep-water stations. There was a moderate difference in intermediate migrators and little difference in the variance and patchiness for poor migrators. Genin *et al.* (1988) suggested that the difference in abundances of strongly migrating plankton was mainly due to the displacement of migrators around the seamount during the day. If predation were the main influence on plankton abundance over sea-mounts then intermediate and weak migrators would be equally affected. Seamounts too shallow, or too deep, for migrating deep scattering layers to cross would have no such effect on migrating plankton. Predation by predators located on seamounts is likely to intensify decreased abund-ances of migrating zooplankton over seamounts. Genin *et al.* (1988) point out that shallow topography has the potential to increase significantly the patchiness of vertically migrating planktonic species.

Micronektonic and nektonic communities over seamounts can show striking differences with those in the surrounding deep water. Boehlert and Seki (1984) used a combination of acoustic observation and mid-water trawls to study scattering layers of vertically migrating organisms both on and around the Southeast Hancock seamount. They found that the very dense scattering layers over the seamount comprised the sternoptychid fish *Maurolicus muelleri*, the lophogastrid mysid *Gnathophausia longispina* and the sepiolid squid *Iridoteuthis iris*. In the less dense scattering layers away from the seamount these taxa were very rare, or even absent, whilst oceanic taxa were much more abundant. During the day the scattering layers remained on the flanks of the seamount, but during the early evening they began to move vertically upwards to a depth of approximately 50 m. This was followed by a consolidation of the shallow layers, a slight sinking of the top layer and an expansion downwards of the scattering layer until it extended from the seamount summit to a depth of approximately 100 m.

The results of Boehlert and Seki (1984) show striking similarities to that of Parin and Prut'ko (1985). This study again combined acoustic observation with trawling over the Equator seamount in the Indian Ocean. Scattering layers above the seamount were mainly composed of the myctophid *Diaphus suborbitalis*, which was found to be absent in deep scattering layers away from the seamount. During daylight hours the scattering layer, mainly comprising *Diaphus suborbitalis*, was located at a depth of 500–600 m on the slopes of the seamount. At night it ascended to become distributed "in dense schools" between 80 and 150 m depth. Gut contents of *D. suborbitalis* indicated that it was feeding on diurnally migrating oceanic plankton, mainly copepods and fish of the genus

Cyclothone (Gorelova and Prut'ko, 1985). Parin and Prut'ko (1985) found that the diversity and abundance of large fish species, including sharks, rays, tuna, swordfish and gempylids, was much higher around the Equator seamount than in the surrounding oceanic waters. The stomach contents of many of these species consisted mainly of *Diaphus suborbitalis*.

D. suborbitalis resides on the slopes of the seamount by day possibly to avoid predation. At night it ascends with diurnally migrating plankton and feeds on it while itself being preyed upon by larger fish species. The ascent of myctophids and sternoptychids around seamounts in the evening (Boehlert and Seki, 1984; Parin and Prut'ko, 1985) contrasts with the behaviour of larger predatory fish, studied by Isaacs and Schwartzlose (1965) and Genin *et al.* (1988), which ascended from the seamount summit in the early morning. These larger fish feed on diurnally migrating plankton trapped on the seamount during their descent. The fact that they are visual predators, probably less well adapted for low light conditions than myctophids and other deeper water micronekton and nekton, probably confines the feeding periods of these fish to the dawn and daytime.

4. BENTHIC BIOLOGY

4.1. Sampling Seamount Benthos

Large depth ranges, hard substrata, steep gradients, cryptic topography, fast and variable impinging currents, clear oceanic water and geographic isolation all combine to make seamounts unique habitats for deep-sea and shallow-water organisms. Though seamounts are very numerous on the ocean floor there have been relatively few studies on their benthic biology. This is partly because of the rugged terrain and strong currents associated with seamounts which make them notoriously difficult to sample (Hughes, 1981; Raymore, 1982; Levin and Nittrouer, 1987). Studies on seamounts have therefore used a variety of sampling strategies. These have included the use of suction dredges (Simpson and Heydorn, 1965), pipe dredges (Rao and Newman, 1972), chain bag dredges (Rao and Newman, 1972), rock dredges (Grigg *et al.*, 1987), tangle dredges (Grigg *et al.*, 1987), beam trawls (Rao and Newman, 1972; Richer de Forges, 1993), otter trawls (Rao and Newman, 1972; Wilson *et al.*, 1985), Sigsbee trawls (Wilson *et al.*, 1985), Agassiz trawls (Wilson *et al.*, 1985), semi-balloon bottom trawl (Kaufmann *et al.*, 1989), box corers (Levin and Thomas, 1989), push corers (Levin *et al.*, 1986; Levin and

Thomas, 1989; Smith *et al.*, 1989) and traps (Hughes, 1981; Raymore, 1982; Wilson *et al.*, 1985). A few surveys have made use of underwater photography taken from remote cameras (Hughes, 1981; Raymore, 1982, Wilson *et al.*, 1985; Grigg *et al.*, 1987) or from submersibles (Keller, 1985; Littler *et al.*, 1985, 1986; Moskalev and Galkin, 1986). Some of the shallower seamounts have been surveyed by divers (Simpson and Heydorn, 1965; Scagel, 1970).

4.2. The Biology of Hard Substrata

One of the first detailed investigations of seamounts was carried out on Vema seamount in the South Atlantic Ocean (Simpson and Heydorn, 1965). Vema seamount was reported to have a fairly shallow flat summit covered in boulders and pebbles. Divers found a rich benthic community composed of sponges, hydroids, ascidians, holothurians, numerous rock lobsters (*Jasus tristani*) and kelp beds of *Ecklonia* spp. (Simpson and Heydorn, 1965; Berrisford, 1969). Extensive collections of the fauna were made and detailed reports on taxa were published in subsequent years (see Millard, 1966; Penrith, 1967; Berrisford, 1969; Levi, 1969; Kensley, 1980). These reports pointed out that many of the species found on Vema appeared to be endemic. Little detail was given about the distribution and ecology of organisms on the seamount.

Subsequently a number of other workers have reported on the communities of hard substrata on various seamounts (Scagel, 1970; Heezen and Hollister, 1971; Hughes, 1981; Littler *et al.*, 1986; Moskalev and Galkin, 1986; Genin *et al.*, 1986; Grigg *et al.*, 1987; Kaufmann *et al.*, 1989). The dominant groups on the hard substrata of the shallower seamounts in these studies are macroalgae and encrusting corallines (Simpson and Heydorn, 1965; Scagel, 1970; Littler *et al.*, 1986) together with sponges, hydroids and ascidians (Simpson and Heydorn, 1965; Littler *et al.*, 1986). The very clear oceanic waters around seamounts can allow benthic macroalgae to extend to great depths. A general zonation of brown algae at the shallowest depths is superseded by green and then red algae at the greatest depths (Littler *et al.*, 1986). The deepest record for macroalgae comes from the San Salvador seamount in the Caribbean on which a crustose coralline algae was found at 268 m (Littler *et al.*, 1985, 1986). This algae appeared to have photosynthetic rates comparable with shallow water species though it existed in conditions of much lower light intensity indicating that it was adapted to utilize very low photon fluxes (Littler *et al.*, 1986).

At greater depths, coelenterates, mainly gorgonians (*Keratoisis* spp., *Callogorgia* spp., *Ellisella* spp.), zoanthids (*Gerardia* spp.), antipatharian corals (*Antipathes* spp., *Stichopathes* spp.), actinarians, pennatulids and hydroids appear to dominate hard substrata along with sponges, ascidians and crinoids (Heezen and Hollister, 1971; Grasshoff, 1972; Genin *et al.*, 1986; Littler *et al.*, 1986; Moskalev and Galkin, 1986; Boehlert and Genin, 1987, Grigg *et al.*; 1987; Kaufmann *et al.*, 1989). Other groups such as echinoderms (asteroids, ophiuroids and holothurians), arthropods (cirripedes and decapods), polychaetes and molluscs are found less frequently (Hughes, 1981; Raymore, 1982; Moskalev and Galkin, 1986; Grigg *et al.*, 1987).

The dominant organisms so far observed on seamounts are suspension feeders (Genin *et al.*, 1986; Boehlert and Genin, 1987). Many gorgonians and black corals require hard substrata, over which flow strong currents, to supply them with food, remove waste products and continuously keep the substratum (and the corals) completely clear of sediment (Grigg, 1974, 1984). Seamounts are one of the few habitats in the deep sea that provide such conditions. They are often located in relatively sediment-free oceanic waters, they have large areas of hard substrata and are often subject to strong currents (see above).

Genin *et al.* (1986) related the abundance of gorgonians, antipatharians and other suspension feeders on Jasper seamount, in the North Pacific, to a local topographically-induced current regime. They found that on a large scale, densities of the antipatharian, *Stichopathes* spp., were higher near peaks than at mid-slope sites of similar depths. On wide peaks highest densities were found along the rims of the summit while on narrow peaks highest densities occurred on the crests. On a smaller scale *Stichopathes* densities were increased on knobs and pinnacles. Direct current measurements showed that mean current speed near peaks was approximately twice that measured on mid-slope sites of corresponding depths (Genin *et al.*, 1986).

Other workers have observed an increased abundance of suspension feeders in areas of topographic current acceleration on seamounts. Submersible observations on Axial seamount on the Juan de Fuca Ridge in the North East Pacific have revealed dense communities of organisms (up to 100 individuals m^{-2}) on the vertical walls of a caldera system along which currents of up to 25 cm s^{-1} were recorded (Tunnicliffe *et al.*, 1985). Ophiuroids, hexactinellid sponges, ascidians and hydroids were the most common groups encountered, with lower numbers of crinoids, brisingids, holothurians, shrimps, actinarians, gorgonians and demosponges also being observed. Moskalev and Galkin (1986) showed an increased abundance of gorgonians, antipatharians and sponges along the edges of terraces on seamounts of the Magellan Range in the Equatorial Pacific. Grigg *et al.* (1987) surveyed Cross seamount, south of the Hawaiian Archipelago, and found highly clumped distributions of the gorgonian

Keratoisis spp. and a black sea anemone both of which were found in an increased abundance near the seamount summit, on basalt dikes and along the rim of the summit. Both the studies of Genin *et al.* (1986) and Grigg *et al.* (1987) showed a strong negative correlation of coral abundance with sediment cover and a decrease in the abundance of suspension-feeding and other organisms with an increasing depth from the summits of Jasper and Cross seamounts. This phenomenon may be related to decreased current velocities at greater depths with a related decrease in food supply and increase in sedimentation.

The topographically induced current regime has a major effect on the distribution and abundance of suspension-feeding organisms on at least some seamounts. Increased mean current speeds, which exist over topographic features at a number of scales, probably bring increased levels of food, prevent sedimentation both on the substratum and on the suspension-feeding organisms themselves (Grigg, 1974, 1984; Genin *et al.*, 1986; Boehlert and Genin, 1987). Increased current will also increase larval supply and therefore recruitment on such areas (Genin *et al.*, 1986; Boehlert and Genin, 1987). The importance of this to populations of corals which are slow growing and often have very low levels of recruitment (Grigg, 1984) is uncertain.

4.3. The Biology of Soft Substrata

Soft substrata on seamounts originate from a number of sources and show a variety of properties that are modified by the complex local current regime and by biological activity (Levin and Nittrouer, 1987; Levin and Thomas, 1989). There have been few extensive studies on the organisms which inhabit these substrata and these are insufficient to draw general conclusions about the ecology of the soft sediment benthos of seamounts.

There have been several studies on the effects of current on soft sediment fauna of seamounts. Levin and Thomas (1989) investigated the relationship between current strength and the abundance of microbes and infauna on two Pacific seamounts, Horizon Guyot and the Magellan Rise. Samples of infauna and microbes were collected at the summit caps of Horizon Guyot and the Magellan Rise and on the perimeter of Horizon Guyot. The Horizon perimeter site was the most exposed to currents and exhibited the coarsest sediments with long crested ripples. The Horizon cap and Magellan cap sites were less exposed with the Magellan site consisting of the finest sediment, poorest sorting and very little current rippling.

Polychaetes were the most common infaunal organisms on the seamounts surveyed, with the majority of them coming from the families

Paraonidae, Cirratulidae, Sabellidae, Syllidae and Ampharetidae. Other common infauna included peracarid crustaceans, aplacophoran, bivalve and gastropod molluscs, sipunculans, nemerteans and oligochaetes. Nematodes and harpacticoid copepods dominated the meiofauna, with lower numbers of ostracods, loriciferans and kinorhynchs. The results of the survey indicated that, in contrast to hard substrata, there was an inverse relationship of current strength to the abundance of bacteria and infauna (Levin and Thomas, 1989). It was suggested that the coarse sediment texture and frequent abrasion by turbation may account for the lowered bacterial abundance and this in turn may have been a contributory factor in decreasing infaunal densities in the more exposed site (Levin and Thomas, 1989). Meiofaunal densities, however, did not appreciably differ between sites.

Surveys of the epibenthic megafauna and lebenspurren on summit and perimeter sites on Horizon Guyot and the Magellan Rise summit (Kaufmann et al., 1989) were confounded by the presence of rocky outcrops on the Horizon perimeter site. This resulted in the inclusion of hard benthos suspension-feeding organisms (i.e. gorgonians and scalpellid barnacles) in a survey comparing the epibenthos of soft sediments. Marked differences were found in the abundance and diversity of epibenthic fauna and lebenspurren between the three sites surveyed and these were related to substratum availability, local current regime, nutrient availability, bioturbation and depth. On all sites, xenophyophores (giant rhizopoid protozoans) were the most abundant epibenthic organisms with densities reaching 492 1000 m^{-2} on the Magellan Rise (Kaufmann et al., 1989). The most abundant megafaunal metazoans included pennatulids (Horizon summit), hexactinellid sponges (Horizon summit, Horizon perimeter), scalpellid barnacles (Horizon perimeter), gorgonians (Horizon perimeter), comatulid crinoids (Horizon perimeter), echinoderms (Magellan Rise) and cerianthid actinarians (Magellan Rise) (Kaufmann et al., 1989).

Xenophyophores are dominant epifaunal organisms on soft substrata on seamounts (Levin et al., 1986; Levin and Thomas, 1988) and common in areas of high surface productivity (Tendal and Gooday, 1981) in the Pacific, Atlantic and Indian Oceans. They agglutinate sand-sized particles of a variety of materials, mainly planktonic foraminiferan tests but also glass chips, sulphides, oxides and black volcanic glass fragments to form a test (Levin and Thomas, 1988) up to 25 cm in diameter (Tendal, 1972). Xenophyophore tests show a great variety of morphologies and include reticulate forms, flattened fanlike forms, flattened oblong forms, rosette forms, flower-like forms and globular forms, and some are naked. Within the tests faecal pellets, known as stercomata, are stored. It has been suggested that bacteria living on stercomata break them down and the resultant products are reabsorbed by the xenophyophore inhabiting the test (Tendal, 1979).

Flattened fan-shaped xenophyophores often orient parallel to the flow of current. They probably feed by trapping particles in pseudopodia stretching between parts of the test (Tendal and Lewis, 1978) and the observed orientation allows exposure of the maximum surface area of xenophyophore to impinging water flow (Levin and Thomas, 1988). Xenophyophores have been observed at several sites with sloping topography and on a smaller scale have been seen to aggregate around basalt outcrops, sediment mounds, steep caldera walls and on small ridges (Tendal and Gooday, 1981; Levin and Thomas, 1988). If they are suspension feeders this may be explained by topography-enhanced currents or by a high concentration of suspended matter associated with these regions.

Xenophyophores apparently modify sediment deposition processes and enhance species diversity on seamounts (Levin et al., 1986; Levin and Thomas, 1988). A variety of taxa have been found living inside or on the external surface of xenophyophore tests (Gooday, 1984; Levin et al., 1986; Levin and Thomas, 1988) and a correlation has been found between the size of the xenophyophore and the number of test inhabitants. The infauna in the sediment beneath xenophyophore tests has also been shown to increase in diversity and abundance with an upward shift in horizontal distribution (Levin et al., 1986; Levin and Thomas, 1988). There is also some evidence that xenophyophores attract errant mega-fauna such as echinoids (Levin and Thomas, 1988).

Xenophyophores may be associated with increased abundances of other organisms for a number of reasons. Both adult and juvenile stages of macro- and meiofauna may use the test as substratum or as a refuge from predation (Gooday, 1984; Levin et al., 1986; Levin and Thomas, 1988). Tests may also be used as breeding grounds and the presence of juvenile and adult specimens of some species within tests suggests that at least some organisms are permanent or semi-permanent residents (Levin and Thomas, 1988). Tests may also be used as feeding grounds for organisms and some may feed upon tissue from the xenophyophore itself or feed upon the stercomata (Gooday, 1984; Levin et al., 1986; Levin and Thomas, 1988). One species of flabelligerid polychaete actually uses the stercomata to make up a sheath in which it lives (Levin et al., 1986). Attraction of megafauna may be stimulated by xenophyophores since they act as irregularities on the ocean bed and as such concentrate suspended particulates in their vicinity. Such particulates may aggregate to form floc-containing biogenic debris attractive to animals such as surface-feeding echinoderms (see Reimers and Wakefield, 1989). Sediment deposition rates calculated using isotopic particle tracers indicate that deposition is enhanced around xenophyophores.

4.4. Seamounts Penetrating Oxygen Minimum Layers

Work on Volcano 7 (Wishner *et al.*, 1990), a seamount located in the tropical East Pacific, the summit of which penetrates an oxygen minimum layer, indicates that bacteria and meiofauna in sediments are relatively unaffected by low oxygen conditions. The hard substratum megafauna, in contrast, shows a marked zonation running from the upper to the lower summit and then down the seamount flanks. On the upper summit (top 20 m), where oxygen concentrations are very low, a few rattails and solitary corals were observed. There was then a transition zone between 750 and 770 m in which increasing densities of sponges, serpulid polychaetes, galatheid crabs and shrimps occurred. Between 770 and 850 m, densities of sea pens, sponges, anemones, serpulids, ophiuroids and shrimps reached a maximum ($2.25\,\mathrm{m^{-2}}$). Faunal maxima were associated with enhanced food supply (degradation of surface water production was inhibited by the oxygen minimum layer) and an increase in oxygen concentrations. Below 1000 m megafaunal abundance was low with xenophyophores dominating the community.

4.5. Hydrothermal Vent Communities

Most seamounts have a volcanic origin (Epp and Smoot, 1989) so it is of no surprise that some are associated with hydrothermal venting. The first hydrothermal vents observed and fully described on a seamount were located on Axial seamount on the Juan de Fuca Ridge in the North East Pacific. Axial seamount is located on the central or Cobb segment of the Juan de Fuca Ridge (Tunnicliffe, 1991) and is the site of four vent fields, CASM (Canadian American Seamount Site), ASHES (Axial Seamount Hydrothermal Emissions Study), East and South vent fields (Massoth *et al.*, 1989). Three of these vent fields, CASM, East and South are low temperature vents; the fourth, ASHES, is a high temperature field which vents clear, brown and black fluids with a temperature measured at up to 328°C (Massoth *et al.*, 1989). All the vent fields are located in the caldera at the seamount summit at a depth of approximately 1500 m (Massoth *et al.*, 1989).

As well as a rich non-hydrothermal vent community growing on the walls of the caldera of Axial seamount (Tunnicliffe *et al.*, 1985; see above) there is a rich hydrothermal vent community growing in and around the four vent fields. The CASM site is located in a fissure in the floor of the caldera which runs from a talus pile in the northern wall of the caldera to a vertical wall 300 m to the south (Canadian American Seamounts Expedition, 1985). The fissure is about 20 m deep and two main vents, together with numerous minor vents, issue from the northern 200 m.

Venting fluids are enriched in helium-3 and carbon dioxide and reach a temperature of up to 35°C (Canadian American Seamounts Expedition, 1985). A vent-specific fauna, consisting of vestimentiferans, was observed to form dense clumps around the main vents while others were surrounded by limpets, alvinellid polychaetes and recumbent vestimentiferans (Tunnicliffe et al., 1985). The vestimentiferan tubes were covered in a thick grey mucus, probably secreted by the alvinellid polychaetes, which contained many small invertebrates. In areas away from the direct influence of the venting fluids, but within 100 m of the vents, white and amber bacterial mats grew along with cobalt blue colonies of colonial ciliates (Tunnicliffe et al., 1985). The biology of the other vent fields on Axial seamount has not been described.

Hydrothermal venting has also been found on a mid-plate seamount, the Loihi seamount off the island of Hawaii in the Pacific (Karl et al., 1988). This vent site was originally detected by a hydrothermal plume over the seamount which was reported to contain elevated levels of methane, helium-3, dissolved inorganic carbon and trace metals (Horibe et al., 1983; Sakai et al., 1987; Gamo et al., 1987). Two hydrothermal fields, Pele's and Kapo's vents, were subsequently located, mapped and sampled using submersibles. These fields were both located near the summit of the seamount on its southern side (south rift zone) on the eastern flanks of pillow lava cones (Karl et al., 1988). Pele's vent is at a depth of approximately 990 m and vents a very clear fluid at a temperature of up to 30°C (Karl et al., 1989). This vent fluid has very high concentrations of dissolved carbon dioxide, trace metals, silicate, ammonium and phosphate ions (Karl et al., 1988, 1989).

The hydrothermal fluids on Loihi seamount support elevated bacterial populations in the vent plumes and extensive red-brown bacterial mats that cover large areas of the hydrothermal fields (Karl et al., 1989). These bacterial mats consist of a filamentous bacteria and very high concentrations of iron, phosphorus and calcium. Despite the presence of a large bacterial population, which often acts as a base of primary productivity on other seamounts, there is a distinct absence of vent endemic megafauna on Loihi. This has been attributed to a number of possible factors including discontinuous venting, geographic isolation, volcanic disturbance, low pH of the venting waters, toxicity of high concentrations of trace metals and the young geological age of Loihi seamount (Grigg, 1987; Karl et al., 1988, 1989). The shallow depth of the Loihi vent system may also be a contributory factor to the differences in biology between it and deeper axis hydrothermal vents.

Low temperature venting has also been discovered on Red Volcano, one of the Larson seamounts, 30 km to the east of the East Pacific Rise. The venting areas are located on small pillow cones, on the summit of

Red Volcano, which are partially mantled in red mud (iron oxide). Chimneys made of minerals rise obliquely from this mud and vent a clear fluid which disperses downhill. Temperatures in the mud appear to reach 10–15°C within 50 cm of the surface (Lonsdale *et al.*, 1982). The surfaces of some of the chimneys were covered in what appear to be bacteria but as with Loihi seamount none of the vent endemic megafauna associated with high temperature sites on the East Pacific Rise (only 30 km away) was present (Lonsdale *et al.*, 1982). The only common epifauna on the iron oxide fields were thin fan-like xenophyophores (Levin and Thomas, 1988).

Most recently hydrothermal vents have been described from submersible observations on the Peepa seamount in the Bering Sea (Sagalevitch *et al.*, 1992). Peepa seamount was found to consist of a northern and southern peak joined at the 700 m isobath. A transect on the northern peak showed a number of distinct faunal zones on this part of the seamount. Fauna included a number of species of sponge, alcyonarians (*Anthomastus* spp.), actinarians, polychaetes, crabs (*Chionoecetes* spp., *Paralithodes* spp.), amphipods and holothurians. The northern peak terminated in a pair of cones between which, at 400 m depth, there were rich bacterial mats, mainly coloured white with patches of black, grey and brown. The mats covered 100% of this area and hydrothermal fluids were seen to be venting through holes approximately 3×7 cm in diameter. The largest bacterial mats were located on the southwestern slope of the northern peak at 650 m and covered an area of approximately 100 m². On the eastern part of the northern peak high temperature hydrothermal fluids (180–360°C) were vented at 50–100 l min^{-1} through aragonite chimneys which were approximately 1.5 m high and 40 cm in diameter. Gas was also seen to be escaping at a rate of approximately 2–5 l min^{-1}. This gas was similar in composition to that released in shallow hydrothermal sites at the Kurile Islands (see Tarasov *et al.*, 1987) but was low in carbon dioxide and hydrogen compared to that released in the Guaymas Basin.

The southern peak is located approximately 2 nautical miles from the northern peak and, like the northern peak, showed a vertical zonation of fauna. Most hydrothermal activity was located in the central part of the summit where hydrothermal fluids were vented from fissures. Fluids were emitted at a lower temperature than on the northern peak and there were chimneys made of aragonite and pyrite, rich in mercury, up to 40 cm in height. Bacterial mats were found from 585 m as a thin covering on stones but were thicker on the fissures close to the summit where active venting was taking place. Bivalves of the genus *Calyptogena* were found in patches at 489 m mainly in cracks and fissures where hydrothermal fluids were being emitted. This represents the most northerly finding of

Calyptogena spp. on vents or seeps (Sagalevitch *et al.*, 1992). No other fauna was found to be associated with the vents though hydrothermal activity with its associated bacteria may have been responsible for large numbers of hydroids, gorgonians and alcyonarians on both peaks.

Hydrothermal deposits have also been observed on Marsili seamount in the Tyrrhenian Basin north of Sicily though no active venting was observed (Uchupi and Ballard, 1989).

The presence of hydrothermal vents on mid-oceanic seamounts and off-axis seamounts may indicate that hydrothermal emissions are more common in the deep sea than previously thought. The biology of vents on off-axis seamounts is markedly different from those located on axis vent systems. The reasons for this are unclear and more data are required on the biology of other seamount vent systems located at different depths and showing different types of venting before any conclusions can be drawn. Deep-water and shallow-water hydrothermal vents show striking differences in their associated fauna. The location of hydrothermal venting at intermediate depths (between 50 and 2000 m) on seamounts may provide a unique insight into the effects of depth (pressure) on hydrothermal vent chemistry and biology.

5. SPECIES DIVERSITY

5.1. The Geographic Affinities of Seamount Organisms

Little is known about the species inhabiting seamounts or the relationships between seamount faunas and those of the surrounding abyssal plain, continental shelf and other seamounts and mid-oceanic islands. Recent studies using submersibles and photography have provided data on factors influencing community structure on seamounts but this has not been accompanied by detailed taxonomic work. Organisms on seamounts are often only referred to by their appearance (i.e. Grigg *et al.*, 1987), with no indication of their systematic status and no details of species biology (i.e. life history, population biology, etc.). Such methods of investigation give little idea of the community dynamics, recruitment, levels of endemism and ultimately the effects of commercial exploitation on seamounts.

Wilson and Kaufmann (1987) reviewed data on the occurrence and geographic affinities of seamount species. They concluded that the seamount biota tended to be dominated by species inhabiting the nearest shelf areas. Shallow seamounts tended to be dominated by regional species with an equal representation of widespread or cosmopolitan species, whilst deeper seamounts tended to be dominated more by the latter category. They also suggested that seamounts appeared to act as stepping stones for transoceanic dispersal of species (see Wilson and Kaufmann, 1987, for discussion).

Another finding by those authors was that seamounts appear to show significant levels of endemism. Their review of current data on seamount invertebrate fauna indicated that over 15% were new species (Wilson and Kaufmann, 1987). For example, studies on the fauna of the Vema seamount, off the coast of South Africa, identified 28% of species as endemic (Berrisford, 1969). Diehl (1970) reported a new species of highly adapted ascidian from the soft sediments of Josephine seamount in the North Atlantic. Two endemic fish species, *Gnathophis codoniphorus* and *Callionymus sousai* and possibly one ray (*Raja* spp.) were identified from the Great Meteor seamount (Maul, 1976; Ehrich, 1977) also in the North Atlantic Ocean. Rice and Williamson (1977) studied crustacean larvae sampled from the Great Meteor and Josephine seamounts and from the coasts of Portugal and Morocco. Of the larvae sampled one type from both seamounts and four from the Great Meteor seamount were not present in the coastal shelf samples and showed "striking characters unknown in any larvae from European or North African waters". The authors suggested that the seamounts may support genera or even higher taxa not represented on the continental shelf.

Other examples of endemism include fish of the Kyushu–Palau ridge (26%) (Okamura *et al.*, 1982), invertebrates from the Horizon Guyot (36%) (Wilson *et al.*, 1985), fish of the Northwest Pacific seamounts (22%) (Borets, 1986). Other examples may be found in the barnacles from the Southeast Pacific (Zullo and Newman, 1964), Southeast Pacific echinoids (Allison *et al.*, 1967), barnacles from the Mid-Pacific Mountains (Rao and Newman, 1972) and in the review on seamount biota by Wilson and Kaufmann (1987).

In view of recent studies on the diversity of deep-sea fauna the results of such studies must be viewed with caution (i.e. Grassle and Maciolek, 1992). In some cases the absence of records of "endemic" seamount species from continental shelf/slope/rise fauna may simply be because they have not yet been captured and described from the limited sampling of these areas. This may be more than compensated for by the limited sampling and description of some groups of seamount organisms (Wilson and Kaufmann, 1987) and the occurrence of cryptic or sibling species (i.e. Grassle and Grassle, 1976).

5.2. Reproductive and Genetic Isolation

Significant levels of endemism in seamount fauna, even amongst groups of well-known and highly mobile organisms such as fish, would suggest that seamounts are sites at which there may be a high incidence of speciation. Such speciation could be the result of reproductive and genetic isolation in seamount populations. A possible cause of such isolation is geographic separation by distance of seamount populations from external populations (other seamounts, continental shelf, etc.). Hydrographic conditions may also isolate seamount populations by trapping larvae originating on a seamount in its immediate vicinity (Shomura and Barkley, 1980) (see Taylor columns, Section 2). In such a case seamount populations would become self-recruiting, allowing adaptation to local environmental conditions. The large variety of substrata and gradients in depth and other physical parameters on seamounts may also enhance speciation.

Evidence for the retention of larvae of seamount species over seamounts by topographically induced currents (such as Taylor columns) is sparse (Boehlert and Mundy, 1993). Over the Great Meteor seamount larvae of oceanic species were found to have a reduced abundance compared to samples taken off the seamount (Nellen, 1973, 1974). It was suggested that predation or physical damage during diurnal descent increased the mortality of vertically migrating planktonic species (Nellen, 1973, 1974) but it is more likely that migrators were displaced by the seamount when below the summit depth (Genin et al., 1988). Larvae of two topographically associated species were more abundant over the summit of Great Meteor than in surrounding waters and it was suggested that the larvae of species spawning on the seamount were retained over it by currents (Nellen, 1973, 1974).

Subsequent sampling at the Great Meteor seamount did not locate elevated populations of topographically associated species (Belyanina, 1984). Decreased abundances of ichthyoplankton were found over Southeast Hancock seamount compared to reference stations at night during the summer (Boehlert, 1985; Boehlert and Genin, 1987). Once again this was probably due to displacement of larvae by the seamount. During winter greater abundances of ichthyoplankton were observed above the seamount both during the day and at night.

Boehlert and Mundy (1993) suggest that supporting data indicating that there are distinct ichthyoplankton assemblages above seamounts are lacking. Furthermore they suggest that in general the dominant larvae at most seamounts are widespread oceanic species. Larvae of topographically associated species are rare, apart from on shallow seamounts, and the abundance of larvae of topographically associated species decreases with increasing distance of the seamount from shelf areas. They also suggest that larvae produced on seamounts are frequently dispersed away from their parent seamount population and that larvae recruiting to seamount populations originate from larger land masses and are transported to seamounts by currents (Boehlert and Mundy, 1993; see also Wilson and Kaufmann, 1987).

In some species, recruitment to seamount populations appears to come from long-lived dispersive larvae that originate from other geographically removed populations. In such cases the seamount populations are described as "dependent populations" (Boehlert *et al.*, 1994). Initial investigations of the Vema seamount revealed a large population of the rock lobster, *Jasus tristani* (Simpson and Heydorn, 1965). It was suggested that recruitment to the rock lobster population on Vema came from long-lived dispersive larvae originating upstream from Tristan da Cunha (Heydorn, 1969). Analysis of the drift of free-floating buoys in this area provided evidence that such a current-borne larval supply was quite feasible (Lutjeharms and Heydorn, 1981a). Immediately after the reports on the Vema seamount lobster population it was severely depleted by fishing (Heydorn, 1969). The population showed no signs of recovery up to 1979 but then made a sudden reappearance, only to be fished again to uneconomic levels (Lutjeharms and Heydorn, 1981b). The variability of recruitment from long-lived larvae originating from a distant population would partially explain the susceptibility of the Vema rock lobster population to overfishing and the unpredictability of its subsequent recovery (Boehlert and Mundy, 1993).

A similar recruitment mechanism was suggested for the most common species of fish at Vema seamount, *Acantholatris monodactylis* (Lutjeharms and Heydorn, 1981a). Larvae of another common seamount species *Pseudopentaceros wheeleri* also appear to disperse away from the seamount by neustonic larvae to recruit back to seamounts after a pelagic life of 1.5–2.5 years (Boehlert and Sasaki, 1988 – see below).

A study of a population of the sternoptychid *Maurolicus muelleri* on the Southeast Hancock seamount concluded that, because of the insufficient number of reproductive adults, it was unlikely that the population was self-sustaining (Boehlert *et al.*, 1994). Recruitment of *Maurolicus muelleri* to the seamount occurs annually, in spring and summer, at sizes greater than 20 mm standard length. On the basis of morphological studies recruits were thought to originate partly from the local population and partly from larger populations located upstream of Southeast Hancock seamount, on the southern Emperor seamounts (Boehlert *et al.*, 1994).

Though some species appear to recruit from geographically separated populations there is evidence that seamount populations show significant differentiation from populations located on other seamounts and on the nearby abyssal plains, continental rise and shelf. Ehrich (1977) compared populations of three species of fish from the Great Meteor seamount and the African shelf using multivariate methods. He demonstrated morphological divergence between the seamount and shelf populations. After relating morphological divergence to genetic divergence, Ehrich (1977) subsequently concluded that demersal fish populations of the seamount and African shelf were separate.

Grasshoff (1972) studied intraspecific variation in populations of the gorgonian *Ellisella flagellum* sampled from the Great Meteor and Josephine seamounts. He found that populations from the two seamounts differed in size and coloration and concluded that the morphological differentiation between them was due to reproductive isolation (as did Ehrich, 1977).

Populations of the frostfish, *Lepidopus caudatus*, sampled from Gettysburg seamount and from the Dacia and Azores seamounts in the Northeast Atlantic, could be differentiated into a southern morphotype (Gettysburg seamount) and a northern morphotype (Dacia and Azores seamounts). Because these populations occurred sympatrically it has been suggested that they are different species (Wilson and Kaufmann, 1987). As with the frostfish, morphologically similar populations of *Maurolicus muelleri* located on the southern Emperor and Southeast Hancock seamounts show significant morphological differentiation to populations located on the continental slope off the coast of Japan (Boehlert *et al.*, 1994). As with the frostfish, it is uncertain whether this morphological variation is intraspecific or interspecific, in which case populations of *Maurolicus muelleri* located on the seamounts and on the continental slope of Japan would represent separate species (Boehlert *et al.*, 1994).

Is the morphological differentiation detected in populations of some species occurring on seamounts indicative of genetic differentiation and therefore reproductive isolation? Morphological differences may simply be phenotypic differences generated by differences in the environmental conditions on seamounts and the adjacent abyssal and shelf areas. There are few studies on the genetics of deep-sea organisms and these are mainly restricted to measures of genetic variability (see Doyle, 1972; Gooch and Schopf, 1972; Ayala and Valentine, 1974; Ayala *et al.*, 1975; Murphy *et al.*, 1976; Siebenaller, 1978; Wilson and Waples, 1984; Bisol *et al.*, 1984; Bucklin, 1985, 1988). Other data on the genetics of deep-sea organisms concern genetic differentiation and molecular taxonomy of hydrothermal vent organisms (France *et al.*, 1992; Denis *et al.*, 1993; Tunnicliffe *et al.*, 1993; Williams *et al.*, 1993). One study at present concerns genetic differentiation of populations of deep-sea amphipods located at different depths on the continental rise (France, 1994) and only one study (Bucklin *et al.*, 1987) concerns organisms found on seamounts.

Bucklin *et al.* (1987) examined differences in allele frequencies between populations of the giant deep-sea scavenging amphipod *Eurythenes gryllus* occurring on the abyssal plain and a population on the summit of the Horizon seamount in the central North Pacific Ocean. They found that abyssal populations, including one sampled at the base of the seamount, were genetically homogeneous, indicating high levels of gene flow between these populations. This was not surprising since *Eurythenes gryllus* is a good swimmer (Smith and Baldwin, 1984) and has a large potential for dispersal. In contrast, there was marked genetic differentiation between amphipod populations at the base of the seamount and those on the summit. Bucklin *et al.* (1987) suggested two possible reasons for this. First the summit population of *Eurythenes gryllus* was subject to strong selection leading to genetic differentiation; alternatively gene flow between the summit and base of the seamount was restricted, probably due to a limited vertical dispersion of *Eurythenes gryllus*. This would indicate a degree of reproductive isolation of the amphipod crest population (a prerequisite for speciation).

In conclusion there is evidence that populations of certain species found on seamounts rely to some degree on an allocthonous input of larvae. Recruitment of larvae to seamounts in such cases may be variable and unpredictable. Models for recruitment to seamounts are shown in Figure 4. There is morphological and genetic evidence that populations of some organisms on seamounts are distinct from surrounding populations located on other seamounts, the abyssal plain and continental shelf. This may indicate a restriction of gene flow between some seamount populations and populations outside the seamount which may be caused by geographic isolation, topographically induced hydrographic conditions or by life-history patterns. This may be a contributory factor in speciation on seamounts.

6. COMMERCIAL EXPLOITATION

6.1. Seamount Fisheries

Seamounts support stocks of commercially valuable fish and shellfish species. Three examples are shown in Figure 5. These include: rock-lobster (*Jasus tristani*) (Simpson and Heydorn, 1965); wreckfish (*Polyprion americanus*) (Penrith, 1967); grouper (*Epinephelus aeneus*) (Penrith, 1967); yellowtail amberjack (*Seriola lalandii*) (Penrith, 1967); mackerel (*Decapterus longimanus*) (Penrith, 1967); (*Plagiogenion rubiginosus*) (Penrith, 1967); biskop (*Mupus imperialis*) (Penrith, 1967); alfonsin (*Beryx splendens*) (Sasaki, 1978) (see Fig. 5); armourhead (*Pentaceros richardsoni*) (Sasaki, 1978); sablefish (*Anoplopoma fimbria*) (Hughes, 1981); deep-sea red king crab (*Lithodes couesi*) (Hughes, 1981); golden

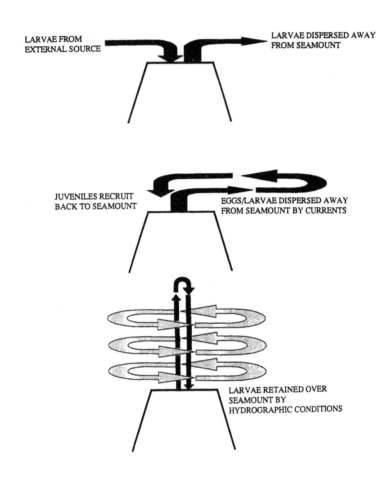

Figure 4   Models for recruitment to populations of species living on sea-mounts.

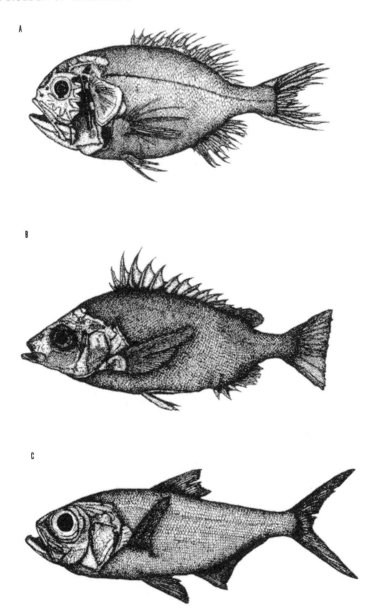

Figure 5 Three species of fish commercially trawled on seamounts. (A) Orange roughy, *Hoplostethus atlanticus*. Standard length of specimen approximately 250 mm (drawn from Nakamura *et al.*, 1986). (B) Pelagic armourhead, *Pseudopentaceros wheeleri*. Standard length of specimen 248 mm (drawn from Okamura *et al.*, 1982). (C) Alfonsin, *Beryx splendens*. Standard length of specimen 387 mm (drawn from Okamura *et al.*, 1982).

king crab (*Lithodes aequispina*) (Hughes, 1981); tanner or snow crab
(*Chionoecetes tanneri*) (Hughes, 1981); caridean shrimp (*Heterocarpus
laevigatus*) (Gooding, 1984); caridean shrimp (*Heterocarpus ensifer*)
(Gooding, 1984); snake mackerel (*Promethichthys prometheus*) (Uchida
and Tagami, 1984); red snapper (*Etelis carbunculus*) (Everson, 1986); red
snapper (*Etelis coruscans*) (Everson, 1986); pink snapper (*Pristipomoides
sieboldii*) (Everson, 1986); Brighams snapper (*Pristipomoides zonatus*)
(Everson, 1986); surgeonfish (*Acanthurus olivaceus*) (Honda, 1986); pink
snapper (*Pristipomoides filamentosus*) (Humphreys, 1986); bluefin tuna
(*Thunnus thynnus*) (Koami, 1986); mackerel (*Scomber japonicus*)
(Koami, 1986); sardine (*Sardinops melanosticta*) (Koami, 1986); blue-
green snapper (*Aprion virescens*) (Kramer, 1986); wahoo (*Acanthocy-
bium solandri*) (Kramer, 1986); pelagic armourhead (*Pseudopentaceros
wheeleri*) (Sasaki, 1986); Japanese beardfish (*Polymixia japonica*) (Sasaki,
1986); broad alfonsin (*Beryx decadactylus*) (Sasaki, 1986); Japanese
butterfish (*Hyperoglyphe japonica*) (Sasaki, 1986); mirror dory (*Zenopsis
nebulosa*) (Sasaki, 1986); skilfish (*Erilepis zonifer*) (Sasaki, 1986); hon-
eycomb rockfish (*Hozukius emblemarius*) (Sasaki, 1986); mackerel scad
(*Decapterus russellii*) (Sasaki, 1986); bigeye driftfish (*Ariomma lurida*)
(Sasaki, 1986); bonnetmouth (*Erythrocles schlegeli*) (Sasaki, 1986); rock-
eye rockfish (*Sebastes aleutianus*) (Sasaki, 1986); shortspine thornyhead
(*Sebatolobus alascanus*) (Sasaki, 1986); red-stripe rockfish (*Sebastes
proriger*) (Sasaki, 1986); rosethorn rockfish (*Sebastes helvomaculatus*)
(Sasaki, 1986); black rockfish (*Sebastes melanops*) (Sasaki, 1986; jack
mackerel (*Trachurus symmetricus*) (Sasaki, 1986); argentine (*Glossano-
don* sp.) (Sasaki, 1986); berycoid fish (*Centroberys affinis*) (Sasaki, 1986);
dory (*Zenion* sp.) (Sasaki, 1986); tarakihi (*Nemadactylus macropterus*)
(Sasaki, 1986); sea bass (*Caprodon longimanus*) (Sasaki, 1986); lizard fish
(*Saurida undosquamis*) (Sasaki, 1986); butterfly bream (*Nemipterus
personii*) (Sasaki, 1986); Antarctic giant fish (*Dissostichus eleginoides*)
(Sasaki, 1986); Antarctic cod (*Notothenia squamifrons, N. rossii*) (Sasaki,
1986); icefish (*Champsocephalus gunnari*) (Sasaki, 1986); icefish
(*Chaenichthys rhinoceratus*) (Sasaki, 1986); sancord (*Helicolenus dacty-
lopterus*) (Sasaki, 1986); rosy snapper (*Lutjanus lutjanus*) (Sasaki, 1986);
butterfish (*Hyperoglyphe* spp.) (Sasaki, 1986); Seales grouper (*Epinephelus quernus*) (Seki, 1986); giant trevally (*Caranx ignobilis*) (Seki, 1986);
white trevally (*Pseudocaranx dentex*) (Seki, 1986); redtail scad (*Decapterus tabl*) (Shiota, 1986); blue-striped snapper (*Lutjanus kasmira*) (Uchida,
1986); little tuna (*Euthynnus affinis*) (Uchiyama, 1986); albacore (*Thunnus alalunga*) (Yasui, 1986); grouper (*Epinephelus* spp.) (Sadakane *et al.*,
1989); orange roughy (*Hoplostethus atlanticus*) (Mace *et al.*, 1990);
muroadsi scad (*Decapterus muroadsi*) (Parin *et al.*, 1990); yellowfin tuna
(*Thunnus albacares*) (Fonteneau, 1991); skipjack tuna (*Katsuwonus*

pelamis) (Fonteneau, 1991); bigeye tuna (*Parathunnus obesus*) (Fonteneau, 1991); Pacific ocean perch (*Sebastes alutus*) (Krieger, 1992); blue whiting (*Micromesistius poutassou*) (Gerber, 1993); widow rockfish (*Sebastes entomelas*) (Pearson *et al.*, 1993); rosy rockfish (*Sebastes rosaceous*) (Pearson *et al.*, 1993); harlequin rockfish (*Sebastes variegatus*) (Pearson *et al.*, 1993); shortbelly rockfish (*Sebastes jordani*) (Pearson *et al.*, 1993).

6.2. The Pelagic Armourhead

Populations of fish species located on seamounts have been over-exploited, an example being the pelagic armourhead fishery over the southern Emperor seamounts and seamounts in the northern Hawaiian Ridge. In the late 1960s Russian trawlers began to exploit fish stocks in the vicinity of these seamounts. The species targeted by this fishery could not be identified since they were gutted and gilled at sea and yet it is estimated that Russian catches reached 133 000 t in 1969 (Sakiura, 1972). The fish was identified by Abe (1972) as the armourhead (*Pseudopentaceros wheeleri*), see Figure 5, considered rare before the development of the fishery (Welander *et al.*, 1957). Catches of pelagic armourhead were maintained at 20 000–30 000 t from 1972 to 1976 but then dropped to 3500 t in 1977. The fishery had been depleted and, though fishing effort decreased thereafter, the fishery did not recover (Sasaki, 1986).

The complex life history of the pelagic armourhead may at least be partially responsible for the rapid decline of the fishery. It has been suggested that *Pseudopentaceros wheeleri* may be semelparous (Humphreys and Tagami, 1986) with adults spawning between November and March over the Southern Emperor and North Hawaiian Ridge seamounts (Boehlert and Sasaki, 1988). The larvae are neustonic for at least the first part of their life and are dispersed away from the seamounts, presumably by ocean currents. They return at an age of between 1.5 and 2.5 years, locating the seamounts by an unknown mechanism, to spawn (Boehlert and Sasaki, 1988). Recruitment to the seamount populations shows a marked interannual variation. This has been attributed to a number of factors from food availability for larval and juvenile armourhead to large-scale variations in the strength and position of oceanic currents possibly involved in the transport of new recruits to the seamount (Boehlert and Sasaki, 1988). As with the rock lobster variability of recruitment is probably a major factor in the rapid decline of pelagic armourhead populations which are heavily exploited. The concentration of these fish around seamounts, with small summit areas for breeding, also makes them vulnerable to an intense and localized fishing strategy (Sasaki, 1986).

6.3. The Orange Roughy

Another fish species which has been subject to increasing levels of fishing effort on seamounts, especially off the coasts of Australia and New Zealand, is the orange roughy, *Hoplostethus atlanticus* (Figure 5). Large-scale trawling for orange roughy around New Zealand commenced in 1978–1979 and has concentrated around the north Chatham Rise, the Ritchie Bank and the Challenger Plateau (Zeldis, 1993). With catches in the Chatham Rise area exceeding 40 000 t, the orange roughy fishery has become one of the most valuable in New Zealand (Smith *et al.*, 1991). More recently, heavy exploitation of orange roughy fisheries has commenced off the south eastern coast of Australia, with approximately 34 000 t of orange roughy being caught in Tasmanian waters in the period from 1989 to 1990 (Smolenski *et al.*, 1993).

The life history of the orange roughy has until recently been very poorly understood and even now important aspects of its biology are still unknown. Orange roughy are mainly found between depths of 700 and 1500 m but occur down to 1800 m (Mace *et al.*, 1990; Fenton *et al.*, 1991; Bell *et al.*, 1992; Smolenski *et al.*, 1993). They are group-synchronous spawners and form large spawning aggregations near banks, pinnacles and canyons during the winter (Pankhurst *et al.*, 1987; Pankhurst, 1988), travelling up to 200 km to reach the spawning grounds (Bell *et al.*, 1992). It is while in these spawning aggregations that they are targeted by trawlers.

The spawning period over the Challenger Plateau and Chatham Rise off the coast of New Zealand appears to be consistent from year to year and there is some evidence that it is regulated by photoperiod (Pankhurst, 1988). Studies using a combination of trawls and acoustic methods indicate that male orange roughy arrive first on the spawning ground and aggregate near the bottom, moving into mid-water when the females arrive. Spawning aggregations do not appear to show diurnal patterns of movement (Pankhurst, 1988). Spawning takes place over approximately 3 weeks and spent males and females disperse in 3–4 weeks (Pankhurst, 1988; Bell *et al.*, 1992). Orange roughy have a low fecundity (Pankhurst and Conroy, 1987; Zeldis, 1993) and studies on populations from Australian waters indicate that not all large females reproduce in a given year (Bell *et al.*, 1992). Failure of individuals to produce eggs may occur as a response to poor food supply and may be a useful spawning strategy in a long-lived species with low natural mortality (see below) such as the orange roughy.

Orange roughy eggs are pelagic but have only been sampled in large numbers in the immediate vicinity of the spawning site. Their distribution is extremely patchy and densities rapidly decline due to dilution moving away from the spawning site. The developmental rates and stages of orange roughy eggs are unknown (Zeldis, 1993) and no larvae of *Hoplostethus* species below 26 mm long have been captured (Jordan and Bruce, 1993). It is thought that the larvae may occur in greater depths than those generally sampled in ichthyoplankton surveys or that the larvae may occur close to the bottom on the continental slope (Jordan and Bruce, 1993). There are therefore no data on larval dispersal in orange roughy though it has been suggested that the pelagic larval/ postlarval stage may last for up to 10 months (Mace *et al.*, 1990). A long larval/juvenile stage may explain to some extent variability in interannual recruitment to orange roughy populations (Mace *et al.*, 1990), in a similar manner to pelagic armourhead.

Original attempts to estimate demographic parameters of orange roughy populations based on studies of otoliths consistently overestimated growth rates and natural mortality and underestimated age of recruitment to the fishery (see Mace *et al.*, 1990). A reinterpretation of otolith and length frequency data by Mace *et al.* (1990) indicated that orange roughy were extremely slow growing with fish reaching an average length at maturity of 30 cm SL in an average time of 20 years. Furthermore it was suggested that orange roughy were extremely long lived (maximum age >50 years) with extremely low rates of natural mortality (M probably <0.08). Subsequent age determination of orange roughy using isotope ratios measured from otoliths (Fenton *et al.*, 1991) indicated that in fact the average age at maturity of orange roughy was 32 years and that the maximum age of adult fish was between 77 and 149 years old.

It has been suggested that, due to the longevity of orange roughy, its low natural rates of mortality, its slow growth rate and variable recruitment, it should be harvested at a rate of less than 10% of the average recruited biomass (Mace *et al.*, 1990). This was based on an underestimate of age of maturity and longevity (Fenton *et al.*, 1991). There is also evidence that orange roughy populations are genetically distinct and should be treated as separate stocks (Ovenden *et al.*, 1989; Smolenski *et al.*, 1993).

Data from the most completely studied orange roughy population, that of the Chatham Rise, indicate that the 1988–1989 quota for orange roughy was 32 637 t, approximately 25% of the estimated total biomass (Smith *et al.*, 1991). Catches of orange roughy from the Chatham Rise have been declining over the last 10 years and without large reductions in allowable catch may soon collapse (Zeldis, 1993). The Challenger Plateau population off New Zealand is also showing signs of overfishing (Zeldis, 1993). Estimations of genetic diversity of the Chatham Rise and Challenger Plateau roughy populations using starch gel electrophoresis have also indicated a decrease in the genetic diversity of these populations (Smith *et al.*, 1991). The exact reason for this is unclear, but a decrease in population size due to overfishing has been implicated (see Smith *et al.*, 1991; Smolenski *et al.*, 1993). It would appear that the orange roughy populations have been subject to unsustainable levels of fishing.

Rapid decline of seamount stocks after the commencement of trawl fisheries has also been reported in rockfish (*Sebastes* spp.) on Cobb seamount in the northeastern Pacific (Sasaki, 1986), though this has been disputed by subsequent workers (Pearson *et al.*, 1993). Depletion of seamount stocks was similarly reported for sea bass (*Caprodon longimanus*) on the Norfolk seamount in the South Pacific Ocean (Sasaki, 1986).

6.4. Precious Corals

Precious corals are highly valued as a raw material for making jewellery and other decorative objects (Grigg, 1984). They have been used as such since ancient times and red coral beads have been recovered in Germany with human remains dating back to about 25 000 BC (Tescione, 1965). Traditionally, the largest fishery for precious corals was in the Mediterranean Sea but in the mid-1960s large precious coral reserves were discovered north of Midway Island on the Milwaukee bank and surrounding Emperor seamounts (Grigg, 1984). The most important precious corals in terms of world harvest and commercial value are the red or pink corals (*Corallium* spp.). Other types of commercially valuable corals are the gold corals (*Primnoa* and *Parazoanthus* spp.), black corals (*Antipathes* spp.) and bamboo corals (*Lepidisis olapa* and *Acanella* spp.) (Grigg, 1974, 1986). In 1983 approximately 70% (about 140 000 kg) of the world's catch of red coral came from the Emperor–Hawaiian seamounts (Grigg, 1986).

Precious corals have fairly precise environmental requirements (see above); they are slow growing and have very low levels of natural mortality and recruitment (Grigg, 1984). Calculations based on the Beverton and Holt yield-per-recruit model indicate that a sustainable level of fishing for a population of black coral (*Antipathes dichotoma*) in Hawaii lies around 3.5% of the standing biomass (Grigg, 1984), a figure similar to that obtained for orange roughy, which also shows slow growth rates and low levels of natural mortality. The history of precious coral fisheries indicates that precious coral beds have frequently been depleted by overfishing (see Grigg, 1984, 1986).

6.5. Overexploitation of Species

It would appear that the biological resources of seamounts have been consistently exploited at unsustainable levels. The reasons for this are complicated but a number of common factors can be identified.

There is often little or no understanding of the biology of the species being fished. Even where quotas have been set, they are often based on inadequate data which have proved to be erroneous (e.g. orange roughy fisheries – Mace et al., 1990).

The life histories of many of the exploited species associated with seamounts are complicated. In several cases the species are very long-lived with slow growth rates, high ages of maturity, low fecundity, low rates of natural mortality and high interannual variability of recruitment (Grigg, 1984, 1986; Boehlert and Sasaki, 1988; Mace et al., 1990; Boehlert and Mundy, 1993; etc.). These factors all combine to give very low levels of sustainable yield for populations of these species.

The use of highly efficient trawl gear on aggregations of individuals around and on seamounts with a relatively small geographic area causes a very intense fishing pressure on the exploited species (Sasaki, 1986).

Many seamounts are located in international waters where no management of fisheries exists. Furthermore where management policies do exist they are often difficult to enforce (Grigg, 1984).

6.6. Exploitation of Geological and Physical Resources

In the future seamounts may be exploited, by ferromanganese crust and polymetallic sulphide mining, for strategic metals such as cobalt (Lonsdale et al., 1982; Welling, 1982; Grigg et al., 1987). Estimating the impact of future mining operations on seamounts is difficult given the current state of knowledge on the population dynamics of seamount organisms. Seamounts also have strategic significance in submarine warfare though data on this are unavailable.

ACKNOWLEDGEMENTS

The author thanks Dr Mike Whitfield, the Director of the Marine Biological Association of the UK, Plymouth for the use of laboratory facilities during the preparation of this manuscript. Furthermore the author thanks Dr Mike Whitfield, Professor Alan Southward and Professor John Blaxter for reviewing and commenting on the present manuscript. Many thanks to Dr Robin Pingree, Plymouth Marine Laboratory, Dr Tony Rice, Institute of Oceanographic Studies and especially Dr John Gage, SAMS, who provided the initial inspiration for this work. Finally I gratefully acknowledge the tireless assistance of the staff at the Marine Biological Association library at Citadel Hill and Neil Bryars, Liverpool John Moores University, and Dave Nichols, Plymouth Marine Laboratory, for assistance in the preparation of figures.

REFERENCES

Abe, T. (1972). Commercial fishing of berycoids and other fishes in previously unfished waters between Japan and Midway Island. In "Kuroshio II. Proceedings of the Second Symposium on the Results of the Cooperative Study of the Kuroshio and the Adjacent Regions. Tokyo, Japan, September 28–October 1", pp. 525–526. Saikon Publishing Co., Tokyo.

Allison, E.C., Durham, J.W. and Mintz, L.W. (1967). New Southeast Pacific echinoids. *Occasional Papers of the California Academy of Science* **62**, 1–23.

Alton, M.S. (1986). Fish and crab populations of Gulf of Alaska seamounts. In "The Environment and Resources of Seamounts in the North Pacific. Proceedings of the Workshop on the Environment and Resources of Seamounts in the North Pacific" (R.N. Uchida, S. Hayasi and G.W. Boehlert, eds), pp. 45–51. *US Department of Commerce, NOAA Technical Report NMFS* **43**.

Ayala, F.J. and Valentine, J.W. (1974). Genetic variability in the cosmopolitan deep-water ophiuran *Ophiomusium lymani*. *Marine Biology* **27**, 51–57.

Ayala, F.J., Valentine, J.W., Hedgecock, D. and Barr, L.G. (1975). Deep-sea asteroids: high genetic variability in a stable environment. *Evolution* **29**, 203–212.

Bell, T.H. (1975). Topographically generated internal waves in the open ocean. *Journal of Geophysical Research* **80**, 320–327.

Bell, J.D., Lyle, J.M., Bulman, C.M., Graham, K.J., Newton, G.M. and Smith, D.C. (1992). Spatial variation in reproduction and occurrence of non-reproductive adults, in orange roughy, *Hoplostethus atlanticus* Collett (Trachichthyidae), from southeastern Australia. *Journal of Fish Biology* **40**, 107–122.

Belyanina, T.N. (1984). Observations on the ichthyofauna in the open waters of the Atlantic near the Great Meteor Seamount. *Journal of Ichthyology* **24**, 127–129.

Belyanina, T.N. (1985). Preliminary data on ichthyoplankton near seamounts of the Northwest Indian Ocean. *Oceanology* **25** (6), 778–780.

Berrisford, C.D. (1969). Biology and zoogeography of the Vema Seamount: a report on the first biological collection made on the seamount. *Transactions of the Royal Society of South Africa* **38** (4), 387–398.

Bezrukov, Y.F. and Natarov, V.V. (1976). Formation of abiotic conditions above submarine elevations of some regions of the Pacific Ocean. *Izvestiya TINRO* **100**, 93–99.

Bisol, P.M., Costa, R. and Sibuet, M. (1984). Ecological and genetical survey on two deep-sea holothurians: *Benthogone rosea* and *Benthodytes typica*. *Marine Ecology Progress Series* **15**, 275–281.

Boehlert, T.H. (1985). Effects of Southeast Hancock Seamount on the pelagic ecosystem. *EOS, Transactions of the American Geophysical Union* **66**, 1336.

Boehlert, T.H. (1986). Productivity and population maintenance of seamount resources and future research directions. *In* "The Environment and Resources of Seamounts in the North Pacific. Proceedings of the Workshop on the Environment and Resources of Seamounts in the North Pacific" (R.N. Uchida, S. Hayasi and G.W. Boehlert, eds), pp. 95–101. *US Department of Commerce, NOAA Technical Report NMFS* **43**.

Boehlert, G.W. and Genin, A. (1987). A review of the effects of seamounts on biological processes. *In* "Seamounts, Islands and Atolls" (B. Keating, P. Fryer, R. Batiza and G. Boehlert, eds), pp. 319–334. *Geophysical Monograph* **43**. American Geophysical Union, Washington.

Boehlert, G.W. and Mundy, B.C. (1993). Ichthyoplankton assemblages at seamounts and oceanic islands. *Bulletin of Marine Science* **53** (2), 336–361.

Boehlert, G.W. and Sasaki, T. (1988). Pelagic biogeography of the armourhead, *Pseudopentaceros wheeleri*, and recruitment to isolated seamounts in the North Pacific Ocean. *Fishery Bulletin US* **86** (3), 453–465.

Boehlert, G.W. and Seki, M.P. (1984). Enhanced micronekton abundance over mid-Pacific seamounts. *EOS, Transactions of the American Geophysical Union* **65**, 928.

Boehlert, G.W., Wilson, C.D. and Mizuno, K. (1994). Populations of the sternoptychid fish *Maurolicus muelleri* on seamounts in the Central North Pacific. *Pacific Science* **48** (1), 57–69.

Borets, L.A. (1986). Ichthyofauna of the Northwestern and Hawaiian submarine ranges. *Journal of Ichthyology* **26** (3), 1–13.

Brink, K.H. (1989). The effect of stratification on seamount-trapped waves. *Deep-Sea Research* **36** (6), 825–844.

Bucklin, A. (1985). Allozymic variability of *Riftia pachyptila* populations from the Galapagos Rift and 21°N hydrothermal vents. *EOS, Transactions of the American Geophysical Union* **66**, 1308.

Bucklin, A. (1988). Allozymic variability of *Riftia pachyptila* populations from the Galapagos Rift and 21°N hydrothermal vents. *Deep-Sea Research* **35**, 1759–1768.

Bucklin, A., Wilson, R.R. and Smith, K.L. (1987). Genetic differentiation of seamount and basin populations of the deep-sea amphipod *Eurythenes gryllus*. *Deep-Sea Research* **34** (11), 1795–1810.

Canadian American Seamounts Expedition (1985). Hydrothermal vents on an axis seamount of the Juan de Fuca ridge. *Nature London* **313**, 212–214.

Chapman, D.C. (1989). Enhanced subinertial diurnal tides over isolated topographic features. *Deep-Sea Research* **36**, 815–824.

Cheney, R.E., Richardson, P.L. and Nagasaka, K. (1980). Tracking a Kuroshio cold ring with a free-drifting surface buoy. *Deep-Sea Research* **27A**, 641–654.

Denis, F., Jollivet, D. and Moraga, D. (1993). Genetic separation of two allopatric populations of hydrothermal snails *Alviniconcha* spp. (Gastropoda) from two south western Pacific back-arc basins. *Biochemical Systematics and Ecology* **21** (4), 431–440.

Diehl, M. (1970). Die neue, okologisch extreme Sand-Ascidie von der Josephine-Bank: *Seriocarpa rhizoides* Diehl 1969 (Ascidiacea, Styelidae). *Meteor Forschungsergebnisse. D* **7**, 43–58.

Dower, J., Freeland, H. and Juniper, K. (1992). A strong biological response to oceanic flow past Cobb Seamount. *Deep-Sea Research* **39** (8), 1139–1145.

Doyle, R.W. (1972). Genetic variation in *Ophiomusium lymani* (Echinodermata) populations in the deep sea. *Deep-Sea Research* **19**, 661–664.

Ehrich, S. (1977). Die Fischfauna der Grossen Meteorbank. *Meteor Forschungsergebnisse. D* **25**, 1–23.

Epp, D. and Smoot, N.C. (1989). Distribution of seamounts in the North Atlantic. *Nature* **337**, 254–257.

Eriksen, C.C. (1982a). An upper ocean moored current and density profiler applied to winter conditions near Bermuda. *Journal of Geophysical Research* **87**, 7879–7902.

Eriksen, C.C. (1982b). Observations of internal wave reflection off sloping bottoms. *Journal of Geophysical Research* **87**, 525–538.

Eriksen, C.C. (1985). Implications of ocean bottom reflection for internal wave spectra and mixing. *Journal of Physical Oceanography* **15**, 1145–1156.

Eriksen, C.C. (1991). Observations of amplified flows atop a large seamount. *Journal of Geophysical Research* **96** (C8), 15,227–15,236.

Everson, A.R. (1986). Lutjanidae: Ehu; Onaga; Kalekale; Gindai. *In* "Fishery Atlas of the Northwestern Hawaiian Islands" (R.N. Uchida and J.H. Uchiyama, eds), pp. 106–109, 114–117. *US Department of Commerce, NOAA Technical Report NMFS* **38**.

Fedosova, R.A. (1974). Distribution of some copepod species in the vicinity of the underwater Hawaiian Ridge. *Oceanology* **14**(5), 724–727.

Fenton, G.E., Short, S.A. and Ritz, D.A. (1991). Age determination of orange roughy, *Hoplostethus atlanticus* (Pisces: Trachichthyidae) using ^{210}Pb: ^{226}Ra disequilibria. *Marine Biology* **109**, 197–202.

Fonteneau, A. (1991). Monts sous-marins et thons dans l'Atlantique tropical est. *Aquatic Living Resources* **4**, 13–25.

France, S.C. (1994). Genetic population structure and gene flow among deep-sea amphipods, *Abyssorchomene* spp., from six California Continental Borderland basins. *Marine Biology* **118**, 67–77.

France, S.C., Hessler, R.R. and Vrijenhoek, R.C. (1992). Genetic differentiation between spatially disjunct populations of the deep-sea, hydrothermal vent-endemic amphipod *Ventiella sulfuris*. *Marine Biology* **114**, 551–559.

Fryer, P. and Fryer, G.J. (1987). Origins of nonvolcanic seamounts in a forearc environment. *In* "Seamounts, Islands and Atolls" (B. Keating, P. Fryer, R. Batiza and G. Boehlert, eds), pp. 61–69, *Geophysical Monograph* **43**. American Geophysical Union, Washington.

Gamo, T., Ishibashi, J., Sakai, H. and Tilbrook, B. (1987). Methane anomalies in seawater above the Loihi submarine summit area, Hawaii. *Geochimica et Cosmochimica Acta* **51**, 2857–2864.

Genin, A. and Boehlert, G.W. (1985). Dynamics of temperature and chlorophyll structures above a seamount: an oceanic experiment. *Journal of Marine Research* **43**, 907–924.

Genin, A., Dayton, P.K., Lonsdale, P.F. and Speiss, F.N. (1986). Corals on seamount peaks provide evidence of current acceleration over deep-sea topography. *Nature London* **322**, 59–61.

Genin, A., Haury, L. and Greenblatt, P. (1988). Interactions of migrating zooplankton with shallow topography: predation by rockfishes and intensification of patchiness. *Deep-Sea Research* **35** (2), 151–175.

Genin, A., Noble, M. and Lonsdale, P.F. (1989). Tidal currents and anticyclonic motions on two North Pacific seamounts. *Deep-Sea Research* **36** (12), 1803–1815.

Gerber, E.M. (1993). Some data on the distribution and biology of the blue whiting, *Micromesistius poutassou*, at the Mid-Atlantic Ridge. *Journal of Ichthyology* **33** (5), 26–34.

Gooch, J.L. and Schopf, T.J.M. (1972). Genetic variability in the deep sea: relation to environmental variability. *Evolution* **26**, 545–552.

Gooday, A. (1984). Records of deep-sea rhizopod tests inhabited by metazoans in the North East Atlantic. *Sarsia* **69**, 45–53.

Gooding, R.M. (1984). Trapping surveys for the deepwater caridean shrimps, *Heterocarpus laevigatus* and *H. ensifer*, in the Northwestern Hawaiian islands. *Marine Fisheries Review* **46** (2), 18–26.

Gorelova, T.A. and Prut'ko, V.G. (1985). Feeding of *Diaphus suborbitalis* (Myctophidae, Pisces) in the equatorial Indian Ocean. *Oceanology* **25** (4), 523–529.

Gould, W.J., Hendry, R. and Huppert, H.E. (1981). An abyssal topographic experiment. *Deep-Sea Research* **28A** (5), 409–440.

Grasshoff, M. (1972) Die Gorgonaria des ostlichen Nordatlantik und des Mittelmeres. I. Die familie Ellisellidae (Cnidaria: Anthozoa). *Meteor Forschungsergerbnisse, D* **10**, 73–87.

Grassle, J.P. and Grassle, J.F. (1976) Sibling species in the marine pollution indicator *Capitella capitata* (Polychaeta). *Science* **192**, 567–569.

Grassle, J.F. and Maciolek, N.J. (1992). Deep-sea species richness: regional and local diversity estimates from quantitative bottom samples. *The American Naturalist* **139** (2), 313–341.

Grigg, R.W. (1974). Distribution and abundance of precious corals in Hawaii. *In* "Proceedings of the 2nd International Coral Reef Symposium" (A.M. Cameron, B.M. Campbell, A.R. Cribb, R. Endean, J.S. Jell, O.A. Jones, P. Mather and F.H. Talbot, eds), Vol. 2, pp. 235–240. The Great Barrier Reef Committee, Brisbane, Australia.

Grigg, R.W. (1984). Resource management of precious corals: a review and application to shallow water reef building corals. *Marine Ecology* **5** (1), 57–74.

Grigg, R.W. (1986). Precious corals: an important seamount fisheries resource. *In* "The Environment and Resources of Seamounts in the North Pacific. Proceedings of the Workshop on the Environment and Resources of Seamounts in the North Pacific" (R.N. Uchida, S. Hayasi and G.W. Boehlert, eds), pp. 43–44. *US Department of Commerce, NOAA Technical Report, NMFS* **43**.

Grigg, R.W. (1987). Loihi Seamount: no macrofauna associated with hydrothermal vents. *EOS, Transactions of the American Geophysical Union* **68**, 1721.

Grigg, R.W., Malahoff, A., Chave, E.H. and Landahl, J. (1987). Seamount benthic ecology and potential environmental impact from manganese crust mining in Hawaii. *In* "Seamounts, Islands and Atolls" (B. Keating, P. Fryer, R. Batiza and G. Boehlert, eds), pp. 379–390. *Geophysical Monograph* **43**. American Geophysical Union, Washington.

Heezen, B.C. and Hollister, C.D. (1971). "The Face of the Deep." Oxford University Press, New York, 659 pp.

Hess, H.H. (1946). Drowned ancient islands of the Pacific basin. *American Journal of Science* **244** (10), 772–791.

Heydorn, S.E. (1969). The South Atlantic rock lobster *Jasus tristani* at Vema Seamount, Gough Island and Tristan Da Cunha. *Republic of South Africa Department of Industries, Division of Sea Fisheries Investigational Series* **73**, 20 pp.

Hinz, K. (1969). The Great Meteor Seamount. Results of seismic reflection measurements with a pneumatic sound source and their geological interpretation. *Meteor Forschungsergebnisse. C* **2**, 63–77.

Honda, V.A. (1986). Acanthuriidae: Naenae. In "Fishery Atlas of the Northwestern Hawaiian Islands" (R.N. Uchida and J.H. Uchiyama, eds), pp. 122–123. US Department of Commerce, NOAA Technical Report NMFS 38.

Horibe, Y., Kim, K. and Craig, H. (1983). Off-ridge submarine hydrothermal vents: Back-arc spreading centers and hotspot seamounts. EOS, Transactions of the American Geophysical Union 64, 724.

Hubbs, C.L. (1959). Initial discoveries of fish faunas on seamounts and offshore banks in the eastern Pacific. Pacific Science 13, 311–316.

Hughes, S.E. (1981). Initial US exploration of nine Gulf of Alaska seamounts and their associated fish and shellfish resources. Marine Fisheries Review 43, 26–33.

Humphreys, R.L. (1986). Lutjanidae: Opakapaka. In "Fishery Atlas of the Northwestern Hawaiian Islands" (R.N. Uchida and J.H. Uchiyama, eds), pp. 112–113. US Department of Commerce. NOAA Technical Report NMFS 38.

Humphreys, R.L. and Tagami, D.T. (1986). Review and current status of research on the biology and ecology of the genus Pseudopentaceros. In "Environment and Resources of Seamounts in the North Pacific. Proceedings of the Workshop on the Environment and Resources of Seamounts in the North Pacific" (R.N. Uchida, S. Hayasi and G.W. Boehlert, eds), pp. 55–62. US Department of Commerce, NOAA Technical Report NMFS 43.

Huppert, H.E. and Bryan, K. (1976). Topographically generated eddies. Deep-Sea Research 23, 655–679.

Isaacs, J.D. and Schwartzlose, R.A. (1965). Migrant sound scatterers: interactions with the sea floor. Science 150, 1810–1813.

Jordan, A.R. and Bruce, B.D. (1993). Larval development of three roughy species complexes (Pisces: Trachichthyidae) from southern Australian waters, with comments on the occurrence of orange roughy Hoplostethus atlanticus. Fishery Bulletin US 91 (1), 76–86.

Kaneko, A., Honji, H., Kawatate, K., Mizuna, S., Masuda, A. and Miita, T. (1986). A note on internal wavetrains and the associated undulation of the sea surface observed upstream of seamounts. Journal of the Oceanographic Society, Japan 42, 75–82.

Karig, D.E., Peterson, M.N.A. and Shor, G.G. (1970). Sediment-capped guyots in the Mid-Pacific Mountains. Deep-Sea Research 17, 373–378.

Karl, D.M., McMurtry, G.M., Malahoff, A. and Garcia, M.O. (1988). Loihi Seamount, Hawaii: a mid-plate volcano with a distinctive hydrothermal system. Nature London 355, 532–535.

Karl, D.M., Brittain, A.M. and Tilbrook, B.D. (1989). Hydrothermal and microbial processes at Loihi Seamount, a mid-plate hot-spot volcano. Deep-Sea Research 36 (11), 1655–1673.

Kaufmann, R.S., Wakefield, W.W. and Genin, A. (1989). Distribution of epibenthic megafauna and lebensspuren on two central North Pacific seamounts. Deep-Sea Research 36 (12), 1863–1896.

Keating, B.H., Fryer, P., Batiza, R. and Boehlert, G.W. (eds) (1987). "Seamounts, Islands and Atolls." Geophysical Monograph 43. The American Geophysical Union, Washington.

Keller, N.B. (1985). Coral populations of underwater ridges in the North Pacific and Atlantic oceans. Oceanology 25(6), 784–786.

Kensley, B. (1980). Decapod and isopod crustaceans from the west coast of southern Africa, including seamounts Vema and Tripp. Annals of the South African Museum 83 (2), 13–32.

Koami, H. (1986). A seamount survey around Izu Islands. In "Environment and Resources of Seamounts in the North Pacific" (R.N. Uchida, S. Hayasi and

G.W. Boehlert, eds), pp. 63–66. *US Department of Commerce, NOAA Technical Report NMFS* **43**.

Kramer, S.H. (1986). Lutjanidae: Uku. Scombridae: Wahoo. *In* "Fishery Atlas of the Northwestern Hawaiian Islands" (R.N. Uchida and J.H. Uchiyama, eds), pp. 104–105, 126–127. *US Department of Commerce, NOAA Technical Report NMFS* **38**.

Krieger, K.J. (1992). Distribution and abundance of rockfish determined from a submersible and by bottom trawling. *Fishery Bulletin US* **91**, 87–96.

Levi, C. (1969). Spongaires du Vema Seamount (Atlantique Sud). *Bulletin du Muséum National D'Histoire Naturelle. 2nd Series.* **41** (4), 952–973.

Levin, L.A. and Nittrouer, C.A. (1987). Textural characteristics of sediments on deep seamounts in the eastern Pacific Ocean between 10°N and 30°N. *In* "Seamounts, Islands and Atolls" (B. Keating, P. Fryer, R. Batiza and G. Boehlert, eds), pp. 187–203. *Geophysical Monograph* **43**. American Geophysical Union, Washington.

Levin, L.A. and Thomas, C.L. (1988). The ecology of xenophyophores (Protista) on eastern Pacific seamounts. *Deep-Sea Research* **12** (12), 2003–2027.

Levin, L.A. and Thomas, C.L. (1989). The influence of hydrodynamic regime on infaunal assemblages inhabiting carbonate sediments on central Pacific Seamounts. *Deep-Sea Research* **36** (12), 1897–1915.

Levin, L.A., DeMaster, D.J., McCann, L.D. and Thomas, C.L. (1986). Effects of giant protozoans (class: Xenophyophorea) on deep-seamount benthos. *Marine Ecology Progress Series* **29**, 99–104.

Littler, M.M., Littler, D.S., Blair, S.M. and Norris, J.N. (1985). Deepest known plant life discovered on an uncharted seamount. *Science* **227**, 57–59.

Littler, M.M., Littler, D.S., Blair, S.M. and Norris, J.N. (1986). Deep-water plant communities from an uncharted seamount off San Salvador Island, Bahamas: distribution, abundance, and primary productivity. *Deep-Sea Research* **33** (7), 881–892.

Lonsdale, P., Batiza, R. and Simkin, T. (1982). Metallogenesis at seamounts on the East Pacific Rise. *Marine Technology Society Journal* **16** (3), 54–60.

Lophukin, A.S. (1986). Distribution of ATP concentration above seamounts in the Atlantic ocean. *Oceanology* **26** (3), 361–365.

Lutjeharms, J.R.E. and Heydorn, A.E.F. (1981a). The rock-lobster *Jasus tristani* on Vema Seamount: drifting buoys suggest a possible recruiting mechanism. *Deep-Sea Research* **28A** (6), 631–636.

Lutjeharms, J.R.E. and Heydorn, A.E.F. (1981b). Recruitment of rock lobster on Vema Seamount from the islands of Tristan da Cunha. *Deep-Sea Research* **28A** (10), 1237.

Mace, P.M., Fenaughty, J.M., Coburn, R.P. and Doonan, I.J. (1990). Growth and productivity of orange roughy (*Hoplostethus atlanticus*) on the north Chatham Rise. *New Zealand Journal of Marine and Freshwater Research* **24**, 105–119.

Marshall, N.B. (1979). "Developments in Deep-sea Biology." Blandford Press, Poole, Dorset, UK, 566 pp.

Massoth, G.J., Butterfield, D.A., Lupton, J.E., McDuff, R.E., Lilley, M.D. and Jonasson, I.R. (1989). Submarine venting of phase separated hydrothermal fluids at Axial Volcano, Juan de Fuca Ridge. *Nature London* **340**, 702–705.

Maul, G.E. (1976). The fishes taken in bottom trawls by R.V. "Meteor" during the 1967 seamounts cruises in the Northeast Atlantic. *Meteor Forschungsergebnisse. D* **22**, 1–69.

Meincke, J. (1971). Observation of an anticyclonic vortex trapped above a seamount. *Journal of Geophysical Research* **76**, 7432–7440.

Menard, H.W. (1964). "Marine Geology of the Pacific." International Series in the Earth Sciences. McGraw-Hill, New York, 271 pp.

Menard, H.W. and Dietz, R.S. (1951). Submarine geology of the Gulf of Alaska. *Bulletin of the Geological Society of America* **62**, 239–253.

Menard, H.W. and Ladd, H.S. (1963). Oceanic islands, seamounts, guyots and atolls. *In* "The Sea", vol. 3 (M.N. Hill, ed.), pp. 365–385. Wiley Interscience, New York and London.

Mikhaylin, S.V. (1977). The intraspecific variability of the frostfish *Lepidopus caudatus*. *Journal of Ichthyology* **17**, 201–210.

Millard, N.A.H. (1966). Hydroids of the Vema Seamount. *Annals of the South African Museum* **48** (19), 489–496.

Moiseyev, Y.V. (1986). Distribution of protozoans near seamounts in the western Indian Ocean. *Oceanology* **26** (1), 86–90.

Moskalev, L.I. and Galkin, S.V. (1986). Investigations of the fauna of submarine upheavals during the 9th trip of the research vessel "Academic Mstislav Keldysh". *Zoologicheskii Zhurnal* **65** (11), 1716–1720. (Russian with English summary.)

Murphy, L.S., Rowe, G.T. and Haedrich, R.L. (1976). Genetic variability in deep-sea echinoderms. *Deep-Sea Research* **23**, 339–348.

Nakamura, I., Inada, T., Takeda, M. and Hatanaka, H. (1986). "Important Fishes Trawled Off Patagonia." Japan Marine Fishery Resource Research Centre. Tosho Printing Co., Tokyo, 369 pp.

Natland, J.J. (1976). Petrology of volcanic rocks dredged from seamounts in the Line Islands. *Initial Reports of the Deep Sea Drilling Project* **33**, 749–777.

Nellen, W. (1973). Untersuchungen zur Verteilung von Fischlarven und Plankton im Gebiet der Grossen Meteorbank. *Meteor Forschungsergebnisse.* D **13**, 47–69.

Nellen, W. (1974). Investigations on the distribution of larvae and plankton above the Great Meteor Seamount. *In* "The Early Life History of Fish" (J.H.S. Blaxter, ed.), pp. 213–214. Springer-Verlag, New York.

Noble, M. and Mullineaux, L.S. (1989). Internal tidal currents over the summit of Cross Seamount. *Deep-Sea Research* **36**, 1791–1802.

Okamura, O., Amakoa, K. and Mitani, F. (1982). "Fishes of the Kyushu-Palau Ridge and Tosa Bay." Japan Fisheries Resource Conservation Association. Tosho Printing Co., Tokyo, 435 pp.

Ovenden, J.R., Smolenski, A.J. and White, R.W.G. (1989). Mitochondrial DNA restriction site variation in Tasmanian populations of orange roughy (*Hoplostethus atlanticus*), a deep-water marine teleost. *Australian Journal of Marine and Freshwater Research* **40**, 1–9.

Owens, W.B. and Hogg, N.G. (1980). Oceanic observations of stratified Taylor columns near a bump. *Deep-Sea Research* **27A**, 1029–1045.

Pankhurst, N.W. (1988). Spawning dynamics of orange roughy, *Hoplostethus atlanticus*, in mid slope waters of New Zealand. *Environmental Biology of Fishes* **21**, 101–116.

Pankhurst, N.W. and Conroy, A.M. (1987). Size—fecundity relationships in the orange roughy *Hoplostethus atlanticus*. *New Zealand Journal of Marine and Freshwater Research* **21**, 295–300.

Pankhurst, N.W., McMillan, P.J. and Tracey, D.M. (1987). Seasonal reproductive cycles in three commercially exploited fishes from the slope waters off New Zealand. *Journal of Fish Biology* **30**, 193–211.

Parin, N.V and Prut'ko, V.G. (1985). The thalassial mesobenthopelagic icthyocoene above the equator seamount in the western tropical Indian Ocean. *Oceanology* **25** (6), 781–783.

Parin, N.V., Konovalenko, I.I. and Nesterov, A.A. (1990). Independent populations of neritic Carangidae above seamounts of the Nazca submarine ridge. *Soviet Journal of Marine Biology* **16** (3), 135–138.

Pearson, D.E., Douglas, D.A. and Barss, B. (1993). Biological observations from the Cobb Seamount rockfish fishery. *Fishery Bulletin US* **91**, 573–576.

Pelaez, J. and McGowan, J.A. (1986). Phytoplankton pigment patterns in the California Current as determined by satellite. *Limnology and Oceanography* **31**, 927–950.

Penrith, M.J. (1967). The fishes of Tristan Da Cuna, Gough Island and the Vema Seamount. *Annals of the South African Museum* **48** (22), 523–548.

Pratt, R.M. (1963). Great Meteor Seamount. *Deep-Sea Research* **10**, 17–25.

Rad, U.von (1974). Great Meteor and Josephine Seamounts (eastern North Atlantic): composition and origin of bioclastic sands, carbonate and pyroclastic rocks. *Meteor Forschungsergebnisse. C* **19**, 1–61.

Rao, M.V. and Newman, W.A. (1972). Thoracic cirripedia from guyots of the Mid-Pacific mountains. *Transactions of the San Diego Society of Natural History* **17** (6), 69–94.

Raymore, P.A. (1982). Photographic investigations on three seamounts in the Gulf of Alaska. *Pacific Science* **36** (1), 15–34.

Reimers, C.E. and Wakefield, W.W. (1989). Flocculation of siliceous detritus on the sea floor of a deep Pacific seamount. *Deep-Sea Research* **36** (12), 1841–1861.

Rice, A.L. and Williamson, D.I. (1977). Planktonic stages of Crustacea Malacostraca from Atlantic Seamounts. *Meteor Forschungsergebnisse. D* **26**, 28–64.

Richardson, P.L. (1980). Anticyclonic eddies generated near Corner Rise seamounts. *Journal of Marine Research* **38**, 673–686.

Richer de Forges, B. (1991). A new species of *Sphenocarcinus* A. Milne Edwards, 1875 from Tasmantid guyots, *S. lowryi* n.sp. (Crustacea, Decapoda, Brachyura). *Records of the Australian Museum* **44** (1), 1–5.

Richer de Forges, B. (1993). Deep sea crabs of the Tasman Seamounts (Crustacea: Decapoda: Brachyura). *Records of the Australian Museum* **45** (1), 11–24.

Roden, G.I. (1987). Effect of seamounts and seamount chains on ocean circulation and thermohaline structure. *In* "Seamounts, Islands and Atolls" (B.H. Keating, P. Fryer, R. Batiza and G.W. Boehlert, eds), pp. 335–354. *Geophysical Monograph* **43**. American Geophysical Union, Washington.

Roden, G.I. (1991). Mesoscale flow and thermohaline structure around Fieberling seamount. *Journal of Geophysical Research* **96**, 16,653–16,672.

Roden, G.I. and Taft, B.A. (1985). Effect of the Emperor seamount on the mesoscale thermohaline structure during the summer of 1982. *Journal of Geophysical Research* **90**, 839–855.

Roden, G.I., Taft, B.A. and Ebbesmeyer, C.C. (1982). Oceanographic aspects of the Emperor seamounts region. *Journal of Geophysical Research* **87**, 9537–9552.

Sadakane, H., Takeda, Y., Tabuchi, K., Tatsumi, S. and Mishiyama, T. (1989). Fishery and echo surveys on the seamounts in the Tonga Ridge using a global positioning system. *The Journal of Shimonoseki University of Fisheries* **38** (1), 1–6.

Sagalevitch, A.M., Torohov, P.V., Matweenkov, V.V., Galkin, S.V. and

Moskalev, L.I. (1992). Hydrothermal activity on the underwater volcano Peepa (Bering Sea). *Izvestiya RAN. Series Biology* **9**, 104–114. (In Russian)

Sakai, H., Tsubota, H., Nakai, T., Ishibashi, J., Akabi, T., Gamo, T., Tilbrook, T., Igarashi, G., Kodera, M., Shitashima, K., Nakamura, S., Fujioka, K., Watanabe, M., McMurtry, G., Malahoff, A. and Ozima, M. (1987). Hydrothermal activity on the summit of Loihi Seamount, Hawaii. *Geochemical Journal* **21**, 11–21.

Sakiura, H. (1972). The pelagic armourhead, *Pseudopentaceros richardsoni*, fishing grounds off the Hawaiian Islands, as viewed by the Soviets. *Suisan Shuho* (The Fishing and Food Industry Weekly) **658**, 28–31.

Sasaki, T. (1978). The progress and current status on exploration of seamounts fishing grounds. *Bulletin of the Japanese Society of Fisheries and Oceanography* **33**, 51–53.

Sasaki, T. (1986). Development and present status of Japanese trawl fisheries in the vicinity of seamounts. *In* "The Environment and Resources of Seamounts in the North Pacific" (R.N. Uchida, S. Hayasi and G.W. Boehlert, eds), pp. 21–30. Proceedings of the Workshop on the Environment and Resources of Seamounts in the North Pacific. US Department of Commerce, *NOAA Technical Report NMFS* **43**.

Scagel, R.F. (1970). Benthic algae of Bowie Seamount. *Syesis* **3**, 15–16.

Seki, M.P. (1986). Serranidae: Hapuupuu. Carangidae: Ulua; Butaguchi. *In* "Fishery Atlas of the Northwestern Hawaiian Islands" (R.N. Uchida and J.H. Uchiyama, eds), pp. 82–83, 86–87, 96–97. *US Department of Commerce, NOAA Technical Report NMFS* **38**.

Seki, M.P. and Tagami, D.T. (1986). Review and present status of handline and bottom longline fisheries for alfonsin. *In* "The Environment and Resources of Seamounts in the North Pacific" (R.N. Uchida, S. Hayasi and G.W. Boehlert, eds), pp. 31–35. Proceedings of the Workshop on the Environment and Resources of Seamounts in the North Pacific. US Department of Commerce, *NOAA Technical Report NMFS* **43**.

Shiota, P.M. (1986). Carangidae: Redtail Scad. "In Fishery Atlas of the Northwestern Hawaiian Islands" (R.N. Uchida and J.H. Uchiyama), pp. 92–93. *US Department of Commerce, NOAA Technical Report NMFS* **38**.

Shomura, R.S. and Barkley, R.A. (1980). Ecosystem dynamics of seamounts – a working hypothesis. *In* "The Kuroshio IV. Proceedings of the Fourth Symposium for the Cooperative Study of the Kuroshio and Adjacent Regions, The Japanese Academy, Tokyo, Japan, 14–17 February 1979." Saikon Publishers, Tokyo. pp. 7789–790.

Siebenaller, J.F. (1978) Genetic variation in deep-sea invertebrate populations: the bathyal gastropod *Bathybembix bairdii*. *Marine Biology* **47**, 265–275.

Simpson, E.S.W. and Heydorn, A.E.F. (1965). Vema Seamount. *Nature London* **207**, 249–251.

Smith, D.K. and Jordan, T.H. (1988). Seamount statistics in the Pacific ocean. *Journal of Geophysical Research* **93**, 2899–2919.

Smith, K.L. and Baldwin, R.J. (1984). Vertical distribution of the necrophagous amphipod, *Eurythenes gryllus*, in the North Pacific: spatial and temporal variation. *Deep-Sea Research* **10**, 1179–1196.

Smith, K.L., Baldwin, R.J. and Edelman, J.L. (1989). Supply of and demand for organic matter by sediment communities on two central North Pacific seamounts. *Deep-Sea Research* **36** (12), 1917–1932.

Smith, P.J., Francis, R.I.C.C. and McVeagh, M. (1991). Loss of genetic diversity due to fishing pressure. *Fisheries Research* **10**, 309–316.

Smolenski, A.J., Ovenden, J.R. and White, R.W.G. (1993). Evidence of stock separation in southern hemisphere orange roughy (*Hoplostethus atlanticus*, Trachichthyidae) from restriction-enzyme analysis of mitochondrial DNA. *Marine Biology* **116**, 219–230.

Tarasov, V.G., Propp, M.V., Propp, L.N., Blinov, S.V. and Kamenev, G.M. (1987). Hydrothermal shallow vents and specific ecosystem in the Kraternaya Caldera (Kurile Islands). *The Soviet Journal of Marine Biology* **12** (2), 122–125.

Taylor, G.I. (1917). Motion of solids in fluids when the flow is not irrotational. *Proceedings of the Royal Society. A* **93**, 99–113.

Tendal, O.S. (1972). A monograph of the Xenophyophoria (Rhizopoda, Protozoa). *Galathea Reports* **12**, 7–99.

Tendal, O.S. (1979). Aspects of the biology of Komokiacea and Xenophyophoria. *Sarsia* **64**, 13–17.

Tendal, O.S. and Gooday, A.J. (1981). Xenophyophoria (Rhizopoda, Protozoa) in bottom photographs from the bathyal and abyssal NE Atlantic. *Oceanologica Acta* **4** (4), 415–422.

Tendal, O.S. and Lewis, A.J. (1978). New Zealand xenophyophores: upper bathyal distribution, photographs of growth position, and a new species. *New Zealand Journal of Marine and Freshwater Research* **12** (2), 197–203.

Tescione, G. (1965). "Il corallo nella storia e nell'arte." Montanino Editore, Napoli, Italy, 405 pp.

Tseytlin, V.B. (1985). Energetics of fish populations inhabiting seamounts. *Oceanology* **25** (2), 237–239.

Tunnicliffe, V. (1991). The biology of hydrothermal vents. *Oceanography and Marine Biology Annual Review* **29**, 319–407.

Tunnicliffe, V., Juniper, S.K. and de Burgh, M.E. (1985). The hydrothermal vent community on Axial Seamount, Juan de Fuca Ridge. *Bulletin of the Biological Society of Washington* **6**, 453–464.

Tunnicliffe, V., Desbruyeres, D., Jollivet, D. and Laubier, L. (1993). Systematic and ecological characteristics of *Paralvinella sulfinicola* Desbruyeres and Laubier, a new polychaete (family Alvinellidae) from northeast Pacific hydrothermal vents. *Canadian Journal of Zoology* **71**, 289–297.

Uchida, R.N. (1986). Lutjanidae: Taape. *In* "Fishery Atlas of the Northwestern Hawaiian Islands" (R.N. Uchida and J.H. Uchiyama, eds), pp. 110–111. *US Department of Commerce, NOAA Technical Report NMFS* **38**.

Uchida, R.N. and Tagami, D.T. (1984). Groundfish fisheries and research in the vicinity of seamounts in the North Pacific Ocean. *Marine Fisheries Review* **46** (2), 1–17.

Uchiyama, J.H. (1986). Scombridae: Kawakawa. *In* "Fishery Atlas of the Northwestern Hawaiian Islands" (R.N. Uchida and J.H. Uchiyama, eds), pp. 128–129. *US Department of Commerce, NOAA Technical Report NMFS* **38**.

Uchupi, E. and Ballard, R.D. (1989). Evidence of hydrothermal activity on Marsili Seamount, Tyrrhenian Basin. *Deep-Sea Research* **36** (9), 1443–1448.

United States Board of Geographic Names (1981). "Gazetteer of Undersea Features", 3rd edn. Defense Mapping Agency, Washington DC.

Vastano, A.C. and Warren, B.A. (1976). Perturbations of the Gulf Stream by Atlantis II seamount. *Deep-Sea Research* **23**, 681–694.

Vastano, A.C., Hagen, D.E. and McNally, G.J. (1985). Lagrangian observations of surface circulation at the Emperor seamount chain. *Journal of Geophysical Research* **90**, 3325–3331.

Voronina, N.M. and Timonin, A.G. (1986). Zooplankton of the region of seamounts in the western Indian ocean. *Oceanology* **26** (6), 745–748.

Welander, A.D., Johnson, R.C. and Hajny, R.A. (1957). Occurrence of the boarfish, *Pseudopentaceros richardsoni*, and the zeid, *Allocyttus verrucosus*, in the North Pacific. *Copeia* **1957** (3), 244–246.

Welling, C.G. (1982). Polymetallic sulfides: an industry viewpoint. *Marine Technology Society Journal* **16** (3), 5–7.

Williams, N., Dixon, D.R., Southward, E.C. and Holland, P.W.H. (1993). Molecular evolution and diversification of the vestimentiferan tube worms. *Journal of the Marine Biological Association of the UK* **73** (2), 437–452.

Wilson, R.R. and Kaufmann, R.S. (1987). Seamount biota and biogeography. In "Seamounts, Islands and Atolls" (B.H. Keating, P. Fryer, R. Batiza and G.W. Boehlert, eds), pp. 319–334. *Geophysical Monograph* **43**. American Geophysical Union, Washington.

Wilson, R.R. and Waples, R.S. (1984). Electrophoretic and biometric variability in the abyssal grenadier *Coryphenoides armatus* of the western North Atlantic, eastern South Pacific and eastern North Pacific Oceans. *Marine Biology* **80**, 227–237.

Wilson, R.R., Smith, K.L., Rosenblatt, R.H. (1985). Megafauna associated with bathyal seamounts in the central North Pacific Ocean. *Deep-Sea Research* **23** (10), 1243–1254.

Wishner, K., Levin, L., Gowing, M. and Mullineaux, L. (1990). Involvement of the oxygen minimum in benthic zonation on a deep seamount. *Nature London* **346**, 57–59.

Wunsch, C. and Webb, S. (1979). The climatology of deep ocean internal waves. *Journal of Physical Oceanography* **9**, 225–243.

Yasui, M. (1986). Albacore, *Thunnus alalunga*, pole-and-line fishery around the Emperor seamounts. In "The Environment and Resources of Seamounts in the North Pacific" (R.N. Uchida, S. Hayasi and G.W. Boehlert, eds), pp. 37–40. *Proceedings of the Workshop on the Environment and Resources of Seamounts in the North Pacific. US Department of Commerce, NOAA Technical Report NMFS* **43**.

Zaika, V.E. and Kovalev, A.V. (1985). Investigation of ecosystems of submarine elevations. *Soviet Journal of Marine Biology* **10** (6), 301–306.

Zeldis, J.R. (1993). Applicability of egg surveys for spawning-stock biomass estimation of snapper, orange roughy, and hoki in New Zealand. *Bulletin of Marine Science* **53** (2), 864–890.

Zhang, X. and Boyer, D.L. (1991). Current deflections in the vicinity of multiple seamounts. *Journal of Physical Oceanography* **21**, 1122–1138.

Zullo, V.A. and Newman, W.A. (1964). Thoracic Cirripedia from a Southeast Pacific guyot. *Pacific Science* **18**, 355–372.

Taxonomic Index

Tables in **bold**, figures in *italic*

Subject Index

Tables in **bold**, figures in *italic*

Cumulative Index of Titles

Alimentary canal and digestion in teleosts, **13**, 109
Antarctic benthos, **10**, 1
Appraisal of condition measures for marine fish larvae, **30**, 217
Artificial propagation of marine fish, **2**, 1
Aspects of stress in the tropical marine environment, **10**, 217
Aspects of the biology of frontal systems, **23**, 163
Aspects of the biology of seaweeds of economic importance, **3**, 105
Assessing the effects of "stress" on reef corals, **22**, 1
Association of copepods with marine invertebrates, **16**, 1
Autrophic and heterotrophic picoplankton in the Baltic Sea, **29**, 73

Behaviour and physiology of herring and other clupeids, **1**, 262
Biological response in the sea to climatic changes, **14**, 1
Biology of ascidians, **9**, 1
Biology of clupeoid fishes, **10**, 1
Biology of coral reefs, **1**, 209
Biology of euphausiids, **7**, 1; **18**, 373
Biology of living brachiopods, **28**, 175
Biology of mysids, **18**, 1
Biology of Oniscid Isopoda of the genus *Tylos*, **30**, 89
Biology of pelagic shrimps in the ocean, **12**, 223
Biology of Phoronida, **19**, 1
Biology of *Pseudomonas*, **15**, 1
Biology of Pycnogonida, **24**, 1
Biology of Seamounts, **30**, 305
Biology of the Atlantic Halibut, *Hippoglossus hippoglossus* (L., 1758), **26**, 1
Biology of the Penaeidae, **27**, 1
Biology of wood-boring teredinid molluscs, **9**, 336
Blood groups of marine animals, **2**, 85
Breeding of the North Atlantic freshwater eels, **1**, 137
Bristol Channel Sole (*Solea solea* (L.)): A fisheries case study, **29**, 215
Burrowing habit of marine gastropods, **28**, 389

Circadian periodicities in natural populations of marine phytoplankton, **12**, 326
Comparative physiology of Antarctic Fishes, **24**, 321
Competition between fisheries and seabird communities, **20**, 225
Coral communities and their modification relative to past and present prospective
  Central American seaways, **19**, 91

Development and application of analytical methods in benthic marine infaunal
  studies, **26**, 169

Cumulative Index of Authors

Printed and bound by CPI Group (UK) Ltd, Croydon, CR0 4YY

03/10/2024

01040415-0001